Uniscience Series on Fine Particle Science and Technology

Editor-in-Chief

John Keith Beddow, Ph.D.
President
The Shapespeare Corporation
Iowa City, Iowa

Advanced Particle Morphology
J.K. Beddow and T.P. Meloy

Separation of Particles From Air and Gases, Volumes I and II
Akira Ogawa

Particle Characterization in Technology
Volume I: Applications and Microanalysis
Volume II: Morphological Analysis
J.K. Beddow, Editor

Theory and Application of Morphological Analysis: Fine Particles and Surfaces
David W. Leurkens

Vortex Flow
Akira Ogawa

Vortex Flow

Author

Akira Ogawa, Dr. Eng.

CRC Series on Fine Particle Science and Technology

Editor-in-Chief

John Keith Beddow

CRC Press

Boca Raton Ann Arbor London Tokyo

Library of Congress Cataloging-in-Publication Data

Ogawa, Akira, 1940–
Vortex flow / author, Akira Ogawa.
p. cm. -- (CRC series on fine particle science and technology)
Includes bibliographical references and index.
ISBN 0-8493-5782-9
1. Vortex-motion. I. Title. II. Series.
QA925.042 1992 92-10982
532′.0595--dc20 CIP

This book represents information obtained from authentic and highly regarded sources. Reprinted material is quoted with permission, and sources are indicated. A wide variety of references are listed. Every reasonable effort has been made to give reliable data and information, but the author and the publisher cannot assume responsibility for the validity of all materials or for the consequences of their use.

Direct all inquiries to CRC Press, Inc., 2000 Corporate Blvd., N.W., Boca Raton, Florida, 33431.

International Standard Book Number 0-8493-5782-9

Library of Congress Card Number 92-10982

Printed in the United States of America 1 2 3 4 5 6 7 8 9 10

Printed on acid-free paper

PREFACE

The subject of vortices is one of general interest in mechanical engineering, chemical engineering, and also in powder technology.

Interest in vortices also has a respectable history. For example, Leonardo da Vinci depicted the very interesting drawings of various kinds of vortex and eddy flows. The fundamental concept of vortex flow which was related to the idea of circulation was developed much later in the form of the lift force of the air plane wing in the theory of Kutta (1910) and Joukowski (1906).

The biggest vortex flows on earth are the cyclones in the Indian Ocean, typhoons in Japan, and also hurricanes. In fiction, there is a most interesting description of the sea wheel by Edgar Allen Poe (1808–1849) in his short story "A Descent into the Maelstrom". The phenomena of vortex flows is also common in daily life although we often don't notice them. For example, vortex flow in the bath tub, Benard-Karman vortex band as a noise which is produced when a strong wind blows across an electric wire, and a vortex ring produced when tobacco smoke is emitted from the smoker's mouth.

Engineering applications of vortex flows include: cyclone dust collectors for separating solid particles from three-dimensional turbulent rotational gas flows by centrifugal force (see "Separation of Particles from Air and Gases", Vols I and II, CRC Press, 1984) and also hydro-cyclones; the vortex tubes for heat separation in three-dimensional turbulent air flow; combustion chambers in the Diesel engine; mixing chambers; and also other technical fields.

However, the internal structure of the mechanisms of vortex flow is rather complicated. The simplest model is the Thomson-Rankine combined vortex model which was defined in a book called *Applied Mechanics* by Rankine (1876). This vortex model is composed of forced and free vortices. This combined vortex model was first proposed by Professor Thomson (Belfast) who studied the hydraulic vortex turbine (see "Hydraulics and its Applications" by Gibson, 1912). In addition, there are many kinds of vortex models which can be used in various technical applications. But, for the present, the more general vortex model which can be applied to air and liquids has not been derived. A key reason for this is that even if many more kinds of detecting methods are developed and further improved, it is very difficult to measure the distributions of three-dimensional Reynolds stresses in turbulent rotational flow.

In this book, the fundamental concepts and applications of vortex flows are described in detail. The phenomena of vortex breakdown, the detailed mathematical descriptions of the stability criterion, and also the thickness of the boundary layer are excluded. For readers who are interested in these and other more advanced studies of vortex flows, they may consult the many references cited at the end of each chapter.

Akira Ogawa

ACKNOWLEDGMENTS

The author wishes to express his gratitude to Dr. John Keith Beddow of The Shapespeare Corporation, Dr. T. Ariman of the University of Tulsa, and also Dr. R. Comolet of the University of Paris VI (Dr. Comolet passed away in 1989. I pray for the repose of his soul.) for their advice, encouragement, and suggestions.

Further, it is a pleasure to acknowledge the contributions of Dr. I. Tani, Tokyo; Dr. K. Iinoya, Nagoya; Dipl.-Ing. R. Pieper of Siemens-Schuckertwerken, Erlangen; Dipl.-Ing. J. Komaroff and K. Ito of Technisches Zentrum; Robert Bosh, Stuttgart; Dr. T. Jaroszczyk of Nelson Industries; Dr. E. Muschelknautz of the University of Stuttgart; Dr. H. Brauer of the University of Berlin; Dr. M. Bohnet of the University of Braunschweig; Dr. D. Etling of the University of Hannover; Dr. F. Löffler of the University of Karlsruhe; Dr. J. C. Rotta of DFVLR of Göttingen; Dr. O. Molerus of the University of Erlangen-Nürnberg; Dr. K. G. Roesner of the University of Darmstadt; Dr. M. P. Escudier of the University of Liverpool; Dr. Camille Bouras of Universite de Techtin, Lattaquié; Ir. J. O. Hinze of Hague, Dr. J. P. Renaudeaux of the University of Paris VI; Dr. A. Fortier of Paris, Dr. C. P. Chen of Paris, Dr. J. J. Bernard of the University of Paris VI; Dr. L. Chincholle of Ecole Supérieure d'Electricité of Paris, Dr. G. Cognet of ENSEM-INPL of Nancy; Dr. F. S. Darozes of the University of Paris VI; Dr. M. H. François of the University of Paris VI; Dr. E. J. Hopfinger of the University of Grenoble; Dr. J. C. Mora of the University of Toulouse; Dr. J. Bataille of the University of Lyon; Dr. D. Bellet of the University of Toulouse; Dr. J. M. Hyun of KAISt of Korea; Dr. J. Abrahamson of the University of Canterbury; Dr. W. Albring of the University of Dresden; Dr. E. Heidenreich of the University of Dresden; Dr. Faume Blasco i Font de Rubinat of l'escole Técnica Superior D'enginyers Industrials de Barcelona; Dr. J. Casal of Universittad Politècnica de Cataluña; Dr. M. G. Velarde of the University of Madrid; Dr. E. València i Leonardo of Escola Tècnica Superior d'Enginyers Industrials; Dr. W. H. Gauvin of Centre de Recherche Noranda; Dr. H. J. Lugt of D. W. Taylor Naval Ship Research and Development Center; Dr. R. H. Page of Texas A&M University; Dr. C. A. Petty of the University of Michigan; Dr. D. L. Soo of the University of Illinois; Dr. J. M. Wu and Dr. H. W. Hsu of the University of Tennessee; Dr. X. C. Yan of Aero-Dynamic Institut of Beijing; Dr. N. X. Chen of Beijing Inst. Eng. Thermo-Physics; Dr. C. T. Wei of Beijing University; Dr. C. K. Lin of Shanhai Technical University; and Dipl.-Ing.-Csc. J. Zahradka of CKD Praha-Lokomativka-Sokolovo; Dr. J. Hanna of University of Alabama; Dr. A. Rayan of Mansoura University of Cairo. The author also thanks the managing editor of Sankaido-Publishing Co., Mr. K. Yagi, who permitted the partial translation of *Vortex Flow.*

The author would like to express his gratitude to the following authors:

Albring, W., *Elementarvorgänge Fluider Wirbelbewegungen,* Akademie-Verlag, Berlin (1981).

Araki, Y. and Nagaoka, H.; *Collected Papers of Famous Scientists,* Helmholts, H. and Thomson, W., Eds., (translated into Japanese) Maruzen-Pub. Co., Tokyo (1913).

Batchelor, G. K., *The Theory of Homogeneous Turbulence,* Cambridge University Press, New York (1953).

Bauer, G., *Die Helmholtzsche Wirbeltheorie über Ingenieure,* Verlag von R. Oldenbourg.

Bjerknes, V., Bjerknes, J., Solberg, H., and Bergeron, T., *Physikalische Hydrodynamik mit Anwendung auf die dynamische Meteorologie,* Springer-Verlag, Berlin (1933).

Bouasse, H., *TOURBILLONS, Tome I et Tome II,* Libraire Delagrave, Paris (1931).

Brauer, H. and Varma, Y. B. G., *Air Pollution Control Equipment,* Springer-Verlag, Berlin (1981).

Chirgwin, B. H. and Plumpton, C., *Elementary Classical Hydrodynamics,* Pergamon Press, New York (1967).

Chorlton, F., *Textbook of Fluid Dynamics,* D. van Nortrand, New York (1967).

Comolet, R., *Biomécanique Circulatoire,* Masson, Paris (1984).

Comolet, R., Mécanique Expérimentale des Fluides, Tome II, Dynamique des Fluides Réels, Turbomachines, 3e. édition, Masson (Paris-1982).

Comolet, R. and Bornin, J., Mécanique Expérimentale des Fluides, Tome III, Recusil de Problèmes, 4e. édition, Masson (Paris-1986).

Cornish, J. J., Vortex Flows, Lockheed Georgia Co. (1982).

Darrozes, J. S. and Francois, C., Mécanique des Fluides Incompressibles, Vol. 163, Lecture Notes in Physics, Springer-Verlag (1982).

Davies, J. T., Turbulence Phenomena, Academia Press (1972).

Ebeling, W. and Klimontovich, Y. L., *Self-Organization and Turbulence in Liquids,* Teubner-Texte zur Physik, Leipzig (1984).

Ekman, V. W., On the influence of the earth's rotation on ocean-currents, Arkiv för Mathematik, Astronomi Och Fysik, Bd. 2, No. 11 (1905), pp. 1–52. (Good will by Prof. Dr. Y. Kikuyama of Nagoya University.)

Eskinazi, S., *Vector Mechanics of Fluids and Magnetofluids,* Academic Press, New York (1967).

Favre, A., *The Mechanics of Turbulence* (Mecanique de la Turbulence) by C.N.R.S. (Paris-1962), Gordon and Breach Science Pub. (1964).

Fortier, A., Mécanique des Fluides et Transferts de Chaleur et de Masse par Convection, Masson, Paris (1975).

Fortier, A., Mécanique des Suspensions, Masson, Paris (1967).

Fruman, D. H. and Darrozes, J. S., *Methodes Optiques Appliquees a la Mécanique des Fluides,* Ecole Nationale Superieure de Techniques Avances, Paris (1980).

Fujita, T., *Tatzumaki* (Tornado, in Japanese) Kyoritzu-Pub. Co., Tokyo (1974).

Granger, R. A. and Wendt, T. F., *Introduction to Vortex Dynamics,* Vols. I and II, Von Karman Institute for Fluid Dynamics (Belgium), Lecture-Series 1986-08 (May 26–30, 1986).

Greenspan, H. P., *The Theory of Rotating Fluids,* Cambridge at the University Press, New York (1968).

Gibson, A. H., *Hydraulics and Its Applications,* London (1912).

Harada, S., *Mechanics of Fluid* (in Japanese), Maki-Pub. Co., Tokyo (1961).

Helmholtz, H., *Über Integrale der Hydrodynamischen Gleichungen Welche den Wirbelbewegungen Entsprechen,* Crelle-Borchardt, Bd. 55 (1858) pp. 25–55.

Holmboe, J. and Forythe, G. E., and Gustin, W., *Dynamic Meteorology,* John Wiley & Sons, New York (1952).

Hunt, J. N., *Incompressible Fluid Dynamics,* Longmans, Green and Co., London (1964).

Kawanischi, K., Oobayashi, K., and Wakatani, M., *Continuous Fluid Physics,* (in Japanese), Asakura-Pub. Co., Tokyo (1981).

Kimura, R., *Introduction to Fluid Mechanics for Meteorology* (in Japanese), Tokyo Do Pub. Co., Tokyo (1985).

Kirchhoff, G., *Vorlesungen über Mathematische Physik,* Mechanik, Verlag von B. G. Teubner (1883).

Lojtchjanskij, L. G., *Mehanika Jidkosti i Gaza,* (Nauka), Moskva (1970).

Lugt, H. J., *Wirbelströmung in Natur und Technik,* G. Braun, Karlsruhe (1979).

Marey, E. J., *La Photographie du Mouvement* (1830/1904), Par Centre Georges Pompidou Musée National D' art Moderne, Paris (1977).

Matzushima, Y., *Mechanics of Deformation and Flow,* (in Japanese), Asakura-Pub. Co., Tokyo (1981).

Matsuura, S., *Guide to the Use of International System of Units(SI),* (in Japanese), J.S.A. (1976).

Mcleod, E. B., *Introduction to Fluid Dynamics,* Pergamon Press, New York (1963).

Milne-Thomson, L. M., *Theoretical Hydrodynamics,* Macmillan, New York (1938).

Miyagi, O., *Hydrodynamics* (in Japanese) ARS-Pub. Co., Tokyo (1935).

Nauman, E. B. and Buffham, B. A., *Mixing in Continuous Flow Systems,* John Wiley & Sons, New York (1983).

Ogawa, A., *Cyclone Dust Collectors* (in Japanese), Earth-Pub. Co., Tokyo (1980).

Ogawa, A., *Vortex Flows* (in Japanese), Sankaido-Pub. Co., Tokyo (1981).

Ogawa, A., *Introduction to Rheology* (in Japanese), Sankaido-Pub. Co., Tokyo (1991).

Ogawa, A., *Separation of Particles from Air and Gases,* Vols. I and II. CRC Press, Boca Raton, FL (1984).

Okada, T., *Fundamental Meteorology,* Vol. I (in Japanese), Iwanami-Pub. Co. (1937).

Panton, R. L., *Incompressible Flow,* John Wiley & Sons, New York (1984).

Peebles, F. N. and Garber, H. J., Studies on the swirling motion of water within a spherical vessel, Contract W-7405 eng., S-370, Oak Ridge National Laboratory, Tennessee (1956), courtesy of Prof. Dr. H. W. Hsu of University of Tennessee.

Poincaré, H., Théorie des Tourbillons, Leçons professées pendant le deuxieme semestre 1891-1892, Georges Carré, Editeur (Paris) (1893) (goodwill by Prof. Dr. R. Comolet, de Univ. Paris VI).

Prandtl, L. and Tietjens, O. G., *Fundamentals of Hydro and Aeromechanics,* Dover-Pub. (1934).

Putterman, S. J., *Superfluid Hydrodynamics,* North-Holland, Amsterdam (1974).

Ramsey, A. S., A Treatise on Hydromechanics, Part II, *Hydrodynamics,* G. Bell and Sons, London (1935).

Ranft, L., Die Bestimung der Strömungsverhältnisse am Eintritt eines Zyklonabscheiders mit Hilfe der Methode der konformen Abbildung, Dissertation (1955), TH-Karlsruhe (goodwill by Prof. Dr. Löffler, F. of TH-Karlsruhe).

Reynolds, A. J., *Turbulent Flows in Engineering,* John Wiley and Sons, New York (1974).

Reydon, R. F., Theoretical and Experimental Studies of Confined Vortex Flow, Thesis, Department of Chemical Eng., Supervision of Dr. W. H. Gauvin, McGill University, Canada (1978).

Scorer, R. S., *Environmental Aerodynamics,* Ellis Horwood (1978).

Silberstein, L., *Vectorial Mechanics,* Macmillan, New York (1913).

Shimanuki, A., *Turbulence and Meteorology* (in Japanese), Tokyo Do Pub., Tokyo (1982).

Soo, S. L., *Fluid Dynamics of Multi-Phase Systems,* Blaisdall Pub. Co. (1967).

Stanisic, M. M., *The Mathematical Theory of Turbulence,* Springer-Verlag, Berlin (1984).

Starr, V. P., *Physics of Negative Viscosity Phenomena,* McGraw-Hill, New York (1968).

Tani, I., *Theory of Turbulence* (in Japanese), Koseido-Pub., Tokyo (1950).

Tennekes, H. and Lumley, J. L., *A First Course in Turbulence,* MIT Press, Cambridge (1972).

Thomson, W., *Mathematical and Physical Papers,* Vol. 4 Cambridge University Press, New York (1910).

Tatsumi, T., Turbulence and chaotic phenomena in fluids, *IUTAM* (September 1983, Kyoto) North-Holland, Amsterdam (1984).

Textor, R. E., A numerical investigation of a confined vortex problem, Union Carbide, Report No. K-1732 (January 1968).

Townsend, A. A., The structure of turbulent-shear flow, Cambridge University Press, New York (1976).
Tzuneto, T., *Elasticity and Fluid* (in Japanese), Iwanami-Pub. Co., Tokyo (1983).
Villamil, R., *Motion of Liquids,* E. & F. N. Spon, London (1914).
Villat, H., Leçons sur la Théorie des Tourbillons, Gauthier-Villars et Cie, Editeurs, Paris (1930).
Walshaw, A. C. and Jobson, D. A., *Mechanics of Fluids,* Longman, London (1972).
Wieghardt, K., *Theoretical Strömungslehre,* B. G. Teubner Verlag, Germany (1965).
Wilson, F., Mansfield, F., Gibson, H., Stobart, R., and Pilsbury, K., *The Weather,* Manyflower Books, New York (1979).
Yamazaki, M., *Typhoon* (in Japanese), Tokyo-do Pub. Co., Tokyo (1983).
Yoshitake, M., *Meteorological Dynamics* (in Japanese), Chijin-Shokan Pub. Co., Tokyo (1956).
Zeytonian, R. K., Notes sur les Ecoulements Rotationnels de Fluides Parfaits, Lecture Notes in Physics-27, Springer-Verlag (1974).
Zierep, J., Ähnlichkeitsgesetze und Modellregeln der Strömungslehre, G. Braun, Karlsruhe, Germany (1972).

Akira Ogawa

THE EDITOR-IN-CHIEF

John Keith Beddow received his Ph.D. in Metallurgy from Cambridge University, Cambridge, England, in 1959. Dr. Beddow is an active lecturer and author. He has also been active as a consultant in particle technology, powder metallurgy, and metallurgy for numerous corporations and governments. He is President of Shapespeare Corporation of Iowa City, Iowa. Professor Beddow has research activities in image analysis and morphological analysis. He is married with four daughters. He and his wife, Joan, reside in Iowa City.

THE AUTHOR

Akira Ogawa, Dr. Eng., was born in 1940 in Tokyo and is currently Full Professor, Department of Mechanical Engineering, College of Engineering, Nihon University, Fukushima, Japan.

He received a Bachelor's degree in Mechanical Engineering from the College of Science and Technology, Nihon University in 1963, a Masters in Engineering from Nihon University in 1965, and a Doctorate in Engineering from Nihon University in 1972.

He worked as an assistant in and as an Assistant Professor at the Department of Mechanical Engineering in the College of Science and Technology, Nihon University from 1968 to 1973. He moved to the Department of Mechanical Engineering, College of Engineering, Nihon University as a Professor in 1975. He became a Full Professor in 1985.

Dr. Ogawa is a member of the Japan Society of Mechanical Engineering, the Japan Society of Chemical Engineering, the Japan Society of Powder Technology, the Japan Society of Precision Engineering, the Japan Society of Air Pollution, the Japan Society of Fluid Mechanics, and the Fine Particle Society (U.S.A.).

Dr. Ogawa is the author of *Cyclone Dust Collectors, Vortex Flow,* and *Rheology* (all in Japanese).

His major research interests incude the turbulent rotational flow in the confined vortex chamber, the Taylor-Görtler vortex flow in the coaxial cylindrical chamber, turbulent swirling air jets, cyclone dust collectors, rotary flow dust collectors, and the behavior of fine solid particles in tuburlent air flow.

ADVISORY BOARD

TABLE OF CONTENTS

Chapter 1

INTRODUCTION

I. NATURAL SPIRAL OF A CHAMBERED NAUTILUS

Figure 1–1 shows a beautiful shell of the chambered nautilus of a diameter about 20 cm which was bought at Carmel near San Francisco. The author was surprised that the inside structure of the chambered nautilus was composed of the simple, solid shell. In addition to this, the natural spiral line of this shell looked like one of the most beautiful formative arts of nature. Figure 1–2 shows the spiral line of this shell of which the author takes an interest in the mathematical description. Therefore, the bold line in Figure 1–3 shows the line of this shell plotted on log-log paper. This figure shows that the natural spiral of this shell does not show an Archimedes spiral which can be represented by $r = k\theta$, or a parabolic spiral which can be represented by $r^2 = k\theta$, or an hyperbolic spiral which can be represented by $r\theta = k$, and/or further an exponential form which can be represented by $r = k\theta^n$, where the symbols r, θ, k, and n denote the radius, the angle (radian), the constant value, and the exponent. However to the author's surprise, as shown by a dashed line in Figure 1–3, a logarithmic spiral line described by

$$r(\text{mm}) = 3 \exp(0.23\theta) \qquad (1\text{–}1)$$

approximates the natural spiral line of this shell.

In general, the logarithmic spiral equation can be expressed by

$$r = k \exp(\beta\theta) \qquad (1\text{–}2)$$

where β is a positive constant value. While the fundamental property of this equation may be represented by

$$dr/rd\,\theta = \beta = \text{constant} \qquad (1\text{–}3)$$

The schematic illustration of Equation 1–3 is shown in Figure 1–4. On the other hand, from the artistic and historical points of view, J. Puree[1] published an interesting book, *The Mystic Spiral-Journey of the Soul.*

II. TORNADO

From the historical point of view, hurricanes, typhoons, and also tornados have inflicted tremendous damage on human societies, and their impacts are well-documented in many kinds of historical record.[2,3]

Here the author would like to introduce an example of the tornado which was written by E. L. Van Tassel.[4] The largest and most devasting tornado in the history of western Nebraska occurred in the North Platte Valley on June 27, 1955. Destruction was spread along the path from Henry, NE, near the Wyoming-Nebraska line, where a tornado struck at 3:30 p.m., MST, down the North Platte Valley to about 12.8 km east of Scottsbluff, NE, where the largest of the observed funnels dissipated about 5:20 p.m., MST. This path was over 48 km in length.

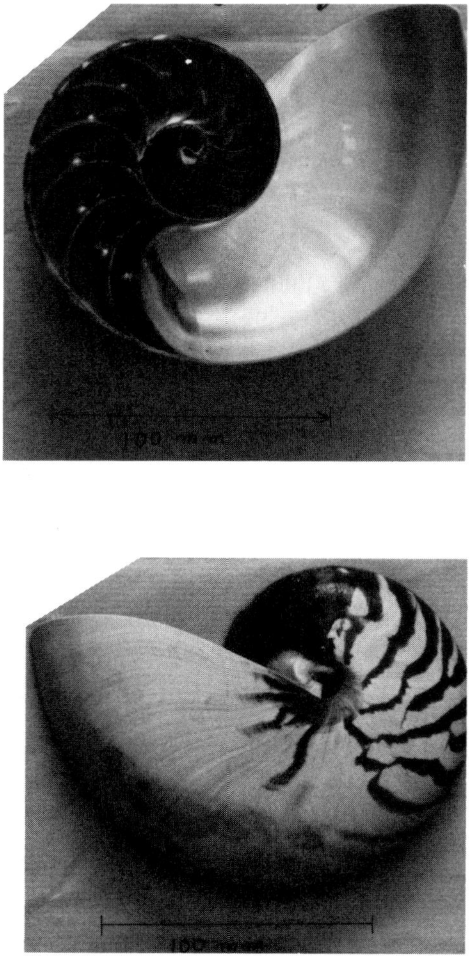

FIGURE 1–1. A shell of the chambered nautilus which was bought at Carmel near San Francisco in August 1986.

Figure 1–5 (a) shows a schematic illustration of this tornado that was measured by the scaled pictures. The view was toward west-northwest from the Scottsbluff Municipal Airport. The following values were estimated as distance to the funnel = 3380 m; width of the funnel = 226 m; height of the funnel = 870 m; and width of dust cloud = 1730 m.

Figure 1–5 (b) shows a schematic illustration of a picture taken from the Scottsbluff Municipal Airport looking southeast. The following values were estimated as distance to the funnel = 3360 m; width of midpoint of funnel = 67 m; width of funnel at top of dust cloud = 32.5 m; height of major dust cloud = 265 m; and width of major dust cloud = 1110 m.

Figure 1–5 (c) shows a schematic illustration of a picture taken from the Scottsbluff Municipal Airport looking east-southeast. The following values were estimated as distance to funnel = 5710 m; width of funnel just about ground = 93.5 m; and height of main cloud deck = 1130 m.

A ground survey did not show a pattern in the markings for obtaining measurements. However, the pictures showed systematic markings and the field was measured in order that

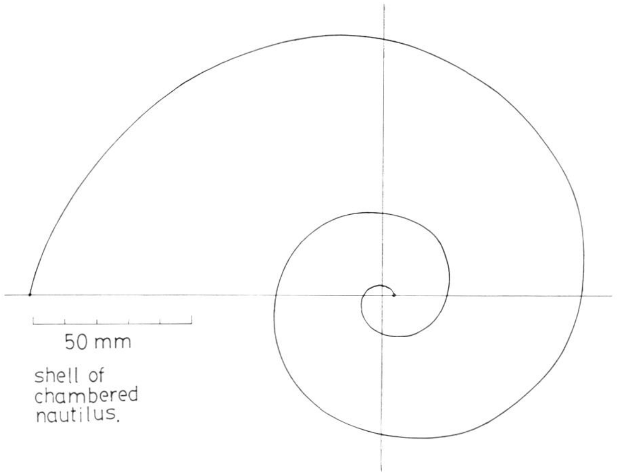

FIGURE 1–2. The shell of the chambered nautilus which was bought at Carmel near San Francisco in August 1986.

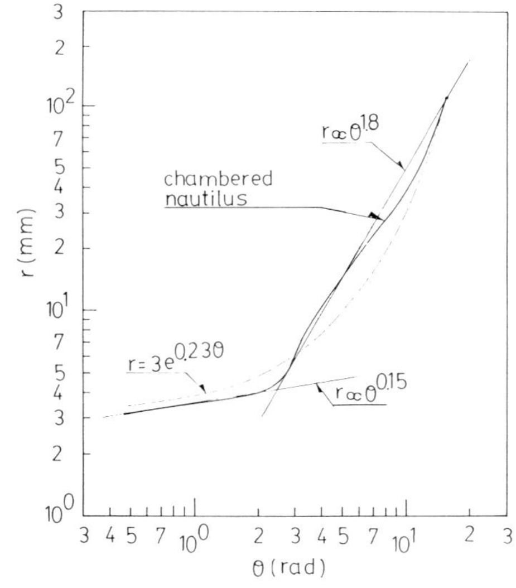

FIGURE 1–3. Development of the natural spiral line of the chambered nautilus.

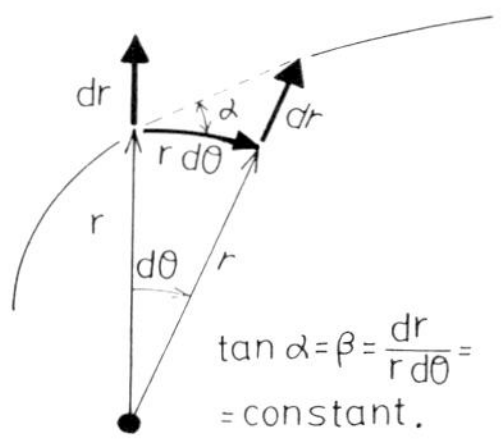

FIGURE 1–4. Schematic illustration of the logarithmic spiral equation.

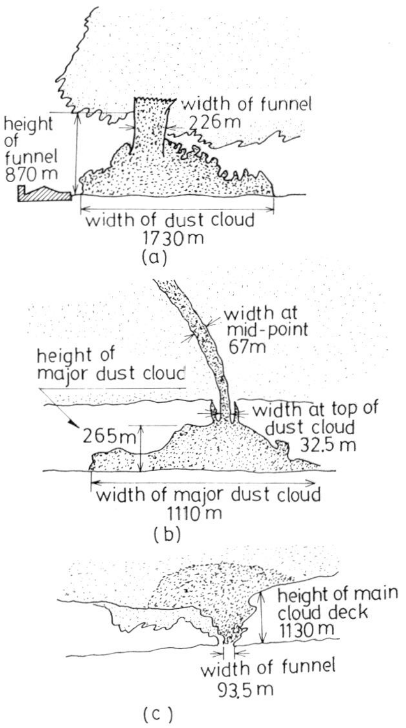

FIGURE 1–5. Tornado of western Nebraska on June 27, 1955.

the pictures could be scaled. Apparently these markings formed a near ellipse. In the following computation it was assumed that they were a true ellipse. The forward speed V_f of the tornado cloud cell as determined by the radar gave an estimated forward speed of the tornado of $V_f = 19.2$ km/h. From the scaled photographs it was determined that the diameter of the major axis of the ellipse was 70 m and that of the minor axis was 46.5 m. By averaging distances between rings at the center of the leading edge over as nearly a uniform portion of the markings as possible it was determined that this average distance between markings was 4.67 m. Then assuming the markings to have been made by something being carried with the speed of the revolving wind within the tornado funnel, the wind speed was determined from the relationship as follows

$$V = CNV_f \tag{1–4}$$

where V was wind speed, $C = 2\pi\sqrt{(a^2 + b^2)/2}$. The approximate circumference of an ellipse, N, was number of rings per unit distance; V_f was forward speed of the tornado; a was one half the major axis of an ellipse; and b was one half the minor axis of an ellipse. With this relationship the numerical values from the scaled photographs gave V = 773 km/h for the speed of the wind in the tornado. It was emphasized that this result was based on an estimated forward speed and on the unconfirmed assumption that the elliptical markings were made by an object that made exactly one revolution about the vortex for each marking.

III. FIRE WHIRLS IN OREGON

Fire whirls are very similar phenomena in appearance to dust devils.[5] Both phenomena are generated by the pronounced heating at the ground surface. Researchers who have studied

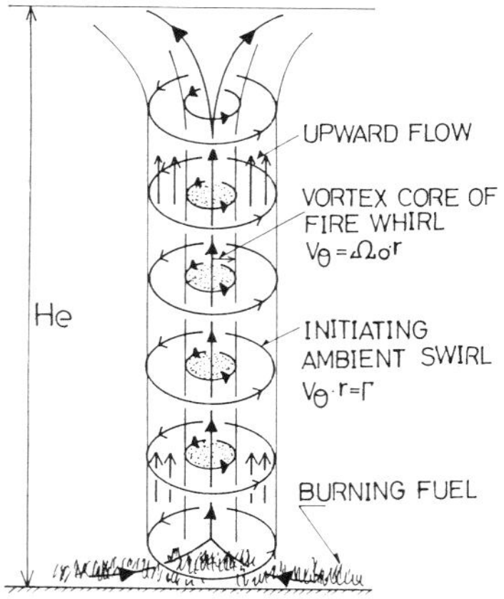

FIGURE 1–6. Flow model of fire whirl.

the theoretical and physical meanings for their development and forming process have found that heat from fires gives rise to the several conditions for the fire whirl growth. Then extremely unstable atmospheric conditions near the ground and the large heat source concentrate and enhance the whirling motion of the three-dimensional vortex. The fire provides the energy source for the continuation of the vortex motion.

For example, Muraszew and Fedele[6] have done a thoretical analysis with experiments applying the flow model of the fire whirl as shown in Figure 1–6. G. P. Hannes and S. M. Hannes observed a large whirl during the postharvest burning of the perennial grass seed fields in the Willamette valley of Oregon. This cultural practice in seed production was originally started for disease control in the Pacific Northwest approximately 36 years ago. These photographs of Figure 1–7 were taken by them at 6:00 p.m. PDT on August 7, 1983, at a postharvest burn of a perennial ryegrass field approximately 9.66 km southeast of downtown Albany, OR. The air temperature was 27.8°C; winds were variable and generally less than 8.05 km/h. The early development of the fire whirl is shown in Figure 1–7 (a). Intense surface heat combined with an atmospheric eddy probably caused it to form. It had a height of about 45.7 m at this time. Figure 1–7 (b) shows the early formation of the fire whirl. In a period of less than three minutes, the whirl intensified as shown in Figure 1–7 (c). The counterclockwise rotation speed around the axis increased as well as the height of the rotary system, which reached over about 106.7 m. And also one of the most interesting phenomena to observe is the Taylor vortices near the ground. Then, the fire whirl migrated from the main fire area and began to dissipate since its major source of heat energy was removed. Figure 1–7 (d) shows a column of the fire whirl before it died out. Finally it became an ordinary whirlwind and died out in a matter of minutes.

The author heartily thanks Gerald P. Hannes and Susan M. Hannes (California State University, Fullerton) for these precious photographs.

IV. HISTORICAL IMAGES OF SPIRALS AND VORTEX FLOW

Figure 1–8 shows a sketch of continuous spirals on the flat bronze sword.[7] This sword was excavated on the ancient Kibi route in Japan. It was presumed to be about five centuries old. The author confirmed that a design of this spiral could be described by the logarithmic spiral as

$$r \fallingdotseq \exp(0.0456 \cdot \theta) \tag{1–5}$$

FIGURE 1–7a

FIGURE 1–7b

FIGURE 1–7. Development processes of the large fire whirl in the Willamette valley of Oregon.

Figure 1–9 shows the photographs of the stone carving of vortex flow and billow in the Yiheyuan garden (Summer palace) which was constructed about 150 years ago near Beijing in China. These are two beautiful sinking vortex flows. These photos were taken by the author in October 1985 in Beijing.

V. VORTEX AND SPIRALS IN FINE ARTS

Figure 1–10 shows a Japanese painting of an ascending dragon shedding the two large strong vortices of clouds which is exhibited in the M. H. de Young Memorial Museum at

FIGURE 1–7c

FIGURE 1–7d

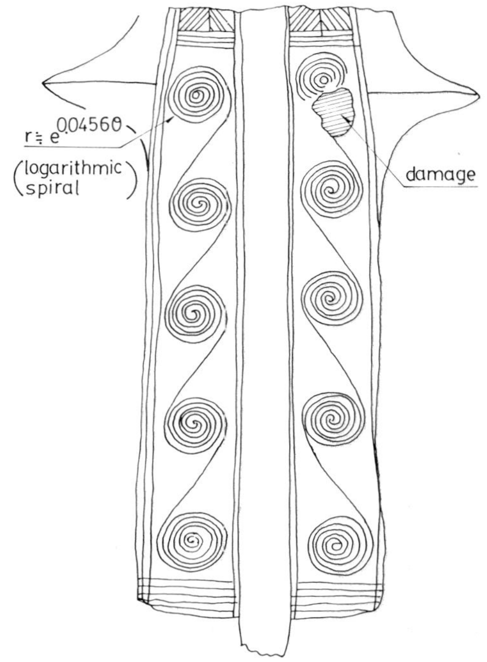

FIGURE 1–8. Sketch of the continuous spirals on the flat bronze sword in Japan.

San Francisco. The artist and the age of painting are unknown. However, it is very interesting to note that this painting of this vortex shedding is very similar to the mechanism of the stable Bénard-Kármán vortex band. Henri Bénard is one of the most famous scholars concerning the study of the Rayleigh-Bénard convection.[8–10] For example, a sketch of the historically interesting Bénard experiment is given in Figure 1–11 which shows the two kinds of the Bénard-Kármán vortex band. The diameter of the obstacles was D = 1.0, 2.0, and 4.0 mm and also the transverse velocity of these obstacles was from $V_\infty = 0.04$ m/s to $V_\infty = 0.22$ m/s in the quiet water or in the metaxylene. This paper was published in November 1909, Paris. On the same figure, a schematic representation of Bénard-Kármán vortex band for the unsymmetrical vortex band is shown in case of the uniform fluid velocity V_∞ around a stationary standing cylinder. A fluid dynamical stable condition on the unsymmetrical vortex is as follows:

$$\cosh(b\pi/a) = \sqrt{2} \quad \text{or} \quad b/a \fallingdotseq 0.281$$

where a is a distance between each vortex of the same rotation and also b is a width of the alternate vortices behind a cylinder of the diameter D. This theory was developed by Kármán. Therefore, these kinds of vortex bands are generally called a Bénard-Kármán vortex band. Concerning the stability of the Bénard-Kármán vortex band, Maue[11] described the detailed analysis of it.

Figure 1–12 shows a photograph of a Maori folkcraft in New Zealand which is exhibited in the M. H. de Young Memorial Museum at San Francisco. The author does not know why this folkcraft was filled with the engraving of these spirals; however, this folkcraft is one of the fascinating fine arts.

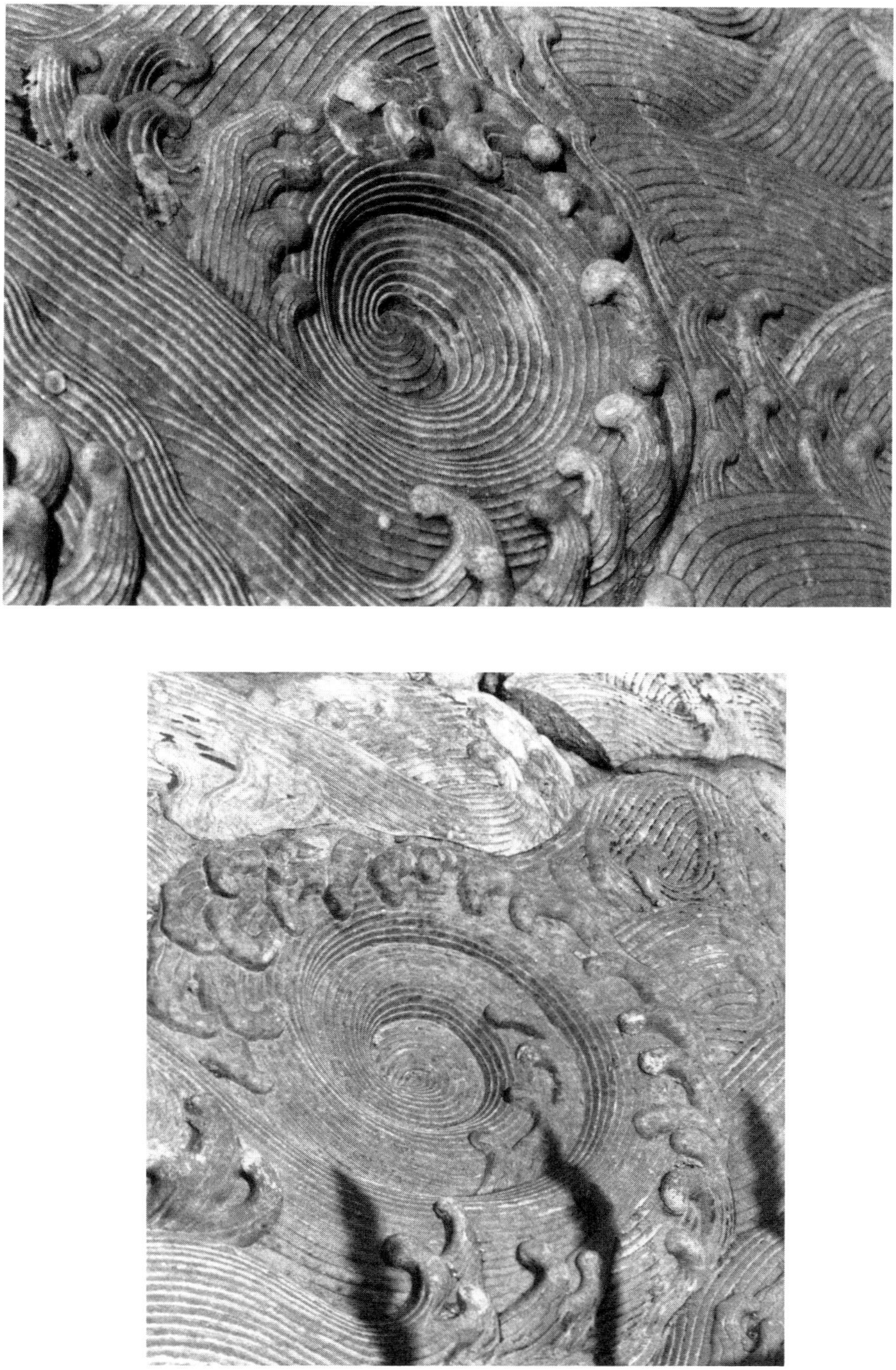

FIGURE 1–9. Stone carving of vortex in Yiheyuan Garden (Summer palace) which was constructed about 150 years ago near Beijing in China.

FIGURE 1–10. Ascending dragon producing the two large vortices which is exhibited in the M. H. de Young Memorial Museum at San Francisco. The artist and the age of painting are unknown.

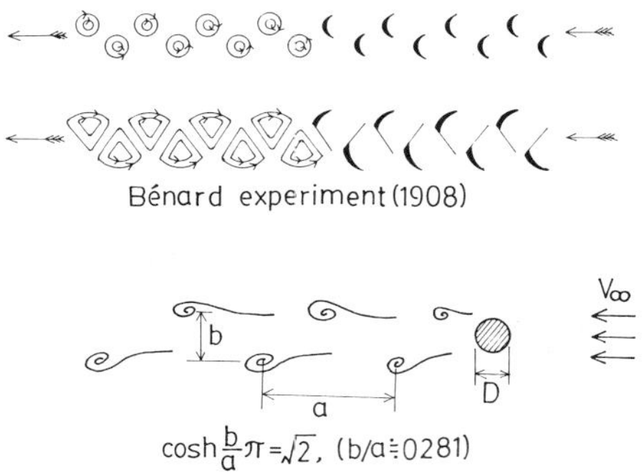

FIGURE 1–11. Bénard-Kármán vortex band and the stable Bénard-Kármán vortex band behind a cylinder.

VI. APPLICATION TO ENGINEERING

A. SEPARATION OF THE SOLID PARTICLES BY LABYRINTH SEPARATOR

Figure 1–13 shows equipment for separating solid particles of diameter X_p = 10 μm ~ 100 μm from dust-laden gas flow. This equipment is called a labyrinth separator.[12] The separation principle is as follows: the dust-laden gas flow collides with the labyrinth separator, then the solid particles of the diameter X_p = 10 μm ~ 100 μm enter the dead space of the vortex flow due to the particle inertia. After that, the rotating solid particles in the scrolls fall due to the gravity force and deposit into the dust bunker. Cleaner air flows through around the labyrinth separator and out into the atmosphere. For the separation of ash of the

FIGURE 1–12. A folkcraft at Maori in New Zealand which is exhibited in the M. H. de Young Memorial Museum at San Francisco.

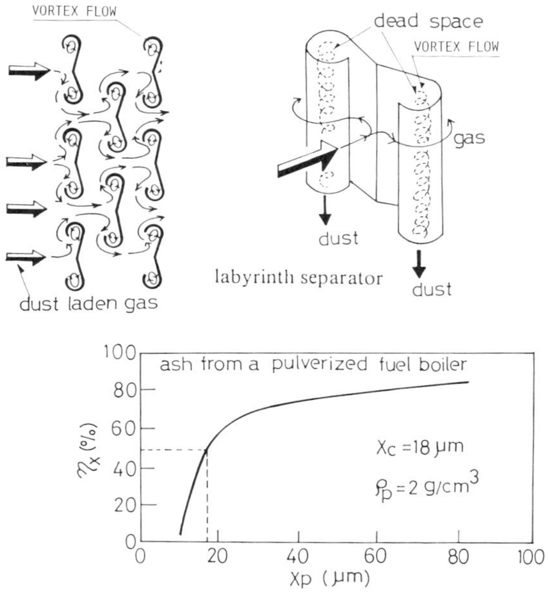

FIGURE 1–13. Separation principle and the fractional collection efficiency of the labyrinth separator.

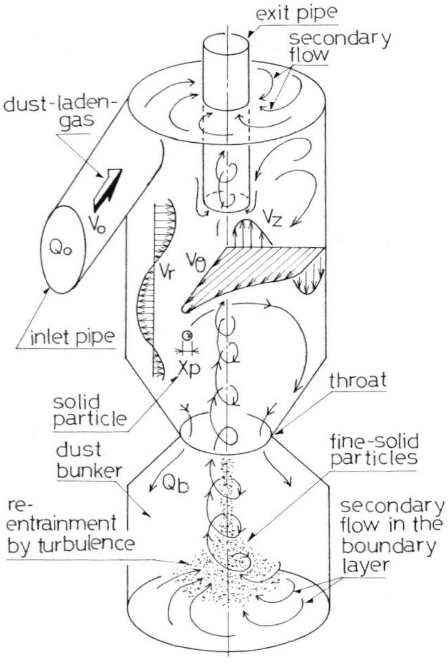

FIGURE 1–14. Schematic representation of the flow pattern in the cyclone dust collector.

density $\rho_p = 1.0$ g/cm^3 from a pulverized fuel boiler by the labyrinth separator, the cut-size (particle size) X_c corresponding to 50% of the fractional collection efficiency $\eta_x(\%)$ was about $X_c = 18$ μm. Therefore, this equipment is generally used for the particle size range $X_p = 10 \sim 100$ μm.

B. CYCLONE DUST COLLECTOR

Figure 1–14 shows a cyclone dust collector, which is one of the most familiar types throughout the world. The cyclone dust collector is an equipment for separating solid particles by centrifugal force in the turbulent rotational air flow.

The particle size X_p which can be separated is about $X_p = 2.0 \sim 50$ μm for the mean inlet velocity of the dust-laden gas of $V_o = 10 \sim 20$ m/s. Then in order to set up a standard for the particle size X_p, we customarily consider the centrifugal effect which can be defined as

$$\text{centrifugal effect} = \text{centrifugal force } (m_p V_\theta^2/r)/\text{gravitational force } (m_p g),$$

where m_p is the particle mass and V_θ is the tangential velocity. When the centrifugal effect reaches a very high value, for example, the centrifugal effect = 6000, then the particle size X_p of the particle density $\rho_p = 2.0$ g/cm^3 for enough separation can be estimated as $X_p = 2 \sim 3$ μm. As shown in Figure 1–15, particles are also carried away due to turbulent diffusion in the three-dimensional turbulent rotational air flow in the cyclone.[13,14] Concerning the historical and engineering point of view, the author has described additional details in the book *Separation of Particles from Air and Gases,* Volume II, pages 1 to 49 (CRC Press, 1984).

C. SEPARATION OF THE FINE SOLID PARTICLES BY THE VORTEX RING BEHIND A SPHERE

Goldshmid and Calvert[15] had considered an interesting separation mechanism of the fine solid particles (submicron particles) by the vortex ring behind a large spherical liquid or

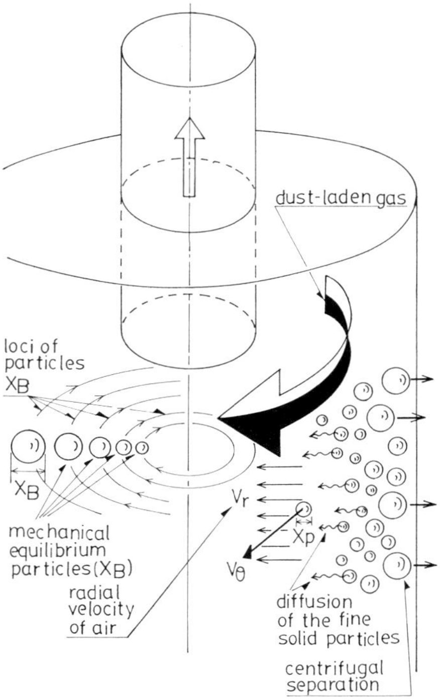

FIGURE 1–15. Schematic illustration of the mechanical equilibrium particle sizes and the separation process for the fine solid particles in the cyclone dust collector.

solid particle which is exposed to the high air velocity. As shown in Figure 1–16, the flow Reynolds number R_{ep} based on the large spherical particle of the diameter $2R_c$ is

$$R_{ep} = \frac{2R_c \cdot V_\infty}{\nu} = 1000 \sim 8000$$

so it is easy to produce the vortex ring behind the large particle. Then it is suggested that the increase in collection efficiency is due to the inertia collection of the fine particles by the back of the large spherical particle. The experimental evidence concerning flow past the large particle describes the flow in the back of the large particle as a shedding vortex loop (vortex ring) in the range of $R_{ep} = 1000 \sim 8000$. Then it is supposed that the curved streamlines act on the fine particles so as to make them change their original course and follow the streamlines. The particles which possess a high momentum do not do so, while the small particles follow the streamlines and are thrown out of the vortex on the back of the collector by the centrifugal force.

Here, in order to derive the fundamental equation of the collection efficiency, the following assumptions are introduced:

1. The concentration of the fine particles in the vortex ring is the same as in the main stream (very rapid transfer of the fine particles into the vortex).
2. The diameter ($2R_v$) of the vortex ring is half the diameter ($2R_c$) of the large particle.
3. The tangential velocity V_θ of the vortex surface at $r = R_v$ is the same as the main free velocity (V_∞).

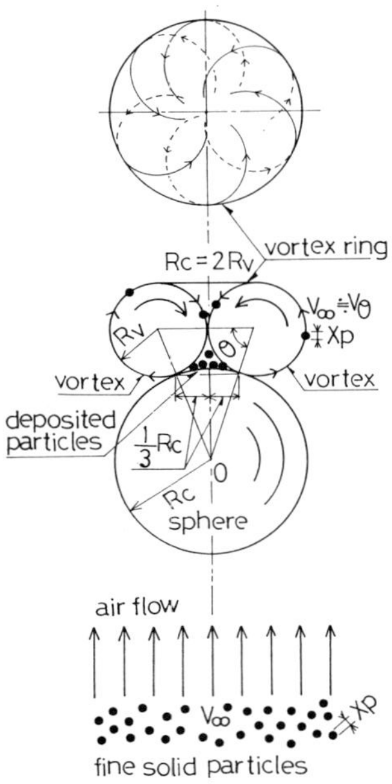

FIGURE 1–16. Separation principle by a vortex ring behind a large sphere.

4. The fine particles obey Stokes law.
5. Every fine particle which hits the surface of the large particle sticks to it.
6. The fine particles start to be deposited only after passing the central tangent point of the two parts of the vortex (deposition occurs over an angle θ).

The terminal velocity W_{sg} of the fine particle obeying Stokes law in the quiet air of the gravitational field is given by

$$W_{sg} = \frac{\rho_p \cdot X_p^2 \cdot g}{18\eta} F \qquad (1\text{–}6)$$

where F is the Cunningham slip correction factor. The radial outward settling velocity U_{pr} of the fine particles in the centrifugal field at the radius $r = R_v$ from the center of the vortex can be calculated by the mechanical equilibrium condition of the centrifugal force and of the Stokes drag force as

$$U_{pr} = \frac{W_{sg} \cdot V_\infty^2}{g \cdot R_v} \qquad (1\text{–}7)$$

At the same time, the lapse of time t in which deposition occurs during the circular distance $R_v \cdot \theta$ is given as

$$t = \frac{\theta \cdot R_v}{V_\infty} \qquad (1\text{–}8)$$

Then from the above assumption and as shown in Figure 1–17, the angle θ can be determined as

$$\theta = \cos^{-1}(1/3) = 1.23 \text{ rad}$$

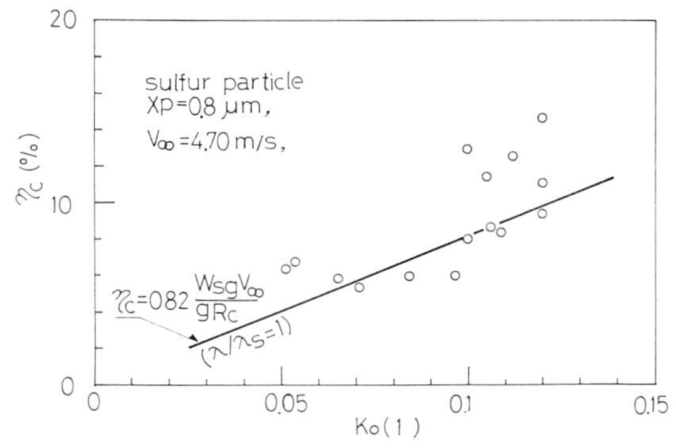

FIGURE 1–17. Collection efficiency at high air velocity for the sulfur particles on a glass sphere.

The radial distance L_v of the fine particles which will be moved in the radial direction in time t is

$$L_v = U_{pr} \cdot t = \frac{W_{sg} \cdot V_\infty \cdot \theta}{g} \tag{1–9}$$

Therefore, the total flow rate Q_v being swallowed into the vortex is

$$Q_v = \pi \cdot \frac{2}{3} \cdot R_c \cdot L_v \cdot V_\infty = \frac{2.46\pi}{3} \frac{R_c \cdot W_{sg} \cdot V_\infty^2}{g} \tag{1–10}$$

Here, using the definition of the collection efficiency

$$\eta_c = \frac{Q_v}{\pi \cdot R_c^2 \cdot v_\infty} \tag{1–11}$$

So, the following equation can be obtained as

$$\eta_c = 0.82 \frac{W_{sg} \cdot V_\infty}{g \cdot R_c} \tag{1–12}$$

This equation shows that the collection efficiency increases with increasing Stokes number $S_t = W_{sg} \cdot V_\infty/g \cdot R_c$. Figure 1–17 shows the experimental result by Goldshmid and Calvert (1963). They used the glass sphere as a large particle and the sulfur particles of $X_p = 0.8$ μm as the fine particles. The free stream velocity of air was $V_\infty = 4.7$ m/s. The ordinate shows the collection efficiency η_c(%) and the abscissa shows the corrected Stokes number K_o defined as

$$K_o = \frac{W_{sg} \cdot V_\infty}{g \cdot R_c} \cdot \frac{\lambda}{\lambda_s} \tag{1–13}$$

where λ/λ_∞ is the ratio of the distance that a particle will travel if projected into still air with velocity V_∞, to the range it would have if the drag force is that of Stokes.

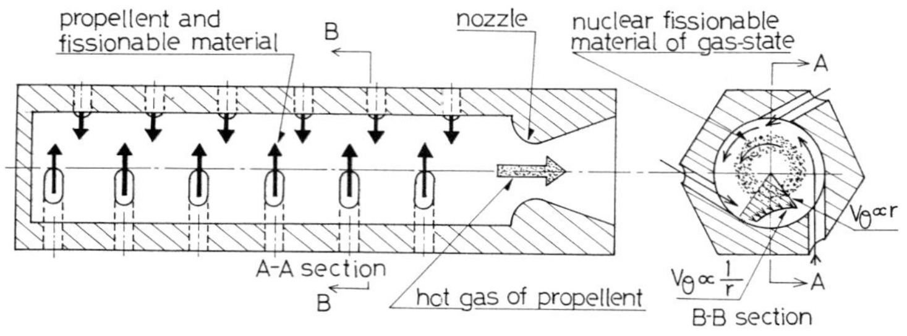

FIGURE 1–18. Schematic illustration of the nuclear rocket engine.

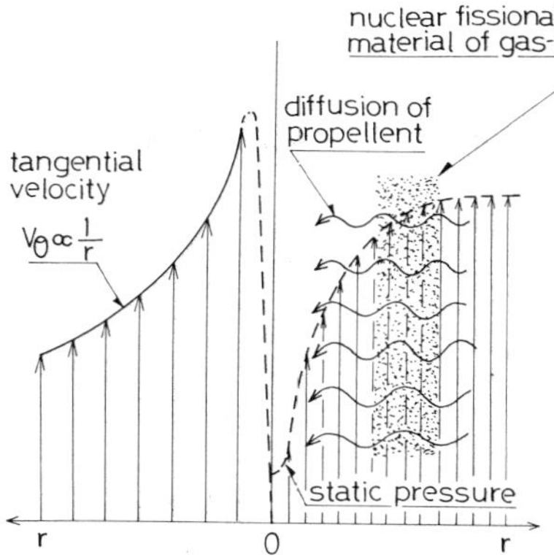

FIGURE 1–19. Mechanism of the nuclear rocket by the strong rotational vortex.

D. NUCLEAR ROCKET

Nuclear rocket contains the possibility of the very high specific impulse in comparison with the specific impulse of the chemical fuel rocket, since the energy content of the nuclear fissionable material per unit mass is very high. Therefore, it is possible to surmise a realization of the holding of the nuclear fissionable material in the gaseous state in the pressure field of the high rotational vortex flow. Kerrebrock and Meghreblian[16] had proposed the application of a vortex tube for suspending an annulus of the gaseous nuclear fuel while hydrogen flows through the tangential inlet pipes and is heated by the nuclear fuel in the nuclear rocket.

Here their idea is explained by a schematic illustration of Figure 1–18. The low molecular propellant and the fissionable material are introduced into the vortex chamber by the tangential inlet pipes. Because the mixed gas of both materials is easily carried away through the convergent-divergent nozzle, so from the economical point of view the loss of the nuclear fissionable material cannot be tolerated. Consequently, it is necessary to contain the expensive fissionable material in the rocket engine. On the other hand, the propellant without mixing with the fissionable material passes through this material and is heated by thermal radiation from the fissionable material. The propellant then flows out through the nozzle. Now, in order to realize this complicated flow mechanism, the high static pressure gradient being produced by the strong rotational vortex flow is necessary to keep the annular fissionable material of gas state with the high concentration in the vortex flow as shown in Figure 1–19. At the same time, the propellant, such as hydrogen, absorbs the energy of the nuclear fission during passing through these heavy nuclear fissionable materials, and the propellant diffuses toward the inward direction in the rocket engine.

MAIN SYMBOLS

m_p (g)	particle mass
N(1)	number of rings per unit distance
n(1)	exponent
R_c(m)	radius of an obstacle
R_{ep}(1)	Reynolds number
r(m)	radius
S_t(1)	Stokes number
V(m/s)	wind velocity
V_f(m/s)	forward speed of the tornado
V_θ(m/s)	tangential velocity
V_∞(m/s)	uniform velocity of fluid
X_c(μm)	cut-size
X_p(μm)	particle size
η_x(%)	fractional collection efficiency
θ(rad)	angle
ν(m^2/s)	kinematic viscosity of fluid
ρ_p(g/cm^3)	density of a particle

REFERENCES

1. **Puree, J.,** *The Mystic Spiral — Journey of the Soul,* Thames and Hudson Ltd., London, 1974.
2. **Watson, L.,** *Heaven's Breath* (Japanese translation by Kobata, K.), Kawaide-Shobo-Pub. Co., Tokyo, 1985.
3. **Taguchi, K.,** *History of the Meteorology on Europe and America* (in Japanese), Kokin-Shoin-Pub. Co., Tokyo, 1942.
4. **Tassel, E. L.,** The north Platte Valley tornado outbreak of June 27, 1955, *Mon. Weather Rev.,* November 1955, 255.
5. **Hannes, G. P. and Hannes, S. M.,** Fire whirls in Oregon, *Weatherwise,* October 1984, 267.
6. **Muraszew, A. and Fedele, J. B.,** The fire whirl phenomenon, *Combust. Flame,* 34, 29, 1979.
7. **Nakamura, A., Makabe, T., and Makabe, Y.,** KIBI-Route Scenery of Ancient History in Japan (in Japanese), Iwanami-Book-Co., Tokyo, 1986.
8. **Bénard, M. H.,** Sur les Tourbillons Alternes Derrière un Obstacle, Extraits des Comptes Rendus de l' Academie des Sciènces, Paris, 1929.
9. **Bénard, M. H.,** Les Tourbillons Cellulaires Dans Une Nappe Liquide Propageant De La Chaleur Par Convection, En Régime Permanent, Théses, A La Faculté des Sciences de Paris, 1901.
10. **Velarde, M. G. and Normand, C., Convection,** *Sci. Am.,* 243 (No. 1) 92, 1980.
11. **Maue, A. W.,** Zur Stabilität der Karmanschen Wirbelstraße, *Z.A.M.M.,* 20 (No. 3), 129, 1940.
12. **Parker, A.,** *Industrial Air Pollution Handbook,* McGraw-Hill, New York, 1978.
13. **Ogawa, A.,** Theoretical analysis of the concentration distributions, preprint of *J. Soc. Mech. Eng.,* No. 64 (October, 1986), No. 860-7, 114.
14. **Fortier, A.,** Méchanique des Suspensions, Masson et Cie, Paris, 1967.
15. **Goldschmid, Y. and Calvert, S.,** Small Particle collection by supported liquid drops, *AIChEJ,* 9 (No. 3), 352, 1963.
16. **Kerrebrock, J. L. and Meghreblian, R. V.,** Vortex containment for the gaseous fission rocket, *J. Aerosp. Sci.,* 28, 710, 1961.

Chapter 2

HYDROSTATICS

I. INTRODUCTION

In this chapter we review hydrostatics. The fundamental concepts concerning potential, equipressure surface, and line of force are described. A practical example for the motion of water in the rotating cylinder with the constant angular velocity around the vertical axis is explained in detail.[1,2]

II. POTENTIAL

We define the following symbols as

X, Y, Z = component of the external forces (N/kg) acting on a unit mass of fluid for the positive directions of x, y, z axis,
ρ = density (kg/m^3) of fluid,
p = pressure (Pa = N/m^2),
g = acceleration (m/s^2) of gravity.

Here, considering a small cube (fluid particle) which has microscopic lengths dx, dy, and dz along the x, y, and z axes in quiet fluid as shown in Figure 2–1, an equation of the mechanical equilibrium of this small cube for the x-axis can be derived. The external force can be written as $\rho \cdot dx \cdot dy \cdot dz \cdot X$. Then the pressure force on a surface BCDG for the positive direction along the x-axis is $dy \cdot dz \cdot p$ and the pressure force on a surface AHEF for the negative direction along the x-axis is

$$dy \cdot dz \cdot \left(p + \frac{dp}{dx} dx\right)$$

Therefore, in order to keep this small cube in a stable condition, the following mechanical equilibrium along the direction of the x-axis must be satisfied as

$$\rho \cdot dx \cdot dy \cdot dz \cdot X + p \cdot dy \cdot dz - \left(p + \frac{\partial p}{\partial x} dx\right) \cdot dy \cdot dz = 0 \qquad (2\text{–}1)$$

Consequently, Equation 2–1 becomes

$$\frac{\partial p}{\partial x} = \rho \cdot X \qquad (2\text{–}2)$$

In the same way for the y- and z- directions of y and z axes, the conditions of the mechanical equlibriums can be written as

$$\frac{\partial p}{\partial y} = \rho \cdot Y \qquad (2\text{–}3)$$

$$\frac{\partial p}{\partial z} = \rho \cdot Z \qquad (2\text{–}4)$$

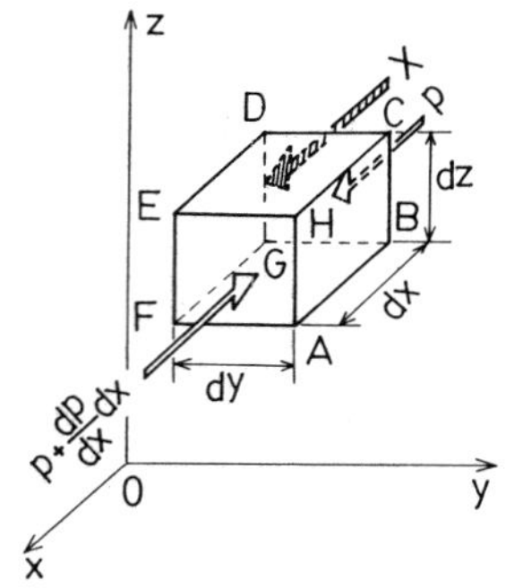

FIGURE 2–1. Schematic illustration of the mechanical equilibrium for the x-direction.

Here $\partial p/\partial x$, $\partial p/\partial y$, and $\partial p/\partial z$ can be represented by the gradient of p(pressure) for the x, y, and z directions, respectively. Therefore the external forces X, Y, and Z are proportional to the gradient of pressure which can be written as

$$\text{grad } p = \rho \mathbf{F} \quad \text{or} \quad (\nabla p = \rho \mathbf{F}) \tag{2–5}$$

where **F** is a vector of the external forces as

$$\mathbf{F} = X\mathbf{i} + Y\mathbf{j} + Z\mathbf{k} \tag{2–6}$$

and also ∇ (nabla) is a vector defined as

$$\nabla = \mathbf{i}\frac{\partial}{\partial x} + \mathbf{j}\frac{\partial}{\partial y} + \mathbf{k}\frac{\partial}{\partial z} \tag{2–7}$$

In this equation, **i**, **j**, and **k** represent unit vectors along the x, y, and z axis. Hereafter, we use the bold faced types for representing the vector quantities. Therefore if we transform tediously Equation 2–5 into a concrete form, Equation 2–5 becomes

$$\mathbf{i}\frac{\partial p}{\partial x} + \mathbf{j}\frac{\partial p}{\partial y} + \mathbf{k}\frac{\partial p}{\partial z} = \rho(X\mathbf{i} + Y\mathbf{j} + Z\mathbf{k}) \tag{2–8}$$

Here assuming that the external forces X, Y, and Z are related to the location, so can they be represented by the partial differential coefficient of the function which depends on x, y, and z, respectively. Then if we represent this function by a symbol Φ, so X, Y, and Z can be written as

$$X = -\frac{\partial \Phi}{\partial x}, \quad Y = -\frac{\partial \Phi}{\partial Y}, \quad Z = -\frac{\partial \Phi}{\partial z} \tag{2–9}$$

This function Φ is usually called a potential function or simply potential. Then from this equation, we can obtain the following relations

$$\frac{\partial X}{\partial y} = \frac{\partial Y}{\partial x}, \quad \frac{\partial Y}{\partial z} = \frac{\partial Z}{\partial y}, \quad \frac{\partial Z}{\partial x} = \frac{\partial X}{\partial z} \tag{2–10}$$

Next, if we assume that the pressure p is a function of the location (x, y, z), so the equation of dp can be written as

$$dp = \frac{\partial p}{\partial x} dx + \frac{\partial p}{\partial y} dy + \frac{\partial p}{\partial z} dz \tag{2–11}$$

Substituting Equations 2–2, 2–3, and 2–4 into Equation 2–11, Equation 2–11 becomes

$$dp = \rho(X\,dx + Y\,dy + Z\,dz) \tag{2–12}$$

Further substituting Equation 2–9 into Equation 2–12, Equation 2–12 becomes

$$dp = -\rho\left(\frac{\partial \Phi}{\partial x} dx + \frac{\partial \Phi}{\partial y} dy + \frac{\partial \Phi}{\partial z} dz\right) \tag{2–13}$$

Here the potential Φ depends on the location of x, y, and z; so the equation in the parentheses of the right hand side is equal to the total differential dp. Therefore Equation 2–13 becomes

$$dp = -\rho\, d\Phi \tag{2–14}$$

Namely, the increment of pressure dp is proportional to the decline of potential.

Further from Equations 2–12 and 2–14 we can obtain

$$X\,dx + Y\,dy + Z\,dz = -d\Phi \tag{2–15}$$

where X dx, Y dy, and Z dz mean work or energy per unit mass of body. Therefore the potential has the same physical meaning as work or energy. Accordingly we can call Φ a potential energy.

In addition to this, X, Y, and Z are equal to the unit mass times d^2x/dt^2, the unit mass times d^2y/dt^2, and the unit mass times d^2z/dt^2, so that the following relation can be obtained

$$X\,dx + Y\,dy + Z\,dz = \frac{d^2x}{dt^2} dx + \frac{d^2y}{dt^2} dy + \frac{d^2z}{dt^2} dz =$$

$$= \frac{dx}{dt} d\left(\frac{dx}{dt}\right) + \frac{dy}{dt} d\left(\frac{dy}{dt}\right) + \frac{dz}{dt} d\left(\frac{dz}{dt}\right) =$$

$$= \frac{1}{2}\left\{\left(\frac{dx}{dt}\right)^2 + \left(\frac{dy}{dt}\right)^2 + \left(\frac{dz}{dt}\right)^2\right\} \tag{2–16}$$

Here a symbol V means a resultant velocity. So we can write

$$V^2 = \left(\frac{dx}{dt}\right)^2 + \left(\frac{dy}{dt}\right)^2 + \left(\frac{dz}{dt}\right)^2 = V_x^2 + V_y^2 + V_z^2 \tag{2–17}$$

where $V_x = dx/dt$, $V_y = dy/dt$, and $V_z = dz/dt$. Consequently, Equation 2–16 becomes

$$X\,dx + Y\,dy + Z\,dz = \frac{1}{2} d(V^2) = d\left(\frac{V^2}{2}\right) \tag{2–18}$$

Therefore Equation 2–18 becomes

$$d\,\Phi = -d\left(\frac{V^2}{2}\right) \tag{2–19}$$

where $V^2/2$ means kinetic energy per unit mass of body. Therefore from Equation 2–19 we can obtain

$$d\Phi + d\left(\frac{V^2}{2}\right) = 0 \tag{2–20}$$

or

$$\Phi + \frac{V^2}{2} = \text{constant} \tag{2–21}$$

Finally, the increment of the potential energy brings on the decline of the kinetic energy and the reverse is also true. Then these phenomena can be called a conservative system.

III. EQUIPRESSURE SURFACE

A surface which consists of an array of points of the pressures of the same values in space is called an equipressure surface. Then the equipressure surface is the surface of the constant pressure, so that the equipressure surface can be represented as $dp = 0$. Therefore, from Equation 2–12, we can obtain

$$X\,dx + Y\,dy + Z\,dz = \mathbf{F} \cdot d\mathbf{L} = 0 \tag{2–22}$$

where $\mathbf{F}$ ($= X\,\mathbf{i} + Y\,\mathbf{j} + Z\,\mathbf{k}$) means the vector of force and $\mathbf{L}$ ($= \mathbf{i}\,dx + \mathbf{j}\,dy + \mathbf{k}\,dz$) means the vector of the line element on the surface of the equipressure. Now the surface of $dp = 0$ is equivalent to $d\Phi = 0$ by Equation 2–14. Therefore the equipressure surface corresponds to the surface of the equipotential, so that the surface of the constant potential can be called an equipotential surface. Consequently, in the conservative system, the equipressure surface is equivalent to the equipotential surface.

IV. LINE OF FORCE

We can imagine an imaginary line which is connected with the directions of the acting forces in order in the given space. This line is generally called a line of force. The direction of force is always tangential to the line of force. Then denoting that the small element of this line of force is ds and the acting force is $\mathbf{F}(= X\,\mathbf{i} + Y\,\mathbf{j} + Z\,\mathbf{k})$, so the force $\mathbf{F}$ is always tangential to ds. Therefore the direction cosines of ds

$$\frac{dx}{ds}, \quad \frac{dy}{ds}, \quad \frac{dz}{ds}$$

are equivalent to the direction cosines of the force $\mathbf{F}$

$$\frac{X}{\mathbf{F}}, \quad \frac{Y}{\mathbf{F}}, \quad \frac{Z}{\mathbf{F}}.$$

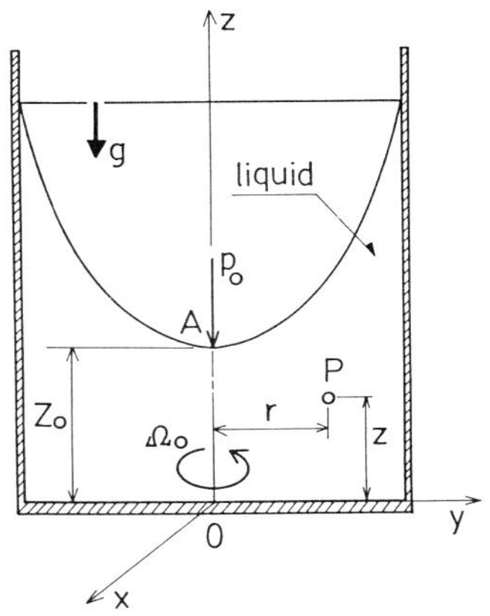

FIGURE 2–2. Illustration of motion of water in the rotating cylinder.

Consequently the following relation can be obtained:

$$\frac{dx}{X} = \frac{dy}{Y} = \frac{dz}{Z} \tag{2–23}$$

This equation is the differential equation of the line of force.

On the other hand, dividing Equation 2–22 by **F** ds, the following equation can be obtained:

$$\frac{X}{\mathbf{F}}\frac{dx}{ds} + \frac{Y}{\mathbf{F}}\frac{dy}{ds} + \frac{Z}{\mathbf{F}}\frac{dz}{ds} = 0 \tag{2–24}$$

Therefore the above equation shows that the direction of the force **F** is perpendicular to the direction of ds. Then we can obtain the following relationships:

1. The external force is acting perpendicular to the equipressure surface.
2. The line of force pierces through the equipressure surface at a right angle.
3. The equipressure surface is a surface which cuts off the line of force at a right angle in order.

V. PRACTICAL EXAMPLE

As shown in Figure 2–2, we consider the motion of water in the rotating cylinder with a constant angular velocity Ω_o around the vertical axis z. The point 0 is the origin and the two axes x and y are laid in the horizontal plane and the positive direction of the z axis is the upward direction. Since the external force acting on the arbitrary point P is the centrifugal force and the gravitational force, we can obtain the following relations

$$X = \Omega_o^2 \cdot x \quad , \quad Y = \Omega_o^2 \cdot y \quad , \quad Z = -g \tag{2–25}$$

These relations satisfy the condition of Equation 2–10 so that this phenomenon shows the effect of the potential. Substituting Equation 2–25 into Equation 2–12, we obtain the equation

$$dp = \rho(\Omega_o^2 x\, dx + \Omega_o^2 y\, dy - g\, dz) \tag{2–26}$$

Integrating Equation 2–26 and giving the boundary condition at a point A that the pressure p at the point x = 0, y = 0, and z = z_o is p = p_0, the following equation can be obtained

$$p - p_0 = \frac{\rho\, \Omega_o^2(x^2 + y^2)}{2} - g\rho(z - z_0) \tag{2–27}$$

This equation gives the pressure at a point P in the rotating water. Here, denoting that the radial distance of a point P is r($r^2 = x^2 + y^2$), Equation 2–27 becomes

$$p - p_0 = \frac{\rho\Omega_o^2 r^2}{2} - g\rho(z - z_0) \tag{2–28}$$

Now the equipressure surface is the surface corresponding to dp = 0; so the differential equation for determining this surface can be written from Equation 2–26 as

$$\Omega_o^2 x\, dx + \Omega_o^2 y\, dy - g\, dz = 0$$

Integrating this equation, we obtain the following equation

$$\frac{\Omega_o^2(x^2 + y^2)}{2} - g\,z = C \quad , \quad \text{or} \quad \frac{\Omega_o^2 r^2}{2} - g\,z = C$$

Consequently the equipressure surface shows the paraboloid of revolution and also the free surface is one of the paraboloid of revolution.

The line of force can be obtained by solving the Equation 2–23 as follows:

$$\frac{dx}{\Omega_o^2 x} = \frac{dy}{\Omega_o^2 \cdot y} = -\frac{dz}{g} \tag{2–29}$$

Hence the equation of the line of force on the y-z plane can be obtained from the above equation as

$$\frac{dy}{y} = -\frac{\Omega_o^2}{g}\, dz$$

The solution of the above equation can be shown to be

$$y = C_2 \exp\left(-\frac{\Omega_o^2}{g} \cdot z\right) \tag{2–30}$$

Therefore Equation 2–30 shows the property of the line of force which depends on the constant of the integration C_2 on the y-z plane. In addition to this, the line of force on the x-y plane can be determined by Equation 2–29 as follows:

$$\frac{dx}{x} = \frac{dy}{y} \tag{2–31}$$

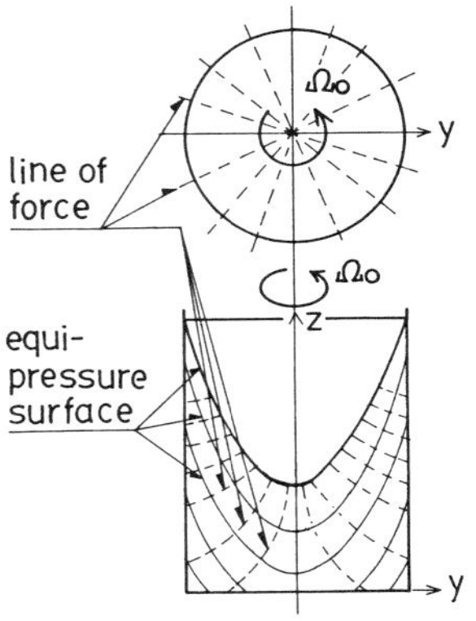

FIGURE 2–3. Schematic illustration of the equipressure surface and of the line of force for the rotating liquid in the rotating cylinder.

Integrating Equation 2–31, we can obtain the following equation:

$$y = C_3 x \tag{2–32}$$

The schematic illustration of the equipressure surface and of the line of force is shown in Figure 2–3.

MAIN SYMBOLS

F(N/kg);	vector of the external force per unit mass
g(m/s^2);	acceleration of gravity
i(l);	unit vector for the x-direction
j(l);	unit vector for the y-direction
k(l);	unit vector for the z-direction
L(m);	vector of a line element
p(Pa);	static pressure
r(m);	radius
V(m/s);	resultant velocity
V_x(m/s);	velocity of x-component
V_y(m/s);	velocity of y-component
V_z(m/s);	velocity of z-component
X(N/kg);	external force per unit mass of fluid
x(m);	distance for the x-direction
Y(N/kg);	external force per unit mass of fluid
y(m);	distance for the y-direction
Z(N/kg);	external force per unit mass of fluid
z(m)	distance for the z-direction
Φ(Nm/kg);	potential
ρ(kg/m^3);	density of fluid
Ω_o(rad/s);	angular velocity

REFERENCES

1. **Miyagi, O.,** Hydrodynamics (in Japanese), ARS Pub. Co., Tokyo, 1935.
2. **Fox, R. W. and Mcdonald, A. T.,** *Introduction to Fluid Mechanics,* John Wiley & Sons, New York, 1973.

Chapter 3

FUNDAMENTAL IDEAS OF FLUID FLOW

I. INTRODUCTION

In this chapter the fundamental concepts of the equation of motion, of the transformation and of the rotation of the small fluid particle, of irrotational flow, of Thomson-Rankine vortex, of the Bernoulli equation, and also of the application of the Bernoulli equation for flow rate of gas flow into the cyclone dust collectors are described in detail.[1]

II. CYLINDRICAL COORDINATES

In the studies of water flow in the hydraulic turbine or in a pump and also in studies of air flow in the turbo-compressor or in the fan, since the fluid flow always rotates around an axis, it is very convenient to use cylindrical coordinates.

As shown in Figure 3–1, we denote one axis which is located on the center of the rotating fluid to be the z-axis. Then, considering one plane perpendicular to the z-axis, we also denote a point of intersection between the z-axis and that plane to be the origin of these coordinates. In addition to this, the positive direction of these axes can be defined for the direction of the right hand rotating screw. The velocity components which are mutually at right angles are denoted as follows:

V_r; radial velocity

V_θ; tangential velocity

V_z; axial velocity

Then the resultant velocity **V** can be obtained as

$$\mathbf{V} = \sqrt{V_r^2 + V_\theta^2 + V_z^2} \tag{3–1}$$

Also, the relationship between the cylindrical and Cartesian coordinates can be defined as

$$x = r \cdot \cos\theta \quad , \quad y = r \cdot \sin\theta \quad \text{and} \quad z = z \tag{3–2}$$

III. EQUATION OF CONTINUITY

As shown in Figure 3–2, in order to calculate the accumulated mass of fluid per unit time in the small volume $r \cdot dr \cdot d\theta \cdot dz$, the accumulated mass into the radial, tangential, and axial directions can be determined, respectively. For the radial direction, the accumulated mass per unit time can be calculated as

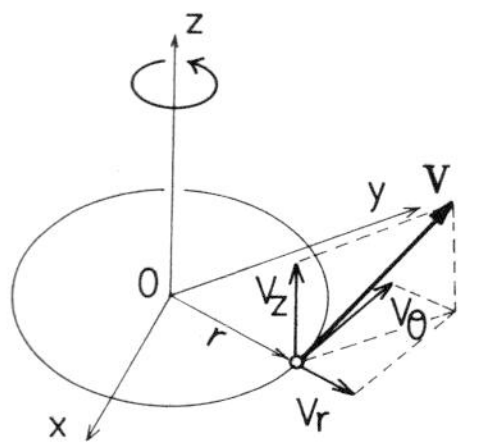

FIGURE 3–1. Cylindrical coordinates and velocity components.

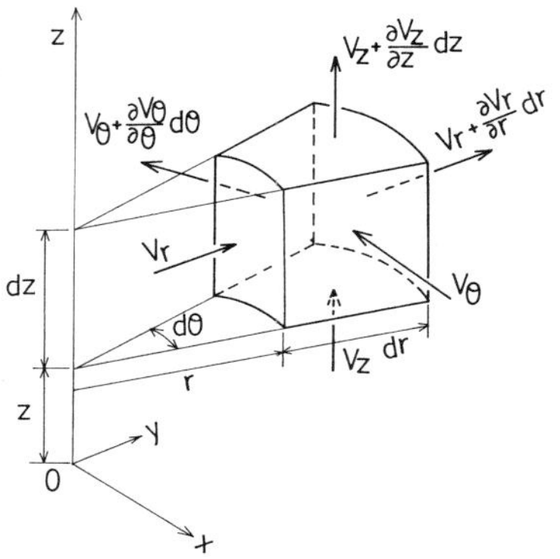

FIGURE 3–2. Schematic illustration of the velocity component on the small fluid particle.

$$\rho V_r \cdot r \cdot d\theta \cdot dz - \left(\rho + \frac{\partial \rho}{\partial r} dr\right)\left(V_r + \frac{\partial V_r}{\partial r} dr\right)(r + dr) \cdot d\theta \cdot dz =$$

$$\doteqdot - \left(V_r \cdot r \cdot \frac{\partial \rho}{\partial r} + \rho \cdot r \cdot \frac{\partial V_r}{\partial r} + \rho V_r\right) dr \cdot d\theta \cdot dz = \tag{3–3}$$

$$- \frac{\partial}{\partial r} (\rho \cdot r \cdot V_r)\, dr \cdot dr\theta \cdot dz$$

and for the circumferential (tangential) direction

$$\rho V_\theta\, dr \cdot dz - \left(\rho + \frac{\partial \rho}{\partial \theta} d\theta\right)\left(V_\theta + \frac{\partial V_\theta}{\partial \theta} d\theta\right) dr \cdot dz =$$

$$\doteqdot - \left(\frac{\partial \rho}{\partial \theta} V_\theta + \rho \frac{\partial V}{\partial \theta} \partial V_\theta\right) d\theta \cdot dr \cdot dz = - \frac{\partial}{\partial \theta} (\rho V_\theta)\, dr \cdot d\theta \cdot dz, \tag{3–4}$$

further for the axial direction

$$\rho V_z\, r \cdot d\theta \cdot dr - \left(\rho + \frac{\partial \rho}{\partial z} dz\right)\left(V_z + \frac{\partial V_z}{\partial z} dz\right) r \cdot d\theta \cdot dr =$$

$$= - \left(\frac{\partial \rho}{\partial z} V_z + \rho \frac{\partial V_z}{\partial z}\right) r \cdot d\theta \cdot dr \cdot dz = - \frac{\partial}{\partial z} (\rho V_z)\, r \cdot dr \cdot d\theta \cdot dz, \tag{3–5}$$

Since the total accumulated fluid mass per unit time flowing into this small volume $r \cdot d\theta \cdot dr \cdot dz$ is the summation of the above-stated three directions, we obtain

$$-\left[\frac{1}{r}\frac{\partial}{\partial r}(\rho \cdot V_r \cdot r) + \frac{1}{r}\frac{\partial}{\partial \theta}(\rho \cdot r \cdot V_\theta) + \frac{\partial}{\partial z}(\rho \cdot V_z)\right] r \cdot dr \cdot d\theta \cdot dz \quad (3\text{–}6)$$

When the density of fluid at the initial state is ρ, the density after the unit time becomes

$$\rho + \frac{\partial \rho}{\partial t} \quad (3\text{–}7)$$

Therefore, the accumulated fluid mass per unit mass per unit time in this small volume can be written as

$$\left[\left(\rho + \frac{\partial \rho}{\partial t}\right) - \rho\right] r \cdot d\theta \cdot dr \cdot dz = \frac{\partial \rho}{\partial t} \cdot r \cdot d\theta \cdot dr \cdot dz \quad (3\text{–}8)$$

Consequently the value of Equation 3–8 must be equal to the value of Equation 3–6. So we can obtain the following equation

$$\frac{\partial \rho}{\partial t} + \frac{1}{r}\frac{\partial}{\partial r}(\rho \cdot r \cdot V_r) + \frac{1}{r}\frac{\partial}{\partial \theta}(\rho \cdot V_\theta) + \frac{\partial}{\partial z}(\rho \cdot V_z) = 0 \quad (3\text{–}9)$$

When the density of fluid is incompressible fluid, the density ρ is constant ($\partial\rho/\partial t = 0$). Accordingly Equation 3–9 becomes

$$\frac{\partial}{\partial r}(r \cdot V_r) + \frac{1}{r}\frac{\partial}{\partial \theta}(r \cdot V_\theta) + \frac{\partial}{\partial z}(r \cdot V_z) = 0 \quad (3\text{–}10)$$

IV. EQUATION OF MOTION

Differentiating the first and second equations of Equation 3–2 by time t, we obtain the following equation

$$U = \frac{dx}{dt} = \frac{\partial r}{\partial t}\cos\theta - r \cdot \sin\theta \cdot \frac{\partial \theta}{\partial t}$$
$$V = \frac{dy}{dt} = \frac{\partial r}{\partial t}\sin\theta + r \cdot \cos\theta \cdot \frac{\partial \theta}{\partial t} \quad (3\text{–}11)$$

Then, as previously stated, $\partial r/\partial t = V_r$ and $r \cdot \partial\theta/\partial t = V_\theta$, so that Equation 3–11 becomes

$$U = V_r \cdot \cos\theta - V_\theta \cdot \sin\theta$$

$$V = V_r \cdot \sin\theta + V_\theta \cdot \cos\theta \quad (3\text{–}12)$$

Again differentiating Equation 3–12 by time t, we obtain the following equation,

$$\frac{dU}{dt} = \frac{\partial V_r}{\partial t} \cdot \cos\theta - V_r \cdot \sin\theta \frac{\partial \theta}{\partial t} - \frac{\partial V_\theta}{\partial t} \cdot \sin\theta - V_\theta \cdot \cos\theta \cdot \frac{\partial \theta}{\partial t} =$$

$$= \frac{\partial V_r}{\partial t} \cdot \cos\theta - \frac{V_r V_\theta}{r} \cdot \sin\theta - \frac{\partial V_\theta}{\partial t} \cdot \sin\theta - \frac{V_\theta^2}{r} \cdot \cos\theta \quad (3\text{–}13)$$

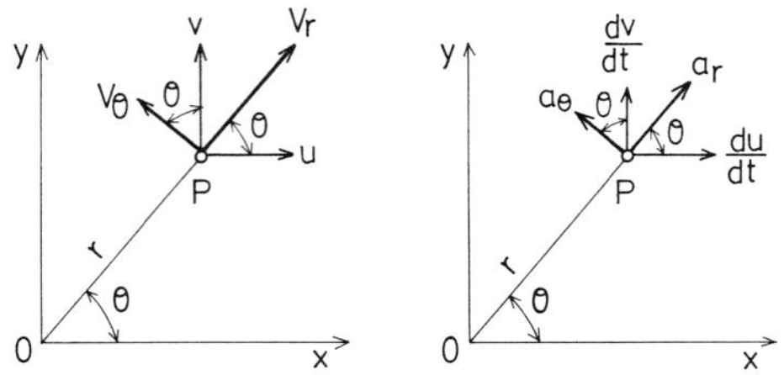

FIGURE 3–3. Schematic illustration of the velocity components and of the acceleration components.

$$\frac{dV}{dt} = \frac{\partial V_r}{\partial t} \cdot \sin\theta + V_r \cdot \cos\theta \cdot \frac{\partial \theta}{\partial t} + \frac{\partial V_\theta}{\partial t} \cdot \cos\theta - V_\theta \cdot \sin\theta \cdot \frac{\partial \theta}{\partial t} =$$

$$= \frac{\partial V_r}{\partial t} \cdot \sin\theta + \frac{V_r V_\theta}{r} \cdot \cos\theta + \frac{\partial V_\theta}{dt} \cdot \cos\theta - \frac{V_\theta^2}{r} \cdot \sin\theta \qquad (3\text{–}14)$$

Further as shown in Figure 3–3, denoting the radial and tangential accelerations as a_r and a_θ at a point P, we obtain the relationship between a_r, a_θ, dU/dt, and dV/dt as follows:

$$a_r = \frac{dU}{dt} \cdot \cos\theta + \frac{dV}{dt} \cdot \sin\theta \qquad (3\text{–}15)$$

$$a_\theta = -\frac{dU}{dt} \cdot \sin\theta + \frac{dV}{dt} \cdot \cos\theta \qquad (3\text{–}16)$$

Here substituting Equations 3–13 and 3–14 into Equations 3–15 and 3–16, we can obtain the equations of the accelerations

$$a_r = V_r - \frac{V_\theta^2}{r} \qquad (3\text{–}17)$$

$$a_\theta = \frac{V_r \cdot V_\theta}{r} + \frac{\partial V_\theta}{\partial t} \qquad (3\text{–}18)$$

Then the equations of motions for radial and tangential directions can be written as

$$\frac{dV_r}{dt} - \frac{V_\theta^2}{r} = -\frac{1}{\rho} \cdot \frac{\partial p}{\partial r} + F_r \qquad (3\text{–}19)$$

$$\frac{dV_\theta}{dt} + \frac{V_r \cdot V_\theta}{r} = -\frac{1}{\rho} \cdot \frac{\partial p}{r \cdot \partial \theta} + F_\theta \qquad (3\text{–}20)$$

where F_r and F_θ are the external force components per unit fluid mass for radial and tangential directions as shown in Figure 3–4. And also for the z-axial direction, the equation of motion can be written as

$$\frac{dV_z}{dt} = -\frac{1}{\rho} \cdot \frac{\partial p}{\partial z} + F_z \qquad (3\text{–}21)$$

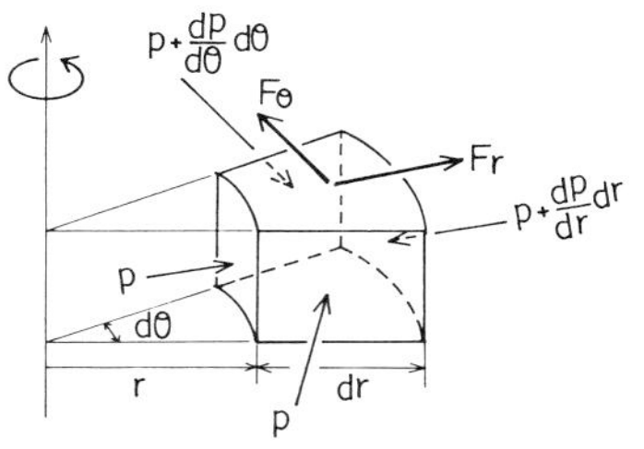

FIGURE 3–4. Schematic illustration of the force components on the small fluid particle.

where F_z is the external force component per unit fluid mass for the axial direction. Further, by multiplying the radius r on both sides of Equation 3–20, the following transformed equation can be obtained

$$\frac{d}{dt}(r \cdot V_\theta) = -\frac{1}{\rho} \cdot \frac{\partial p}{\partial \theta} + r \cdot F_\theta \tag{3–22}$$

The terms V_r, $r \cdot V_\theta$, and V_z are, in general, dependent on the radius r, the circumferential angle θ, the axial distance z, and the time t; so, the above terms can be transformed as

$$\frac{dV_r}{dt} = \frac{\partial V_r}{\partial t} + V_r \frac{\partial V_r}{\partial r} + V_\theta \frac{\partial V_r}{r \cdot \partial \theta} + V_z \frac{\partial V_r}{\partial z}$$

$$\frac{d(r \cdot V_\theta)}{dt} = \frac{\partial (r \cdot V_\theta)}{\partial t} + V_r \frac{\partial (r \cdot V_\theta)}{\partial r} + V_\theta \frac{\partial (r \cdot V_\theta)}{r \cdot \partial \theta} + V_z \frac{\partial (r \cdot V_\theta)}{\partial z}$$

$$\frac{dV_z}{dt} = \frac{\partial V_z}{\partial t} + V_r \frac{\partial V_z}{\partial r} + V_\theta \frac{\partial V_z}{r \cdot \partial \theta} + V_z \frac{\partial V_z}{\partial z}$$

Consequently the above-stated equation of motions for each direction can be rewritten as

$$\frac{\partial V_r}{\partial t} + V_r \frac{\partial V_r}{\partial r} + V_\theta \frac{\partial V_r}{r \cdot \partial \theta} + V_z \frac{\partial V_r}{\partial z} = -\frac{1}{\rho}\frac{\partial p}{\partial r} + \frac{V_\theta^2}{r} + F_r \tag{3–23}$$

$$\frac{\partial (r \cdot V_\theta)}{\partial t} + V_r \frac{\partial (r \cdot V_\theta)}{\partial r} + V_\theta \frac{\partial (r \cdot V_\theta)}{r \cdot \partial \theta} + V_z \frac{\partial (r \cdot V_\theta)}{\partial z} = -\frac{1}{\rho}\frac{\partial p}{\partial \theta} + F_\theta \cdot r \tag{3–24}$$

$$\frac{\partial V_z}{\partial t} + V_r \frac{\partial V_z}{\partial r} + V_\theta \frac{\partial V_z}{r \cdot \partial \theta} + V_z \frac{\partial V_z}{\partial z} = -\frac{1}{\rho}\frac{\partial p}{\partial z} + F_z \tag{3–25}$$

In these equations, when the flow is a steady-state condition,

$$\frac{\partial V_r}{\partial t} = 0 \quad , \quad \frac{\partial (r \cdot V_\theta)}{\partial t} = 0 \quad , \quad \frac{\partial V_z}{\partial t} = 0 \tag{3–26}$$

In addition to this, when flow is axis-symmetric for the z-axis,

$$\frac{\partial p}{\partial \theta} = 0 \quad , \quad \frac{\partial V_r}{r \cdot \partial \theta} = 0 \quad , \quad \frac{\partial (r \cdot V_\theta)}{r \cdot \partial \theta} = 0 \quad , \quad \frac{\partial V_z}{r \cdot \partial \theta} = 0 \tag{3–27}$$

V. VELOCITY POTENTIAL

On the irrotational flow, that is to say, the flow which has the velocity potential Φ, the velocity components for the arbitrary directions are equal to the negative gradients of the potentials, so that the following relationships can be obtained

$$V_r = -\frac{\partial\Phi}{\partial r}, \quad V_\theta = -\frac{\partial\Phi}{r \cdot \partial\theta}, \quad V_z = -\frac{\partial\Phi}{\partial z} \tag{3–28}$$

Next, substituting the terms of Equation 3–28 into the equation of continuity for steady and incompressible flow of Equation 3–10, the following equation is obtained

$$r\frac{\partial^2\Phi}{\partial r^2} + \frac{\partial\Phi}{\partial r} + \frac{1}{r}\frac{\partial^2\Phi}{\partial\theta^2} + r\frac{\partial^2\Phi}{\partial z^2} = 0 \tag{3–29}$$

Then the velocity potential Φ, depending on r, θ, and z, can be written as

$$d\Phi = \frac{\partial\Phi}{\partial r}dr + \frac{\partial\Phi}{\partial\theta}d\theta + \frac{\partial\Phi}{\partial z}dz \tag{3–30}$$

Substituting Equation 3–28 into Equation 3–30,

$$d\Phi = -(V_r \cdot dr + V_\theta \cdot r \cdot d\theta + V_z \cdot dz) \tag{3–31}$$

The equipotential surface is the surface corresponding to $d\Phi = 0$, so that the differential equation of the equipotential surface becomes

$$V_r \cdot dr + V_\theta \cdot r \cdot d\theta + V_z \cdot dz = 0 \tag{3–32}$$

And also the equation of the stream line becomes

$$\frac{dr}{V_r} = \frac{r \cdot d\theta}{V_\theta} = \frac{dz}{V_z} \tag{3–33}$$

Further, comparing Equation 3–33 with Equation 3–32, it is clear that the stream line pierces perpendicularly through the equipotential surface.

VI. TRANSFORMATION AND ROTATION OF THE SMALL FLUID PARTICLE

A. TRANSFORMATION

In order to consider the quantities of the small translations of the small fluid volume ABCDEFGH during the time dt as shown in Figure 3–5, in the first place we must estimate the quantities of the transformation of the small cylindrical surface AHEF. Therefore we may apply approximately the Cartesian coordinates as shown in Figure 3–6, in which the transformation of the small surface AHEF during the time dt can be calculated as

$$\theta_1 + \theta_2 = \frac{\dfrac{\partial V_z}{r \cdot \partial\theta} r \cdot d\theta \cdot dt}{r \cdot d\theta} + \frac{\dfrac{\partial V_\theta}{\partial z} \cdot dz \cdot dt}{dz} = \frac{1}{r}\left[\frac{\partial V_z}{\partial\theta} + \frac{\partial(r \cdot V_\theta)}{dz}\right] dt \tag{3–34}$$

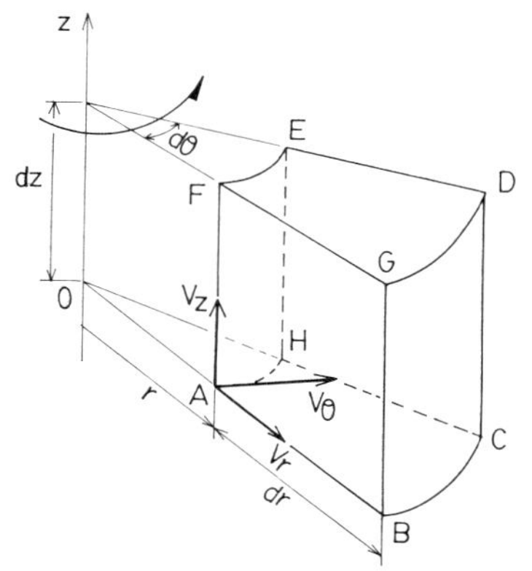

FIGURE 3–5. Schematic illustration of the velocity components.

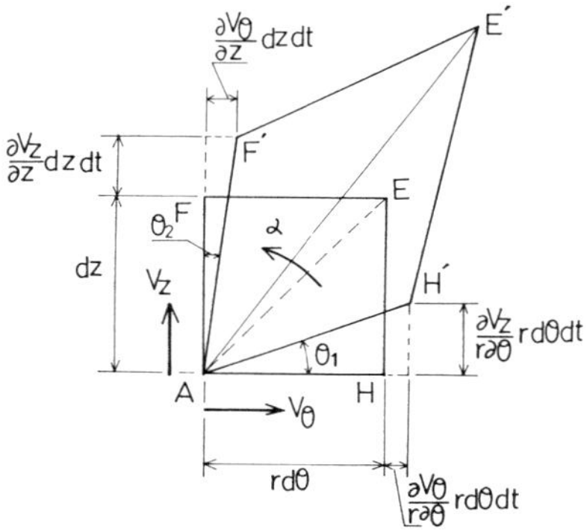

FIGURE 3–6. Schematic illustration of the transformation of the small surface.

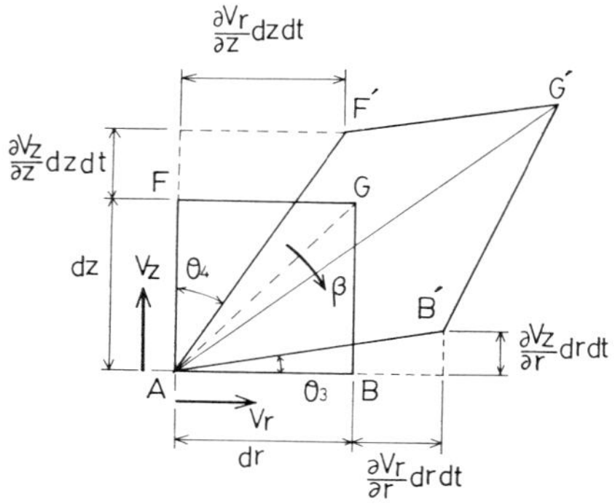

FIGURE 3–7. Schematic illustration of the transformation of the small surface.

and also the transformation of the small surface ABGF during the time dt as shown in Figure 3–7 can be calculated as

$$\theta_3 + \theta_4 = \frac{\dfrac{\partial V_r}{\partial z} \cdot dz \cdot dt}{dz} + \frac{\dfrac{\partial V_z}{\partial r} \cdot dr \cdot dt}{dr} = \left(\frac{\partial V_r}{\partial z} + \frac{\partial V_z}{\partial r}\right) dt \qquad (3\text{–}35)$$

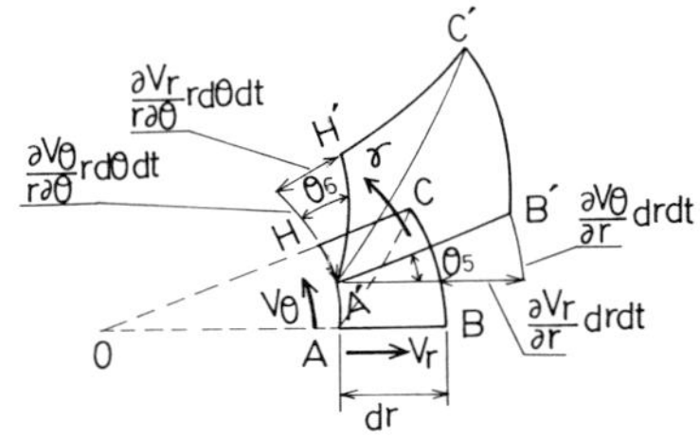

FIGURE 3–8. Schematic illustration of the transformation of the small surface.

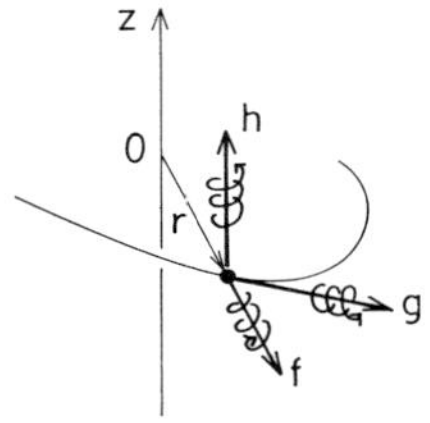

FIGURE 3–9. Schematic illustration of the components of the transformations for the radial, tangential, and axial directions.

In addition to this, the transformation of the small surface ABCH during the time dt as shown in Figure 3–8 can be calculated as

$$\theta_5 + \theta_6 = \frac{\dfrac{\partial V_\theta}{\partial r} \cdot dr \cdot dt}{dr} + \frac{\dfrac{\partial V_r}{r \cdot \partial\theta} \cdot r \cdot d\theta \cdot dt}{r \cdot d\theta} = \left(\frac{\partial V_\theta}{\partial r} + \frac{\partial V_r}{r \cdot \partial\theta}\right) dt \qquad (3\text{–}36)$$

Consequently the quantities of the transformation for each small surface per unit time can be obtained with Equations 3–34, 3–35, and 3–36 by dividing the time dt. Therefore as shown in Figure 3–9, the quantities of the transformations of the small surface for radial (AHEF), tangential (ABGF), and axial (ABCH) directions per unit time can be defined as

$$f = \frac{1}{r}\left[\frac{\partial V_z}{\partial\theta} + \frac{\partial(r \cdot V_\theta)}{\partial z}\right] \qquad (3\text{–}37)$$

$$g = \frac{\partial V_r}{\partial z} + \frac{\partial V_z}{\partial r} \qquad (3\text{–}38)$$

$$h = \frac{\partial V_\theta}{\partial r} + \frac{\partial V_r}{r \cdot \partial\theta} \qquad (3\text{–}39)$$

B. ROTATION

As shown in Figure 3–6, the rotating angle of the small surface AHEF during the time is denoted as dt α, so that the angle α can be calculated as

$$\alpha = \frac{\theta_1 - \theta_2}{2} = \frac{1}{2r}\left[\frac{\partial V_z}{\partial\theta} - \frac{\partial(r \cdot V_\theta)}{\partial z}\right] dt \qquad (3\text{–}40)$$

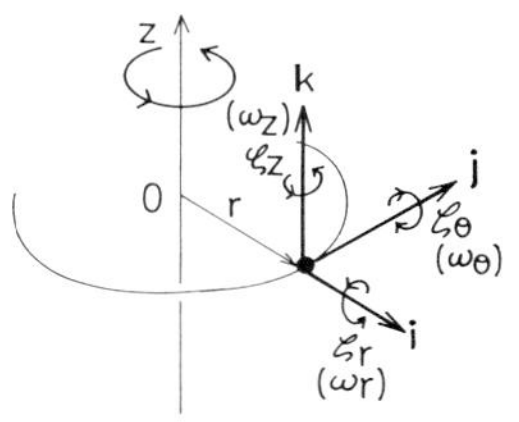

FIGURE 3–10. Schematic illustration of the vorticity components for the radial, tangential, and axial directions.

and also the rotating angle of the small surface ABGF during the time dt is β as shown in Figure 3–7, so that

$$\beta = \frac{1}{2}(\theta_4 + \theta_3) = \frac{1}{2}\left(\frac{\partial V_r}{\partial z} - \frac{\partial V_z}{\partial r}\right) dt \qquad (3\text{–}41)$$

In addition, the rotating angle of the small surface ABCH during the time dt is γ as shown in Figure 3–8, so that γ must be the half of the angle difference between the angle between AB and A′B′ and θ_6, that is to say,

$$\gamma = \frac{1}{2}[(\angle BOA' + \theta_5) - \theta_6] \qquad (3\text{–}42)$$

where $\angle BOA'$ is equal to $V_\theta \cdot dt/r$. Therefore,

$$\gamma = \frac{1}{2}\left(\frac{V_\theta}{r} + \frac{\partial V_\theta}{\partial r} - \frac{\partial V_r}{r \cdot \partial \theta}\right) dt = \frac{1}{2r}\left[\frac{\partial (r \cdot V_\theta)}{\partial r} - \frac{\partial V_r}{\partial \theta}\right] dt \qquad (3\text{–}43)$$

As shown in Figure 3–10, denoting the angular velocity components for radial, tangential, and axial directions as ω_r, ω_θ, and ω_z, respectively, each component can be represented by

$$\omega_r = \frac{1}{2r}\left[\frac{\partial V_z}{\partial \theta} - \frac{\partial (r \cdot V_\theta)}{\partial z}\right] \qquad (3\text{–}44)$$

$$\omega_\theta = \frac{1}{2}\left(\frac{\partial V_r}{\partial z} - \frac{\partial V_z}{\partial r}\right) \qquad (3\text{–}45)$$

$$\omega_z = \frac{1}{2r}\left[\frac{\partial (r \cdot V_\theta)}{\partial r} - \frac{\partial V_r}{\partial \theta}\right] \qquad (3\text{–}46)$$

Further we must define the vorticity components for radial (ζ_r), tangential (ζ_θ), and axial (ζ_z) directions as follows:

$$\zeta_r = \frac{1}{r}\left[\frac{\partial V_z}{\partial \theta} - \frac{\partial (r \cdot V_\theta)}{\partial z}\right] \qquad (3\text{–}47)$$

$$\zeta_\theta = \frac{\partial V_r}{\partial z} - \frac{\partial V_z}{\partial r} \qquad (3\text{–}48)$$

$$\zeta_z = \frac{1}{r}\left[\frac{\partial(r \cdot V_\theta)}{\partial r} - \frac{\partial V_r}{\partial \theta}\right] \tag{3–49}$$

Therefore applying these equations, we obtain the following equation

$$\frac{\partial(r \cdot \zeta_r)}{\partial r} + \frac{\partial \zeta_\theta}{\partial \theta} + \frac{\partial(r \cdot \zeta_z)}{\partial z} = 0 \tag{3–50}$$

This equation represents the equation of continuity of the vorticity. If we want to represent these equations by the vector forms, we can write the following vorticity equation

$$\boldsymbol{\Omega}\ (= \zeta_r\mathbf{i} + \zeta_\theta\mathbf{j} + \zeta_z\mathbf{k}) = \text{rot}\ \mathbf{V} = \nabla \times \mathbf{V} \tag{3–51}$$

and as the equation of continuity of vorticity

$$\text{div}\ \boldsymbol{\Omega} = \text{div rot}\ \mathbf{V} = 0 \tag{3–52}$$

where $\boldsymbol{\Omega}$ represents the vorticity vector and $\mathbf{V}(= V_r\mathbf{i} + V_\theta\mathbf{j} + V_z\mathbf{k})$ the velocity vector. Then Equation 3–52 means that the rotation of the small fluid element keeps continuation always.

On the other hand, where the fluid flow has symmetry around the z-axis, Equation 3–50 becomes

$$\frac{\partial(r \cdot \zeta_r)}{\partial r} + \frac{\partial(r \cdot \zeta_z)}{\partial z} = 0 \tag{3–53}$$

Now the function Ψ which satisfies the above differential equation depends on r and z, and also the following relationships can be derived

$$r \cdot \zeta_r = -\frac{\partial \Psi}{\partial z}\ , \quad r \cdot \zeta_z = \frac{\partial \Psi}{\partial r} \tag{3–54}$$

VII. VORTEX LINE

We can surmise that one curve connects the rotating axes one after another in the rotating fluid as shown in Figure 3–11. This curve is called a vortex line. Then the closed pipe which is composed of these vortex lines is called a vortex tube. Further in the case of the tube which is infinitesimally contracted, the sectional area of this tube is called a vortex filament.

VIII. IRROTATIONAL FLOW

In the fluid motion of the irrotational flow, the vorticity components are zero; so, we obtain the following equation

$$\zeta_r = 0\ , \quad \zeta_\theta = 0\ , \quad \zeta_z = 0 \tag{3–55}$$

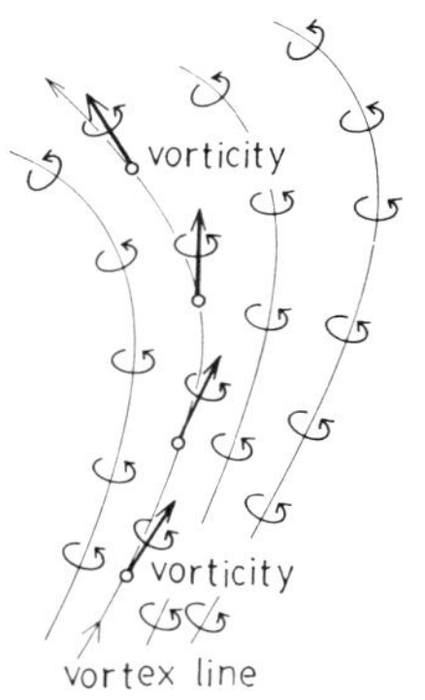

FIGURE 3–11. Schematic illustration of the vorticity and of the vortex lines.

Then from Equations 3–47, 3–48, and 3–49, the concrete equations can be represented as

$$\frac{\partial V_z}{\partial \theta} = \frac{\partial (r \cdot V_\theta)}{\partial z} \quad , \quad \frac{\partial V_r}{\partial z} = \frac{\partial V_z}{\partial r} \quad , \quad \frac{\partial (r \cdot V_\theta)}{\partial r} = \frac{\partial V_r}{\partial \theta} \tag{3–56}$$

Here substituting Equation 3–28 of the velocity potential into Equation 3–56, the velocity potential Φ satisfies the condition of Equation 3–56. Therefore the fluid flow which can be represented by the velocity potential is corresponding to the motion of the irrotational flow. Further, assuming that fluid flow is the axis-symmetry, the terms including the differentiation with angle θ all vanish. Then Equation 3–56 becomes

$$\frac{\partial (r \cdot V_\theta)}{\partial z} = 0 \quad , \quad \frac{\partial V_r}{\partial z} = \frac{\partial V_z}{\partial r} \quad , \quad \frac{\partial (r \cdot V_\theta)}{\partial r} = 0 \tag{3–57}$$

In this equation, on the first and third terms the product $r \cdot V_\theta$ is independent of r and z, so that the value of $r \cdot V_\theta$ in all points of the flow space shows the constant value, that is to say,

$$r \cdot V_\theta = \text{constant} \tag{3–58}$$

Therefore in the above stated flow conditions, the tangential velocity is always inversely proportional to the radius r.

IX. THOMSON-RANKINE VORTEX MODEL

A. FREE VORTEX (IRROTATIONAL VORTEX)

When the external force is only the gravitational force (for example, the fluid is liquid) and the fluid is rotating about the z-axis as shown in Figure 3–12, we can put $F_r = 0$, $F_\theta = 0$, and $F_z = -g$ and also $\partial/\partial\theta = 0$ for Equations 3–23, 3–24, and 3–25, respectively. And further, assuming that $V_r = 0$ and $V_z = 0$,

$$\frac{\partial (r \cdot V_\theta)}{\partial t} = 0 \tag{3–59}$$

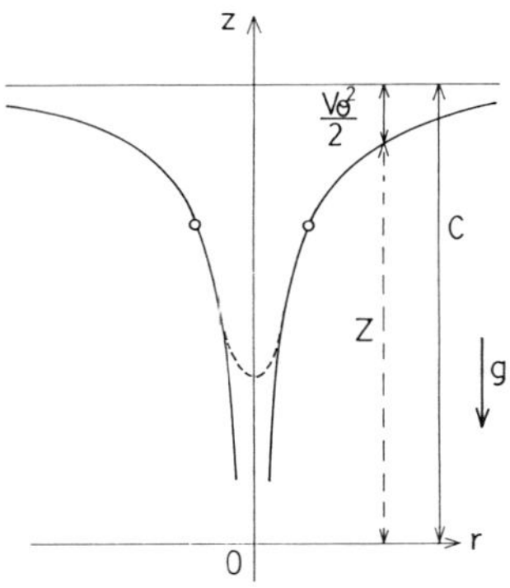

FIGURE 3–12. Schematic illustration of the static pressure in the free vortex region.

Therefore, Equation 3–59 becomes

$$V_\theta \cdot r = \text{constant} \tag{3–60}$$

This equation satisfies the first and third terms of Equation 3–57. Consequently when the external force is only the gravitational force, the fluid flow becomes irrotational motion, that is to say, the flow has the velocity potential. Such a kind of rotational motion is in general called a free vortex motion or natural vortex motion.

Then we can obtain the form of the equipressure surface or the free surface of the free vortex motion by Equations 3–23 and 3–25 as

$$\frac{1}{\rho}\frac{\partial p}{\partial r} = \frac{V_\theta^2}{r} \tag{3–61}$$

$$g + \frac{1}{\rho}\frac{\partial p}{\partial z} = 0 \tag{3–62}$$

Then combining Equations 3–61 and 3–62

$$dp = \frac{\partial p}{\partial r}dr + \frac{\partial p}{\partial z}dz = \rho\left(\frac{V_\theta^2}{r}\cdot dr - g \cdot dz\right) \tag{3–63}$$

and also denoting the constant value of $r \cdot V_\theta$ as

$$V_\theta \cdot r = \Gamma \tag{3–64}$$

Equation 3–63 becomes

$$dp = \rho\left(\Gamma^2 \frac{1}{r^3}dr - g \cdot dz\right) \tag{3–65}$$

Since the equipressure surface is the surface corresponding to dp = 0, the equation of the equipressure surface must satisfy the following equation

$$\frac{\Gamma^2}{r^3}dr - g \cdot dz = 0 \tag{3–66}$$

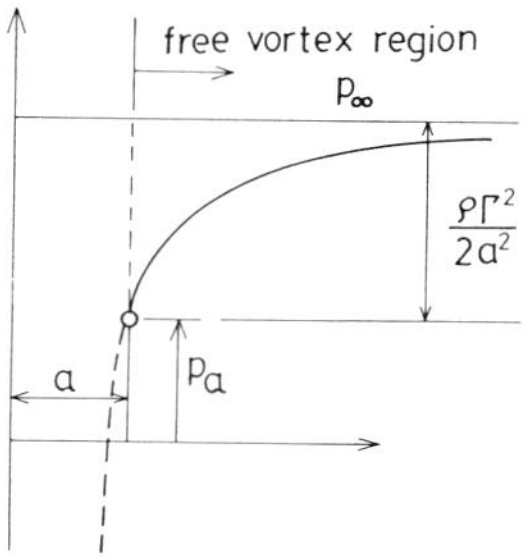

FIGURE 3–13. Schematic illustration of the static pressure depression in the free vortex region.

Here integrating this equation once, and putting C as the constant of integration, we can obtain the following equation of the equipressure surface as

$$\frac{\Gamma^2}{2r^2} + g \cdot z = \frac{V_\theta^2}{2} + g \cdot z = C \qquad (3\text{–}67)$$

On the other hand, when we can neglect the gravitational force (for example, the fluid is gas), Equations 3–23 and 3–25 become

$$\frac{1}{\rho}\frac{\partial p}{\partial r} = \frac{V_\theta^2}{r} = \frac{\Gamma^2}{r^3} \qquad (3\text{–}68)$$

$$\frac{\partial p}{\partial z} = 0 \qquad (3\text{–}69)$$

Then from Equation 3–68 we can obtain the following equation of the static pressure as

$$\int_{p_a}^{p} dp = \rho\Gamma^2 \int_a^r \frac{dr}{r^3}$$

$$p - p_a = \frac{\rho\Gamma^2}{2}\left(\frac{1}{a^2} - \frac{1}{r^2}\right) \qquad (3\text{–}70)$$

Further when the radius r increases to infinity and at the same time the static pressure p can be written as $p = p_\infty$, Equation 3–70 becomes

$$p_\infty - p_a = \frac{\rho\Gamma^2}{2 \cdot a} \qquad (3\text{–}71)$$

This physical illustration is shown in Figure 3–13. From the equation we may suppose that the static pressure p_a becomes negative when the radius approaches $a \rightarrow 0$. But this kind of flow condition never occurs in the natural phenomena on earth.

B. FORCED VORTEX

When we consider the motion of a viscous fluid, since the viscosity gives the resistance force (frictional force) to the motion of fluid, the external force is the gravitational force.

Therefore when there is the very high deformation, the frictional force becomes very high. Then even if we assume the fluid to be a nearly perfect fluid, it is impossible to create the free vortex motion near the center axis due to the high deformation of the fluid particle. Therefore we assume that the tangential velocity near the center axis can be written as

$$V_\theta = \Omega_o \cdot r \tag{3–72}$$

Then the equation of the equipressure surface under the gravitational force can be written as

$$dp = \frac{\partial p}{\partial r} dr + \frac{\partial p}{\partial z} dz = \rho\left(\frac{V_\theta^2}{r} dr - g \cdot dz\right) = (\Omega_o^2 \cdot r \cdot dr - g \cdot dz) \tag{3–73}$$

Since the equation of the equipressure surface must satisfy the equation $dp = 0$, we obtain the following equation

$$\Omega_o^2 \cdot r \cdot dr - g \cdot dz = 0 \tag{3–74}$$

Integrating this equation, we obtain

$$\frac{\Omega_o^2}{2} \cdot r^2 - g \cdot z = C \tag{3–75}$$

This equation was already derived in Chapter 2 (Hydrostatics). On the other hand, when we consider the motion of gas, the relationship between the static pressure p and the centrifugal force can be written as

$$\frac{dp}{dr} = \rho \frac{V_\theta^2}{r} = \rho\, \Omega_o^2 \cdot r \tag{3–76}$$

Integrating this equation once and denoting that the static pressure at the center ($r = 0$) is p_o,

$$p = p_o + \frac{\rho}{2} \Omega_o^2 \cdot r^2 \tag{3–77}$$

Then the static pressure difference Δp_{oa} can be written as

$$\Delta P_{oa} = p_a - p_o = \frac{\rho\, \Omega_o^2}{2} \cdot a^2 \tag{3–78}$$

C. THOMSON-RANKINE VORTEX

Here we consider a combined vortex model which is composed of the free vortex and the forced vortex as shown in Figure 3–14. First of all, the tangential velocity V_θ at the boundary radius $r = a$ is the same value; therefore, at the radius $r = a$, the following relation can be obtained

$$V_{\theta a} = \Omega_o \cdot a = \Gamma/a \tag{3–79}$$

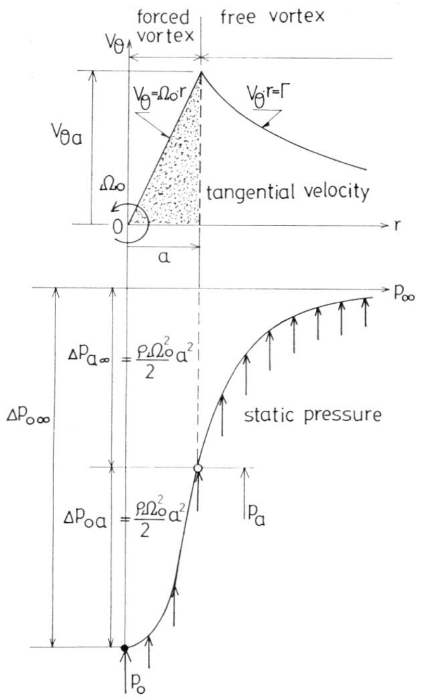

FIGURE 3–14. Schematic illustration of the distributions of the velocity and of the static pressure of the Thomson-Rankine vortex.

Next the distributions of the static pressure on the free vortex and on the forced vortex can be represented as

$$p - p_a = \frac{\rho \Gamma^2}{2} \left(\frac{1}{a^2} - \frac{1}{r^2} \right) \tag{3–70}$$

$$p_\infty - p_a = \frac{\rho \Gamma^2}{2 \cdot a} \tag{3–71}$$

$$p = p_o + \frac{\rho}{2} \Omega_o^2 \cdot r^2 \tag{3–77}$$

$$\Delta P_{oa} = p_a - p_o = \frac{\rho\ \Omega_o^2}{2} \cdot a^2 \tag{3–78}$$

Now substituting Equation 3–79 into Equation 3–71, so Equation 3–71 becomes

$$\Delta p_{a\infty} = \frac{\rho\ \Omega_o^2}{2} \cdot a^2 \tag{3–80}$$

Therefore we can find that the static pressure difference $\Delta p_{a\infty}$ on the free vortex is equal to that Δp_{oa} on the forced vortex. These relationships can be applied to the vortex motion of liquid. This combined vortex model was created by Professor James Thomson of Belfast, and the principle of this model was applied to his turbine or vortex water-wheel. This description was written in the book *A Manual of Applied Mechanics* by W. J. M. Rankine.[2] Therefore this combined vortex model can be called the Thomson-Rankine Vortex Model in his honor.

X. BERNOULLI EQUATION

The equation of motion of Euler can be represented as

$$\frac{\partial \mathbf{V}}{\partial t} + \mathbf{V} \cdot \nabla \mathbf{V} = -\frac{1}{\rho} \nabla p + \mathbf{F} \tag{3–81}$$

where **F** is the external force per unit mass of fluid. And the continuity equation can be written as

$$\frac{\partial \rho}{\partial t} + \nabla \cdot (\rho \, \mathbf{V}) = 0 \tag{3–82}$$

When we consider the compressibility of fluid, the density is the function of the position (x, y, z). Then the change of state during the movement of fluid may be assumed to be adiabatic; so, entropy is constant due to the combined effect of incoming and outgoing heat on the mutual relation of fluid particles. This flow is generally called an isentropic flow.

Here assuming that a relationship between pressure p and density ρ may be written as

$$\rho = f(p) \tag{3–83}$$

where a medium which satisfies the above equation is said to be barotropic; for example, an ideal fluid of a constant temperature can be represented as

$$\rho \propto p \tag{3–84}$$

and also for an incompressible fluid, the density is constant. These examples are special cases of barotropic fluid.

For barotropic flow, Equation 3–82 can be applied, and the following equation can be obtained

$$P = \int \frac{dp}{\rho} \tag{3–85}$$

Therefore the pressure term of Equation 3–81 can be represented as

$$-\frac{1}{\rho} \cdot \nabla p = \nabla P \tag{3–86}$$

Further, assuming that the external force **F** per unit mass of fluid has potential (potential energy per unit mass of fluid) Φ, so **F** can be rewritten as

$$\mathbf{F} = -\nabla \Phi \tag{3–87}$$

And also applying vector analysis, the second term of the left-hand side of Equation 3–81 can be transformed

$$\mathbf{V} \cdot \nabla \mathbf{V} = \frac{1}{2} \nabla \mathbf{V}^2 - \mathbf{V} \times \boldsymbol{\Omega} \quad (\boldsymbol{\Omega} = \nabla \times \mathbf{V}) \tag{3–88}$$

Substituting Equations 3–86, 3–87, and 3–88 into Equation 3–81, Equation 3–81 can be transformed

$$\frac{\partial \mathbf{V}}{\partial t} + \nabla \frac{\mathbf{V}^2}{2} - \mathbf{V} \times \boldsymbol{\Omega} = -\nabla P - \nabla \Phi$$

or

$$\frac{\partial \mathbf{V}}{\partial t} + \nabla\left(\frac{\mathbf{V}^2}{2} + P + \Phi\right) = \mathbf{V} \times \boldsymbol{\Omega} \tag{3–89}$$

$$H = \frac{V^2}{2} + P + \Phi \quad , \quad V = |\mathbf{V}| \tag{3–90}$$

where H is called a Bernoulli function.

A. IRROTATIONAL FLOW (FREE VORTEX)

Assuming that the vorticity $\boldsymbol{\Omega}$ is zero, the velocity potential Φ is

$$\mathbf{V} = \nabla \Phi \tag{3–91}$$

Substituting Equation 3–91 into Equation 3–90, Equation 3–90 becomes

$$\nabla\left(\frac{\partial \Phi}{\partial t} + H\right) = 0 \tag{3–92}$$

Then integrating Equation 3–92, we can obtain the following equation

$$\frac{\partial \Phi}{\partial t} + \frac{1}{2} V^2 + P + \Phi = \text{constant} \tag{3–93}$$

where a constant value of the right-hand side of Equation 3–93 may be regarded as a function of time t.

B. STEADY FLOW

In the case of steady flow, Equation 3–89 becomes

$$\nabla\left(\frac{V^2}{2} + P + \Phi\right) = \mathbf{V} \times \boldsymbol{\Omega} \tag{3–94}$$

Here taking an inner product by $\mathbf{V}$ and by $\boldsymbol{\Omega}$ to both sides of Equation 3–94,

$$\mathbf{V} \cdot \nabla\left(\frac{V^2}{2} + P + \Phi\right) = 0 \quad , \quad \boldsymbol{\Omega} \cdot \nabla\left(\frac{V^2}{2} + P + \Phi\right) = 0 \tag{3–95}$$

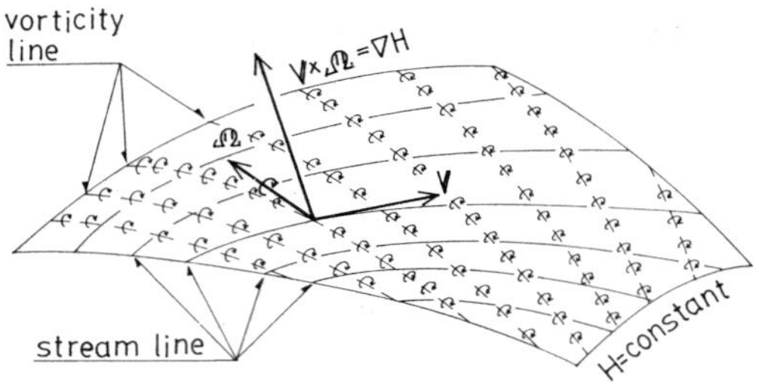

FIGURE 3–15. Schematic illustration of the Bernoulli surface.

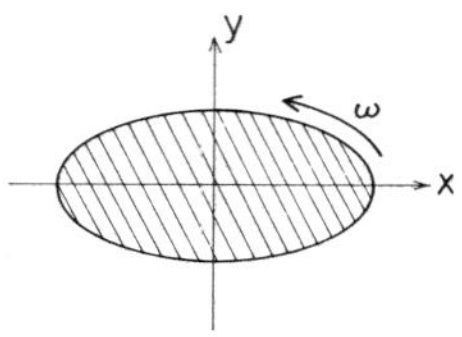

FIGURE 3–16. Schematic illustration of the elliptic section.

since by the vector analysis

$$\mathbf{V} \cdot (\mathbf{V} \times \mathbf{\Omega}) = \mathbf{\Omega} \cdot (\mathbf{V} \times \mathbf{\Omega}) = 0$$

This equation has the following geometrical meaning. As shown in Figure 3–15, assuming that the curved surface (Bernoulli surface) of H is constant, then taking a normal line at an arbitrary point on this surface, this normal direction coincides with the direction of ∇H. And also Equation 3–95 shows that the vectors **V** and $\mathbf{\Omega}$ both make a right angle with the direction of ∇H. Consequently when we make a reticulated surface which is composed of a group joining the stream lines and the vorticity lines, this surface is a Bernoulli surface. Therefore the value of H is constant along a stream line or along a vorticity line. Further, generally speaking, the value of H being a constant value along each stream line is called the Bernoulli theorem.

XI. VORTEX OF THE ELLIPTIC SECTION

A. ELLIPTIC CYLINDER ROTATING IN AN INFINITE MASS OF LIQUID AT REST AT INFINITY

Here ω denotes the angular velocity as shown in Figure 3–16. The boundary conditions can be written as[3,4]

$$\Psi = \frac{1}{2}\omega(x^2 + y^2) + C \tag{3–96}$$

or

$$x + iy = C\cosh(\xi + i\eta) = C[\cosh\xi \cdot \cos\eta + i\sinh\xi \cdot \sin\eta]$$

therefore

$$x = C\cosh\xi \cdot \cos\eta \quad , \quad y = C\sinh\xi \cdot \sin\eta \tag{3–97}$$

So the stream function Ψ can be represented as

$$\Psi = \frac{1}{2}\omega\,[(c\cosh\xi \cdot \cos\eta)^2 + (c\sinh\xi \cdot \sin\eta)^2] + C =$$

$$= \frac{1}{4}\omega \cdot c^2 \cdot (\cosh 2\xi + \cos 2\eta) + C \tag{3–98}$$

where i is an imaginary number.

Since the effect vanishes at infinity, the only variable term in the boundary condition is $\cos 2\eta$. Therefore, we must assume a complex relation which gives to Ψ of the form of

$$\exp(-2\xi) \cdot \cos 2\eta$$

Further assume that

$$\Phi + i\Psi = Ai\exp[-2(\xi + i\eta)] \tag{3–99}$$

so that the stream function becomes

$$\Psi = A \cdot e^{-2\xi} \cdot e^{-2i\eta} = A \cdot e^{-2\xi} \cdot (\cos 2\eta - i\sin 2\eta) = A \cdot e^{-2\xi} \cdot \cos 2\eta \tag{3–100}$$

Hence at the boundary $\xi = \alpha$, we must have

$$A \cdot e^{-2\alpha} \cdot \cos 2\eta = \frac{1}{4}\omega \cdot c^2\,(\cosh 2\alpha + \cos 2\eta) + C \tag{3–101}$$

for all values of η. In this case A and C become

$$A = \frac{1}{4}\omega \cdot c^2 \cdot e^{2\alpha} \quad , \quad C = -\frac{1}{4}\omega \cdot c^2 \cdot \cosh 2\alpha \tag{3–102}$$

Therefore when the cylinder rotates with angular velocity ω, the stream function

$$\Psi = \frac{1}{4}\omega \cdot c^2 \cdot e^{(2\alpha - 2\xi)} \cdot \cos 2\eta \tag{3–103}$$

gives a stream function which makes the boundary of an elliptical stream line. Further

$$c^2 \cdot e^{2\alpha} = (a + b)^2 \tag{3–104}$$

so we may write the results as

$$\Psi = \frac{1}{4}\omega \cdot (a + b)^2 \cdot e^{-2\xi} \cdot \cos 2\eta \tag{3–105}$$

$$\Phi = \frac{1}{4}\omega \cdot (a + b)^2 \cdot e^{-2\xi} \cdot \sin 2\eta \tag{3–106}$$

Then from Equation 3–105, the velocity vanishes at infinity.

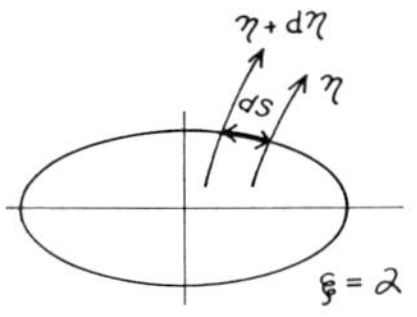

FIGURE 3–17. Schematic illustration of ds and dη on the elliptic section.

B. CIRCULATION ABOUT AN ELLIPTIC CYLINDER

When the irrotational motion is cyclic with circulation round an elliptic cylinder, we can take this into account by means of the function

$$\Phi + i\eta = \frac{i\Gamma}{2\pi} \cdot (\xi + i\eta) \tag{3–107}$$

The circulation defined as

$$\Gamma = \int - \frac{\partial \Phi}{\partial s} \cdot ds \tag{3–108}$$

taken round the elliptic cylinder of $\xi = \alpha$ as shown in Figure 3–17 can be calculated as

$$\Gamma = \int_0^{2\pi} - \frac{\partial \Phi}{\partial \eta} \cdot d\eta = \int_0^{2\pi} \frac{\Gamma}{2\pi} \cdot d\eta = \Gamma \tag{3–109}$$

Hence in addition to the rotation with angular velocity ω, there is a circulation Γ about the elliptic cylinder, so we have

$$\Psi = \frac{1}{4} \omega \cdot (a + b)^2 \cdot e^{-2\xi} \cdot \cos 2\eta + \frac{\Gamma \xi}{2\pi} \tag{3–110}$$

$$\Phi = \frac{1}{4} \omega \cdot (a + b)^2 \cdot e^{-2\xi} \cdot \sin 2\eta - \frac{\Gamma \eta}{2\pi} \tag{3–111}$$

XII. APPLICATION OF BERNOULLI EQUATION FOR FLOW RATE OF GAS FLOW INTO THE DUST BUNKER OF CYCLONE DUST COLLECTORS

A. INTRODUCTION

Cyclone dust collectors are one of the most simple, fixed constructions of dust separators. The air flow pattern in the cyclone dust collector is three-dimensional turbulent rotational air flow, and it is possible to separate the fine solid particles of diameter about three micron meters.

The fine solid particles with air flow entering into the cyclone separation chamber are thrown out near the region of the outer concave surface wall due to the centrifugal force acting on the solid particles by the tangential air velocity V_θ, and these separated particles are transported into the dust bunker of cyclone by the downward axial air flow near the

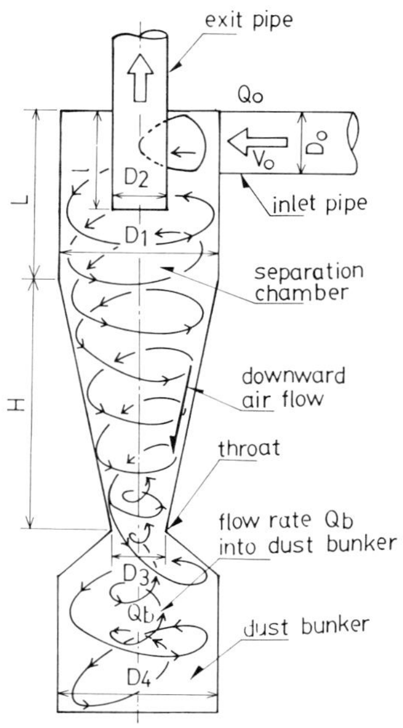

FIGURE 3–18. Schematic illustration of the flow pattern in the cyclone dust collector.

concave surface. Finally these transported particles are collected in the dust bunker as shown in Figure 3–18. Hence it is very important to confirm the flow rate Q_b of gas flow into the dust bunker.[5–7] In order to simplify the mathematical analysis, assuming that the tangential velocity distribution can be represented by free vortex, an equation of the flow rate Q_b can be obtained by the idea of Binnie et al.[8–10]

B. FUNDAMENTAL EQUATION

As shown in Figure 3–19, in the downward flow region of the axial flow velocity V_z for radius $r \geq$ radius b, if we assume that the distribution of the rotational flow is an irrotational flow (free vortex), then a relationship between the tangential velocity V_θ and the radius r can be written as

$$r \cdot V_\theta = \Gamma = \text{constant} \tag{3–112}$$

where Γ is a constant.

Here let us denote that p is static pressure, and neglecting the radial velocity V_r in comparison with V_θ and V_z, the Bernoulli equation can be represented as

$$p + \frac{\rho}{2}(V_\theta^2 + V_z^2) = P \tag{3–113}$$

where P is the total pressure. The distribution of the static pressure p may be written as

$$p - p_b = \rho \frac{\Gamma^2}{2}\left(\frac{1}{b^2} - \frac{1}{r^2}\right) \tag{3–114}$$

Where p_b is static pressure at the radius $r = b$, then substituting Equation 3–114 into Equation 3–113, we obtain the following equation:

$$\frac{\Gamma^2}{2}\left(\frac{1}{b^2} - \frac{1}{r^2}\right) + \frac{1}{2}\left(V_z^2 + \frac{\Gamma^2}{r^2}\right) = \frac{P + p_b}{\rho} = \frac{P'}{\rho} \qquad (3\text{–}115)$$

Therefore the magnitude of the axial velocity V_z can be obtained from Equation 3–115 as

$$V_z = \left(\frac{2P'}{\rho} - \frac{\Gamma^2}{b^2}\right)^{1/2} \qquad (3\text{–}116)$$

Thereupon the equation of flow rate at an arbitrary z-axis can be written as

$$Q = \pi(R^2 - b^2)\, V_z \qquad (3\text{–}117)$$

Further differentiating Equations 3–116 and 3–117 with z and then eliminating a term dV_z/dz, the following equation is obtained

$$\frac{R\Gamma^2}{b^4} \cdot \frac{dR}{dz} = \frac{V_z^2 - V_{zt}^2}{V_z} \cdot \frac{dV_z}{dz} \qquad (3\text{–}118)$$

where V_{zt} can be represented as

$$V_{zt} = \left[\frac{\Gamma^2(R^2 - b^2)}{2b^4}\right]^{1/2} \qquad (3\text{–}119)$$

Now assuming that $dR/dz = 0$ and $dV_z/dz \neq 0$ at the throat on the model flow of Figure 3–19, the axial velocity at the throat plane becomes $V_z = V_{zt}$. Consequently the radius $b = b_t$ at the throat plane can be represented with Equations 3–116 and 3–119 as

$$\left(\frac{b_t}{R_3}\right)^2 = \frac{\rho\Gamma^2}{8P'\, R_3^2}\left\{1 + \left[1 + \left(\frac{16P'\, R_3^2}{\rho\Gamma^2}\right)\right]^{1/2}\right\} \qquad (3\text{–}120)$$

Finally a theoretical equation of flow rate Q_b from the throat plane into dust bunker can be obtained as

$$\frac{Q_b}{\Gamma R_3} = \frac{\pi}{\sqrt{2}} \cdot \frac{[1 - (b_t/R_3)^2]^{3/2}}{(b_t/R_3)^2} \qquad (3\text{–}121)$$

Theoretically speaking, the flow rate Q_b depends on the terms (b_t/R_3) and ΓR_3, respectively.

C. EXPERIMENTAL RESULTS

Here in order to recognize the relation of Equation 3–121, the main size of cyclone dust collectors, which were made of transparent celluloid, are shown in Figure 3–20. The diameters of throat are $D_3 = 50$, 80, and 100 mm, respectively. Figure 3–21 and Figure 3–22 show the experimental results of the tangential V_θ and axial V_z velocities at the throat planes. In the quasi-free vortex region, a relationship between the tangential velocity V_θ and the radius r can be represented as

$$V_\theta \cdot r^n = \Gamma_n = \text{constant} \qquad (3\text{–}122)$$

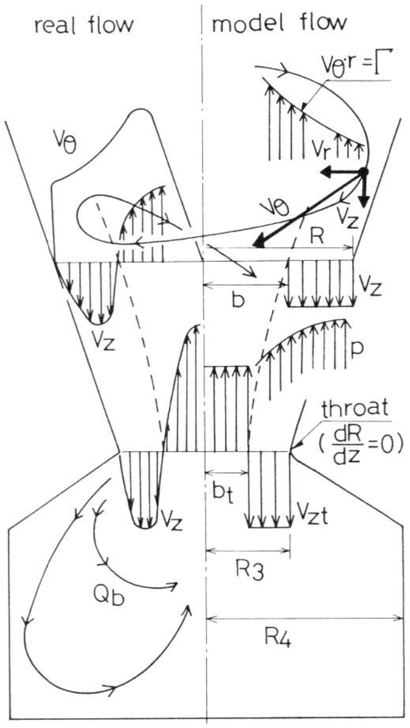

FIGURE 3–19. Fundamental idea of the velocity distributions for real flow and for model flow in the cyclone dust collector.

The velocity exponential n (dimensionless number) has a tendency to increase with increasing the mean inlet velocity V_o — the exception to this is a cyclone of $D_3 = 50$ mm (as will be described in a later section) in which the velocity exponential n will be increased with increasing Reynolds number R_{ec}. On the other hand from the experimental results of the axial velocities V_z, the radial position $r = b_t$, which corresponds to $V_z = 0$, is nearly independent of the inlet velocity V_o in the cyclone dust collectors.[11] Therefore Figure 3–23 shows the experimental results of how to relate b_t/R_3 with D_3/D_1. A relationship between b_t/R_3 and D_3/D_1 can be obtained as

$$\frac{b_t}{R_3} = 0.6\left(\frac{D_3}{D_1}\right)^{-0.28} \tag{3–123}$$

Next a relationship between circulation Γ and the throat radius R_3 can be represented by the following equation

$$\Gamma R_3 = k \cdot Q_o \tag{3–124}$$

where the magnitude of circulation Γ is defined at $r = 20$ mm in the throat plane, and Q_o is the flow rate of air flow into the cyclone dust collector. A proportional factor k can be determined as shown in Figure 3–23 as

$$k = 1.14\left(\frac{D_3}{D_1}\right) \tag{3–125}$$

Further tentatively let us denote that Q_{bm} is the measured flow rate of air flow into the dust bunker and Q_b is the theroetical flow rate of that so, as shown in Figure 3–23, a relationship between Q_b and Q_{bm} can be written as

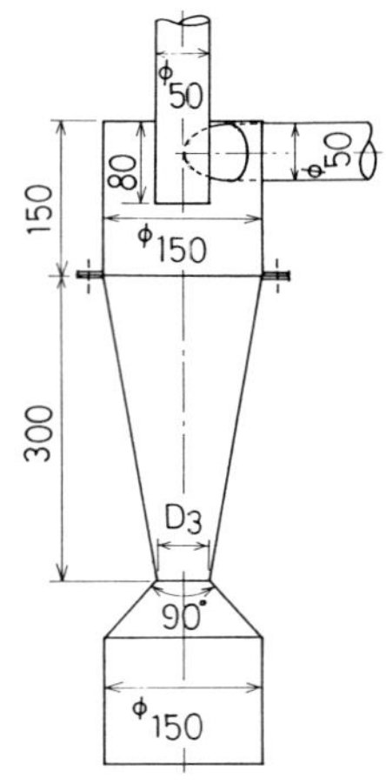

FIGURE 3–20. Main sizes of the test cyclone.

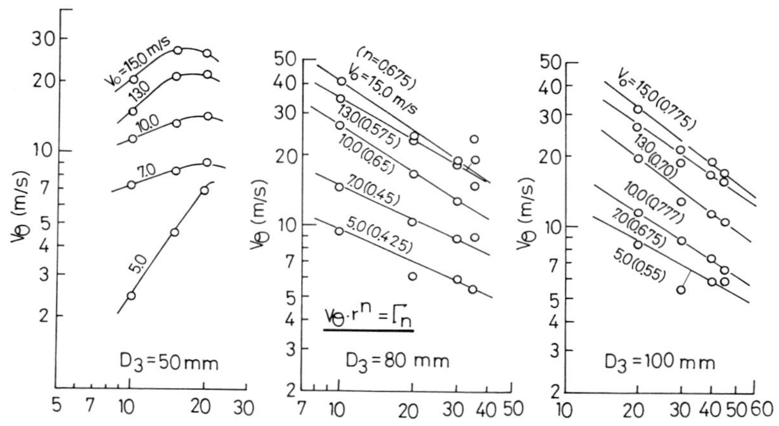

FIGURE 3–21. Distribution of the measured tangential velocities in the cyclone dust collectors.

$$\frac{Q_b}{Q_{bm}} = 2.56 \left(\frac{R_3 \cdot \Gamma}{Q_o}\right)^{0.4} \tag{3–126}$$

Consequently substituting Equations 3–123, 3–125, and 3–126 into Equation 3–121, we can obtain an empirical equation of flow rate Q_b of air flow into the dust bunker as follows:

$$\frac{Q_b}{Q_o} = 2.62 \left[1 - 0.36 \left(\frac{D_3}{D_1}\right)^{-0.56}\right]^{3/2} \cdot \left(\frac{D_3}{D_1}\right)^{1.16} \tag{3–127}$$

Figure 3–24 shows the relationship between the experimentally determined flow rate Q_b and the empirical equation 3–127.

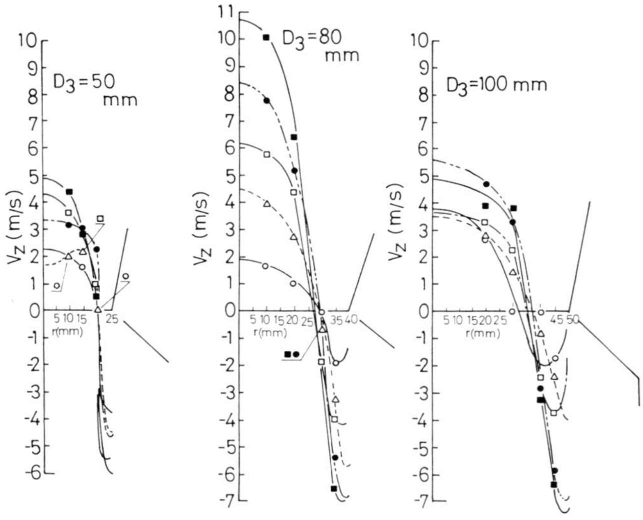

FIGURE 3–22. Distributions of the measured axial velocities at the throat of the cyclone dust collectors.

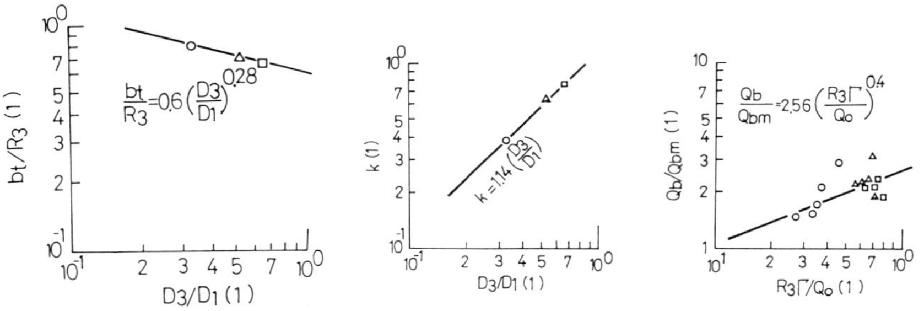

FIGURE 3–23. Experimental results of b_t/R_3, k, and Q_b/Q_{bm} for D_3/D_1.

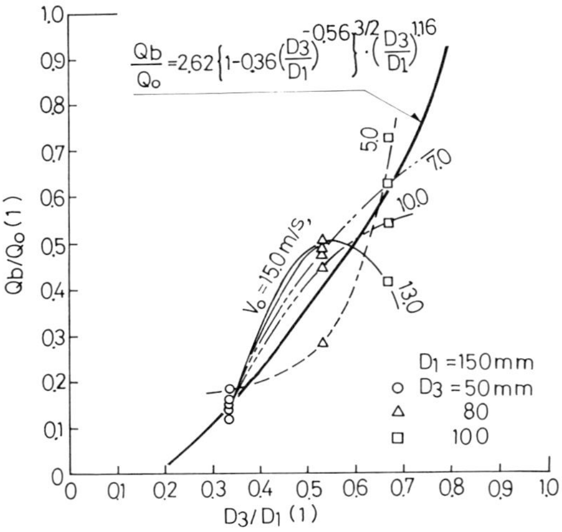

FIGURE 3–24. Relationship between the empirical equation and the experimental results of Q_b/Q_o for D_3/D_1.

MAIN SYMBOLS

a(m);	boundary radius between forced and free vortices
a_r(m/s^2);	radial acceleration
b(m);	radius corresponding to the axial velocity to be zero
a_θ(m/s^2);	tangential acceleration
D_1(m);	diameter of the cyclone dust collector
D_3(m);	diameter of the throat of the cyclone dust collector
F(N/kg);	external force per unit mass
F_r(N/kg);	external force per unit mass
F_θ(N/kg);	external force per unit mass
F_z(N/kg);	external force per unit mass
p(Pa);	static pressure
Q(m^3/s);	flow rate
Q_b(m^3/s);	flow rate into the dust bunker
u(m/s);	velocity for x-direction
V(m/s);	resultant velocity vector
V_r(m/s);	radial velocity
V_θ(m/s);	tangential velocity
V_z(m/s);	axial velocity
v(m/s);	velocity for y-direction
Γ(m^2/s);	circulation
ζ_r(1/r);	vorticity component
ζ_θ(1/s)	vorticity component
ζ_z(1/s);	vorticity component
θ(rad);	angle
ρ(kg/m^3);	density
Φ(N·m/kg);	potential energy per unit mass
Φ(m^2/s);	potential
Ψ(m^2/s);	stream function
ω_r(rad/s);	angular velocity
ω_θ(rad/s);	angular velocity
ω_z(rad/s);	angular velocity
$\mathbf{\Omega}$(1/s);	vorticity vector

REFERENCES

1. **Miyagi, O.,** *Hydrodynamics* (in Japanese), ARS-Pub. Co., Tokyo, 1935.
2. **Rankine, W. J. M.,** *Manual of Applied Mechanics,* Charles Griffin and Co., London, 1876.
3. **Ramsey, A. S.,** *A Treatise on Hydromechanics,* Part II, G. Bell & Sons, Ltd., London, 1935.
4. **Villat, H.,** Leçons sur la Théorie des Tourbillons, Gauthier-Villars et Cie, Paris, 1930.
5. **Ogawa, A.,** Flow in the dust bunker of cyclone dust collectors, Kikai No Kenkyu (in Japanese), 27(10), 1253, 1975.
6. **Ogawa, A.,** Flow rate of air flow entering into cyclone dust bunker, Preprint of J. Soc. Mech. Eng., (Nagano-prefecture), October 1974.
7. **Ogawa, A.,** Re-Entrainment of dusts from the dust layer on the deposited dusts in the cyclone dust collectors, *Particulate Sci. Technol.,* 2, 305, 1984.
8. **Binnie, A. M. and Hookings, G. A.,** Laboratory experiments on whirlpools, *Proc. R. Soc. London,* A-194, 398, 1948.

9. **Binnie, A. M., Hookings, G. A., and Kamel, M. Y. M.,** The flow of swirling water through a convergent-divergent nozzle, *J. Fluid Mech.*, 3, 261, 1958.
10. **Binnie, A. M. and Harris, D. P.,** The application of boundary layer theory to swirling liquid flow through a nozzle, *Q. J. Mech. Appl. Math.*, 3(1), 89, 1950.
11. **Ogawa, A.,** Sur la distribution de la vitesse tangentielle et ce rayon spécifique dans le cyclone dépoussiéreur, *Nihon Univ. J. Coll.*, A-24, 93, 1983.

Chapter 4

STREAM FUNCTION

I. INTRODUCTION

In this chapter, we introduce the stream function and potential of the axis-symmetrical flow in cylindrical coordinates. We also show applications of these.[1,2] One example is the stream lines of the logarithmic spiral and another example is the stream lines of the hyperbolic curves.

II. TWO-DIMENSIONAL FLOW

When the fluid flow is limited on the surface of r, θ plane, the equation of continuity can be written as

$$\frac{\partial(r \cdot V_r)}{\partial r} + \frac{\partial(r \cdot V_\theta)}{r\,\partial\theta} = 0 \tag{4–1}$$

Therefore the relationship between the stream function Ψ and the velocity components V_r, V_θ can be written as

$$V_r = -\frac{\partial\Psi}{r \cdot \partial\theta} \tag{4–2}$$

$$V_\theta = \frac{\partial\Psi}{\partial r} \tag{4–3}$$

Then concerning the rotation of the fluid particle on this plane, the following terms vanish

$$\frac{\partial V_z}{\partial\theta} = 0 \quad , \quad \frac{\partial V_z}{\partial r} = 0 \quad , \quad \frac{\partial(r \cdot V_\theta)}{\partial z} = 0 \quad , \quad \frac{\partial V_r}{\partial z} = 0 \tag{4–4}$$

so only the ζ component along the z-axis has the component of rotation

$$\zeta_z = \frac{1}{r}\left\{\frac{\partial(r \cdot V_\theta)}{\partial r} - \frac{\partial V_r}{\partial\theta}\right\} \tag{4–5}$$

Here when we consider the irrotational motion (or potential flow), Equation 4–5 becomes

$$\frac{\partial(r \cdot V_\theta)}{\partial r} - \frac{\partial V_r}{\partial\theta} = 0 \tag{4–6}$$

Then substituting Equations 4–2 and 4–3 into Equation 4–6, Equation 4–6 becomes

$$r\frac{\partial^2\Psi}{\partial r^2} + \frac{\partial\Psi}{\partial r} + \frac{\partial^2\Psi}{r \cdot \partial\theta^2} = 0 \tag{4–7}$$

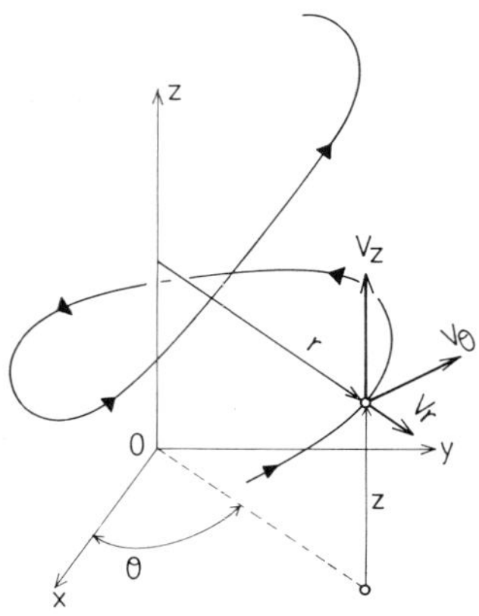

FIGURE 4–1. Cylindrical coordinates and the velocity components of fluid flow.

In this connection the interesting theoretical analysis of steady axis-symmetric flow with swirls for flow of effectively inviscid fluid with vorticity, which was analyzed with Bessel functions, was described by Professor G. K. Batchelor[3,4] and Professor C. S. Yih.[5]

On the other hand, the continuity equation for the two-dimensional flow ($\partial^2\Phi/\partial z^2 = 0$) by Equation 3–29 can be written as

$$r\frac{\partial^2\Phi}{\partial r^2} + \frac{\partial\Phi}{\partial r} + \frac{1}{r}\frac{\partial^2\Phi}{\partial\theta^2} = 0 \qquad (4\text{–}8)$$

Comparing Equations 4–7 and 4–8, the reader will recognize that Equations 4–7 and 4–8 are the equations of the same form. Here the equation of the stream line on this plane can be written from Equation 3–33 as

$$\frac{dr}{V_r} = \frac{r \cdot d\theta}{V_\theta} \qquad (4\text{–}9)$$

Therefore

$$V_\theta \cdot dr - r \cdot d\theta \cdot V_r = 0 \qquad (4\text{–}10)$$

Now substituting Equations 4–2 and 4–3 into Equation 4–10, we obtain

$$\frac{\partial\Psi}{\partial r}dr + \frac{\partial\Psi}{r \cdot \partial\theta}r \cdot d\theta = \frac{\partial\Psi}{\partial r} \cdot dr + \frac{\partial\Psi}{\partial\theta} \cdot d\theta = d\Psi = 0 \qquad (4\text{–}11)$$

Consequently the line of Ψ = constant is equivalent to the stream line.

III. AXIS-SYMMETRICAL FLOW

As shown in Figure 4–1, when the fluid flow is symmetrical with respect to the z-axis, the fluid flow can be described by the function of r and z. In this case the equation of continuity can be written as

$$\frac{\partial(r \cdot V_r)}{\partial r} + \frac{\partial(r \cdot V_z)}{\partial z} = 0 \qquad (4\text{–}12)$$

Therefore the relationship between Ψ and V_r, V_z can be represented by

$$V_r = -\frac{\partial \Psi}{r \cdot \partial z} \tag{4–13}$$

$$V_z = \frac{\partial \Psi}{r \cdot \partial r} \tag{4–14}$$

And concerning the rotation of the fluid particle, when the fluid flow is the irrotational motion, the vorticity components $\zeta_r = 0$ of Equation 3–47, $\zeta_\theta = 0$ of Equation 3–48, and $\zeta_z = 0$ of Equation 3–49 satisfy

$$r \cdot V_\theta = \text{constant} \tag{4–15}$$

$$\frac{\partial V_r}{\partial z} = \frac{\partial V_z}{\partial r} \tag{4–16}$$

Then substituting Equations 4–13 and 4–14 into Equation 4–16, the following equation can be obtained

$$r\frac{\partial^2 \Psi}{\partial r^2} - \frac{\partial \Psi}{\partial r} + r\frac{\partial^2 \Psi}{\partial z^2} = 0 \tag{4–17}$$

The equation of continuity of Equation 3–29 can be written as

$$r\frac{\partial^2 \Phi}{\partial r^2} + \frac{\partial \Phi}{\partial r} + r\frac{\partial^2 \Phi}{\partial z^2} = 0 \tag{4–18}$$

Here the equation of the stream line for the axis-symmetrical flow can be written from Equation 3–33 as

$$\frac{dr}{V_r} = \frac{dz}{V_z} \tag{4–19}$$

Then substituting Equations 4–13 and 4–14 into Equation 4–19, we can obtain the following

$$\frac{\partial \Psi}{\partial r} dr + \frac{\partial \Psi}{\partial z} dz = d\Psi = 0 \text{ or } \Psi = \text{constant} \tag{4–20}$$

Therefore for two-dimensional flow, the line of Ψ = constant is equivalent to the stream line.

IV. APPLICATION OF STREAM FUNCTION AND POTENTIAL

A. STREAM LINE OF LOGARITHMIC SPIRAL

When we consider the fluid motion of a free vortex on the r, θ plane, the product of the tangential velocity V_θ and radius r is constant

$$V_\theta \cdot r = \Gamma = \text{constant} \tag{4–21}$$

Then the continuity equation for the axis-symmetrical motion becomes

$$\frac{\partial(r \cdot V_r)}{\partial r} = 0 \tag{4–22}$$

Equation 4–22 becomes

$$V_r \cdot r = K = \text{constant}, \tag{4–23}$$

Combining Equations 4–21 and 4–23

$$\frac{r \cdot V_\theta}{r \cdot V_r} = \frac{V_\theta}{V_r} = \frac{\Gamma}{K} = M = \text{constant}, \tag{4–24}$$

Equation 4–24 becomes

$$\frac{r \cdot d\theta}{dr} = M \tag{4–25}$$

since $V_\theta = r \cdot d\theta/dt$ and $V_r = dr/dt$.

Now integrating Equation 4–25, we obtain

$$\theta = M \ln r + C \tag{4–26}$$

In order to determine the constant of integration, putting $\theta = \theta_o$ at $r = r_o$, Equation 4–26 becomes

$$\theta - \theta_o = M \ln \frac{r}{r_o} \tag{4–27}$$

or

$$r = r_o \exp\left(\frac{\theta - \theta_o}{M}\right) \tag{4–28}$$

These equations indicate the equation of stream line which is called a logarithmic spiral.

The equation of potential can be written as

$$d\Phi = \frac{\partial\Phi}{\partial r} dr + \frac{\partial\Phi}{\partial\theta} d\theta \tag{4–29}$$

and also the components $\partial\Phi/\partial r$ and $\partial\Phi/\partial\theta$ can be represented as

$$\frac{\partial\Phi}{\partial r} = -V_r = -\frac{K}{r}, \quad \frac{\partial\Phi}{\partial\theta} = -r \cdot V_\theta = -\Gamma \tag{4–30}$$

So equation 4–29 becomes

$$d\Phi = -\frac{K}{r} dr - \Gamma\, d\theta \tag{4–31}$$

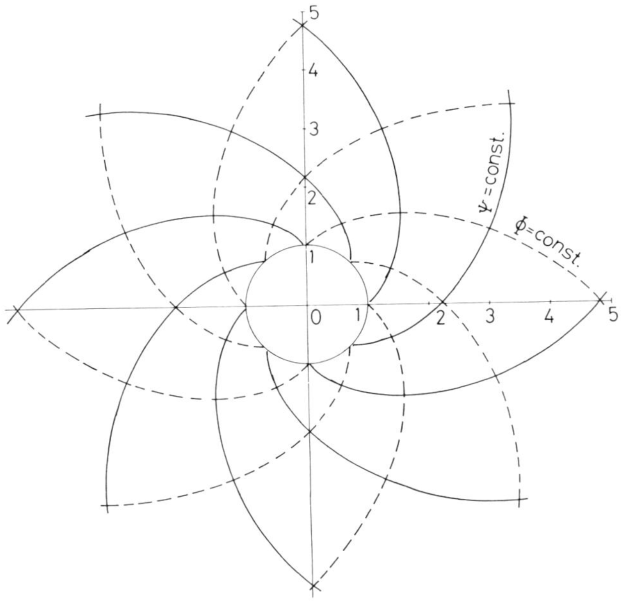

FIGURE 4–2. Numerical example of the stream function and potential for M = 1.0.

Since the value of $d\Phi$ is $d\Phi = 0$ along the equipotential line, Equation 4–31 can be transformed

$$\frac{dr}{r} = -\frac{\Gamma}{K} d\theta = -M\, d\theta \tag{4–32}$$

Integrating Equation 4–32 on the same boundary condition, we obtain

$$r = r_o \exp[-M(\theta - \theta_o)] \tag{4–33}$$

This equation is an equipotential line which is perpendicular to the stream line. Figure 4–2 shows one numerical example for $M = 1$ on the boundary condition $r_o = 1$ at $\theta_o = 0$.

B. AXIS-SYMMETRICAL FLOW

Consider the axis-symmetrical irrotational flow and assume that the radial V_r and axial V_z velocities can be written as

$$V_r = a \cdot r \quad , \quad V_z = -b \cdot z \tag{4–34}$$

where the constant values a and b are positive values. This flow model must satisfy the continuity equation

$$\frac{\partial(r \cdot V_r)}{\partial r} + \frac{\partial(r \cdot V_z)}{\partial z} = \frac{\partial(a \cdot r^2)}{\partial r} - \frac{\partial(b \cdot r \cdot z)}{\partial z} = 2 \cdot a \cdot r - b \cdot r = 0 \tag{4–35}$$

Then a relationship between a and b can be written as

$$2 \cdot a = b \tag{4–36}$$

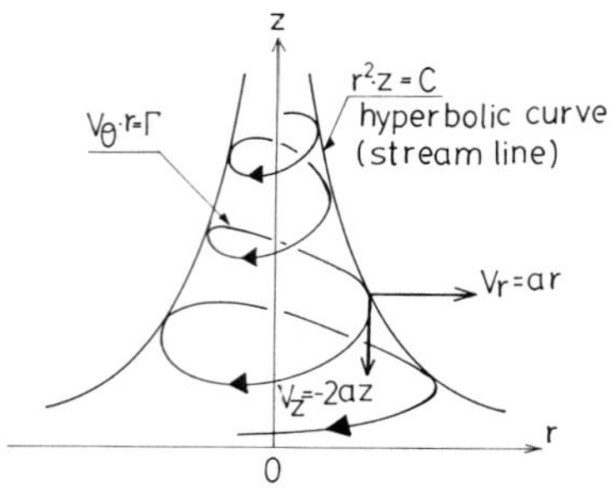

FIGURE 4–3. Schematic illustration of the hyperbolic curve on the axis-symmetric flow.

Therefore V_r and V_z can be described as

$$V_r = a \cdot r \quad , \quad V_z = -2 \cdot a \cdot z \tag{4–37}$$

From Equations 4–13 and 4–14 we obtain the following

$$V_r = a \cdot r = -\frac{\partial \Psi}{r \cdot \partial z} \tag{4–38}$$

$$V_z = -2az = \frac{\partial \Psi}{r \cdot \partial r} \tag{4–39}$$

Consequently equation $d\Psi$ can be transformed

$$d\Psi = \frac{\partial \Psi}{\partial r} \cdot dr + \frac{\partial \Psi}{\partial z} \cdot dz = -2 \cdot a \cdot z \cdot r \cdot dr - a \cdot r^2 \cdot dz \tag{4–40}$$

Here the equation of the stream line for $d\Psi = 0$ can be written as

$$2 \cdot r \cdot z \cdot dr + r^2 \cdot dz = 0 \tag{4–41}$$

or

$$\frac{2 \cdot dr}{r} + \frac{dz}{z} = 0 \tag{4–42}$$

Integrating Equation 4–42, we obtain the equation of the stream line

$$r^2 \cdot z = C \tag{4–43}$$

This equation shows the hyperbolic curves which have r- and z-axes as the asymptotes. A physical illustration is shown in Figure 4–3.

As shown in Figure 4–4, assuming that the flow rate flowing between the stream function Ψ and the stream function $\Psi + d\Psi$ is dQ, so

$$dQ = -2\pi \cdot r \cdot dr \cdot V_z \tag{4–44}$$

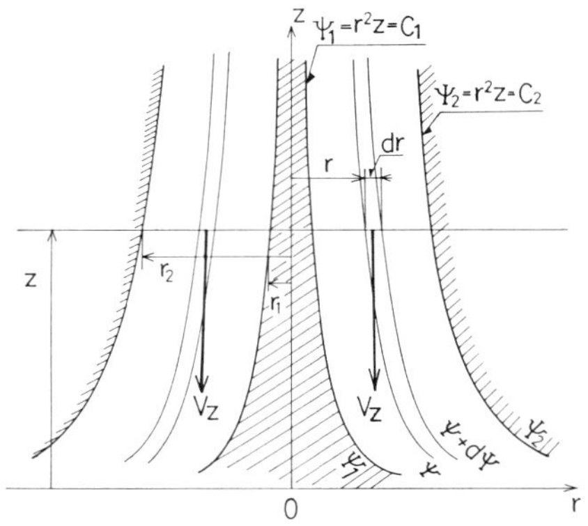

FIGURE 4–4. Relationship between the stream function and the flow rate.

Substituting Equation 4–39 into Equation 4–44, Equation 4–44 becomes

$$dQ = -2\pi \cdot r \cdot dr \frac{\partial\Psi}{r \cdot \partial r} = -2\pi \frac{\partial\Psi}{\partial r} \cdot dr \tag{4–45}$$

For a given value of z, Ψ is a function of r. Then Equation 4–45 becomes

$$dQ = -2\pi \cdot d\Psi \tag{4–46}$$

Denoting the total flow rate flowing between the stream lines Ψ_1 and Ψ_2 as Q, Equation 4–46 becomes

$$Q = -2\pi \int_{\Psi_1}^{\Psi_2} d\Psi = 2\pi(\Psi_1 - \Psi_2) \tag{4–47}$$

Substituting Equation 4–39 into Equation 4–44 yields

$$dQ = -2\pi r \cdot dr \cdot V_z = -2\pi r \cdot dr(-2 \cdot a \cdot z) = 4\pi \cdot a \cdot z \cdot r \cdot dr \tag{4–48}$$

Now fixing one plane of z and integrating Equation 4–44 between $r = r_1$ and $r = r_2$, we obtain

$$Q = 4\pi a \cdot z \int_{r_1}^{r_2} r \cdot dr = 2\pi a \cdot z \cdot (r_2^2 - r_1^2) \tag{4–49}$$

As shown in Figure 4–4, writing the stream functions at radii $r = r_1$ and $r = r_2$ as follows:

$$r_1^2 \cdot z = C_1 \quad , \quad r_2^2 \cdot z = C_2 \tag{4–50}$$

so Equation 4–49 becomes

$$Q = 2\pi \cdot a(C_2 - C_1) \tag{4–51}$$

Here assuming that the z-axis is one of the stream lines, C_1 becomes zero due to $r = r_1 = 0$. Therefore a relationship between a and Q can be written as

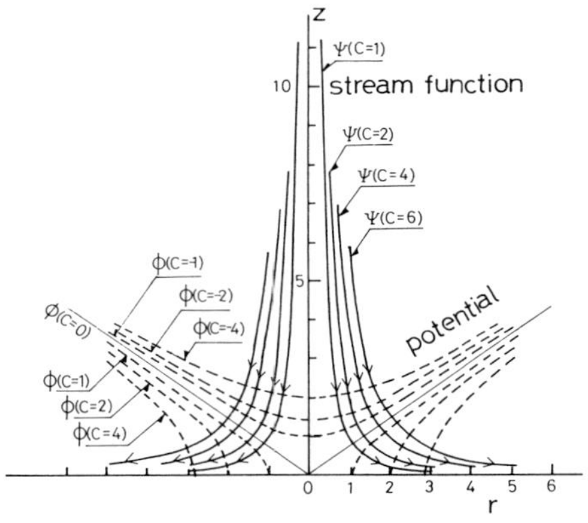

FIGURE 4–5. Numerical example of the stream function and potential.

$$a = \frac{Q}{2\pi \cdot C_2} = \frac{Q}{2\pi \cdot r_2^2 \cdot z} \tag{4–52}$$

The constant a can be estimated by Equation 4–52 of the flow rate Q.

By definition the velocity potential Φ, V_r, and V_z can be written as

$$V_r = -\frac{\partial \Phi}{\partial r} = a \cdot r \quad , \quad V_z = -\frac{\partial \Phi}{\partial z} = -2 \cdot a \cdot z \tag{4–53}$$

And then

$$d\Phi = \frac{\partial \Phi}{\partial r} \cdot dr + \frac{\partial \Phi}{\partial z} \cdot dz = -a \cdot r \cdot dr + 2 \cdot a \cdot z \cdot dz \tag{4–54}$$

Therefore the equation of the equipotential line for $d\Phi = 0$ can be written as

$$r \cdot dr - 2 \cdot z \cdot dz = 0 \tag{4–55}$$

Integrating Equation 4–55, we can obtain the following equation as

$$\frac{r^2}{2 \cdot C} - \frac{z^2}{C} = 1 \tag{4–56}$$

This equation is the equipotential line on the r-z plane and is the hyperbolic equation with semi-axes $\sqrt{2C}$ and $\sqrt{C}$. Now the asymptote can be obtained as

$$r = \pm \sqrt{\frac{2C}{C}}\, z = \pm \sqrt{2}\, z \tag{4–57}$$

Figure 4–5 shows the numerical example of the stream function and the potential Φ.

MAIN SYMBOLS

a(1/s);	constant value
b(1/s);	constant value
$K(m^2/s)$;	constant value
M(1);	constant value
$Q(m^3/s)$;	flow rate
r(m);	radius
$V_r(m/s)$;	radial velocity
$V_\theta(m/s)$;	tangential velocity
$V_z(m/s)$;	axial velocity
z(m);	axial velocity
$\Gamma(m^2/s)$;	circulation
$\zeta(1/s)$;	vorticity
$\zeta_z(1/s)$;	vorticity component along the z-axis
θ(rad);	angle
$\Phi(m^2/s)$;	potential
$\Psi(m^2/s)$;	stream function

REFERENCES

1. **Miyagi, O.,** *Hydrodynamics* (in Japanese), ARS-Pub. Co., Tokyo, 1935.
2. **Albring, W.,** Elementarvorgänge fluider Wirbelbewegungen, Akademie-Verlag, Berlin, 1981.
3. **Batchelor, G. K.,** *An Introduction to Fluid Dynamics,* Cambridge, 1967.
4. **Batchelor, G. K.,** Note of a class of solutions of the Navier-Stokes equation representing steady rotationally symmetric flow, *Q. J. Mech. Appl. Math.,* 4(1) 29, 1951.
5. **Yih, C. S.,** Two solutions for inviscid rotational flow with corner eddies, *J. Fluid Mech.,* 5, 36, 1959.

Chapter 5

CIRCULATION

I. INTRODUCTION

In this chapter, from the historical point of view, the fundamental ideas concerning the fluid mechanics of vortex flows are described in the following sections. The important theorems are illustrated by the figures in detail.

II. GAUSS THEOREM

As shown in Figure 5–1, the volume of fluid accumulating in the small volume (dx dy dz) per unit time can be written as

$$\left[\frac{\partial(\rho V_x)}{\partial x} + \frac{\partial(\rho V_y)}{\partial y} + \frac{\partial(\rho V_z)}{\partial z}\right] dx\, dy\, dz \qquad (5\text{–}1)$$

Therefore the total volume of fluid accumulating in the total volume of space per unit time is

$$\iiint\left[\frac{\partial(\rho V_x)}{\partial x} + \frac{\partial(\rho V_y)}{\partial y} + \frac{\partial(\rho V_z)}{\partial z}\right] dx\, dy\, dz \qquad (5\text{–}2)$$

Then as shown in Figure 5–2, denoting that ds is the small surface area of the enveloping surface on the given volume, and that **V** is the velocity of fluid flowing through this small surface from inside to outside, and further that θ is the angle between **V** and the vertical line N on the small surface ds, the mass of fluid flowing through this small surface ds per unit time is

$$\rho \mathbf{V} \cos\theta\, ds \qquad (5\text{–}3)$$

However the total mass of fluid penetrating in the given space per unit time must be the difference between the mass flowing into this space per unit time and the mass flowing out per unit time. While we define the positive value of N having the direction from inside to outside of this space, and we also measure the angle θ from this positive value of N to the velocity **V**, the difference of the mass between the mass of fluid flowing into this space per unit time and the mass of fluid flowing out from this space per unit time is equal to

$$\int \rho \mathbf{V} \cos\theta\, ds \qquad (5\text{–}4)$$

Consequently we can obtain the following equation

$$\iiint\left[\frac{\partial(\rho V_x)}{\partial x} + \frac{\partial(\rho V_y)}{\partial y} + \frac{\partial(\rho V_z)}{\partial z}\right] dx\, dy\, dz = \int \rho\, \mathbf{V} \cos\theta\, ds \qquad (5\text{–}5)$$

The left-hand side of this equation gives the volume integral in the total volume of the given space and the right-hand side of this equation gives the surface integral on the surrounding

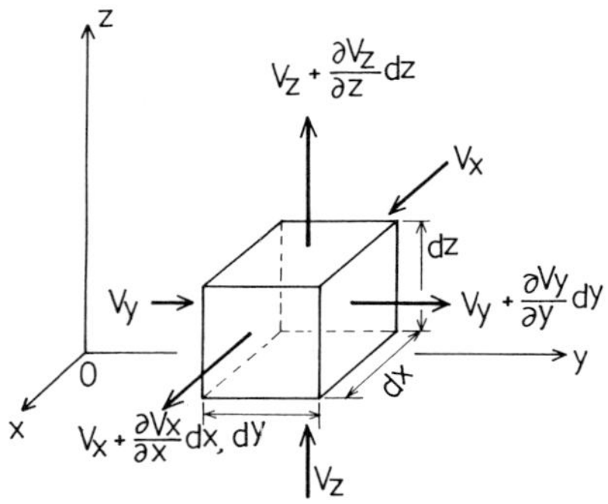

FIGURE 5–1. x, y, z-coordinate system and the velocity components.

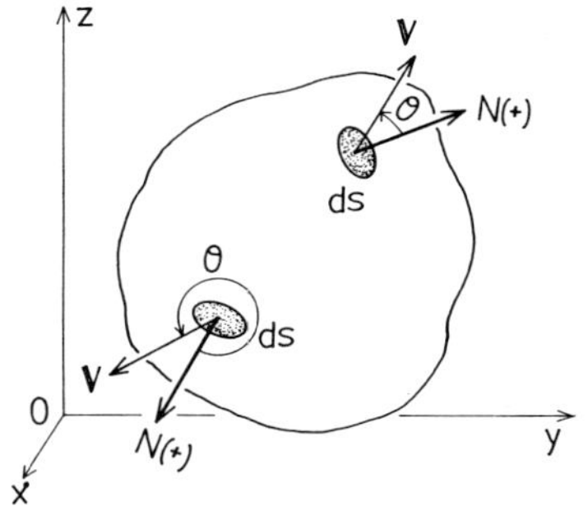

FIGURE 5–2. Schematic illustration of the closed surface and the velocity vector.

surface of this space. Therefore this equation shows a relationship between the volume integral and the surface integral, and in general this equation is called a Gauss integral. This equation can be written by the vector notation as

$$\int \mathrm{div}(\rho \mathbf{V})\ \mathrm{dv} = \int \rho \mathbf{V} \cdot \mathbf{n}\ \mathrm{ds} \tag{5–6}$$

where dv is the small volume element (dx dy dz) and **V** is the velocity vector and **n** is the unit vector of the vertical line N, and the positive value of **n** looks to the outward direction of this surface. Further assuming that the fluid is incompressible (ρ = constant), so Equation 5–6 becomes

$$\iiint \left(\frac{\partial V_x}{\partial x} + \frac{\partial V_y}{\partial y} + \frac{\partial V_z}{\partial z} \right) \mathrm{dx\ dy\ dz} = \int \mathbf{V} \cos\theta\ \mathrm{ds} \tag{5–7}$$

and in addition to this, Equation 5–6 becomes

$$\int \mathrm{div}\ \mathbf{V}\ \mathrm{dv} = \int \mathbf{V} \cdot \mathbf{n}\ \mathrm{ds} \tag{5–8}$$

The continuity equation of the incompressible fluid can be written as

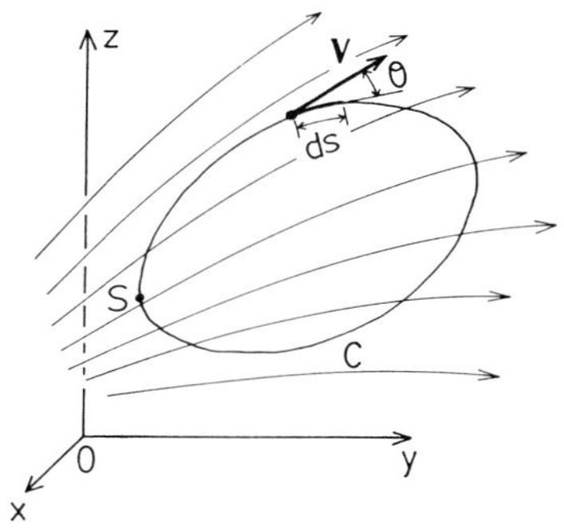

FIGURE 5–3. Arbitrary closed circuit and the velocity vector in fluid flow.

$$\frac{\partial V_x}{\partial x} + \frac{\partial V_y}{\partial y} + \frac{\partial V_z}{\partial z} = 0 \qquad (5\text{–}9)$$

so Equation 5–7 becomes

$$\int \mathbf{V} \cos\theta \, ds = 0 \qquad (5\text{–}10)$$

III. DEFINITION OF CIRCULATION

As shown in Figure 5–3, forming an arbitrary closed circuit C in the fluid flow, the circulation Γ can be defined as

$$\int_C V \cos\theta \, ds \qquad (5\text{–}11)$$

where Equation 5–11 means that $\int_C$ integrates one revolution along the closed circuit C. Hence Equation 5–11 can be transformed

$$\Gamma = \int_C V \cos\theta \, ds = \int_C (V_x \, dx + V_y \, dy + V_z \, dz) \qquad (5\text{–}12)$$

This equation can be written by vector notation as

$$\Gamma = \int_C \mathbf{V} \cdot \mathbf{ds} \qquad (5\text{–}13)$$

where

$$\mathbf{V} = V_x \, \mathbf{i} + V_y \, \mathbf{j} + V_z \, \mathbf{k}$$

$$\mathbf{ds} = dx \, \mathbf{i} + dy \, \mathbf{j} + dz \, \mathbf{k}$$

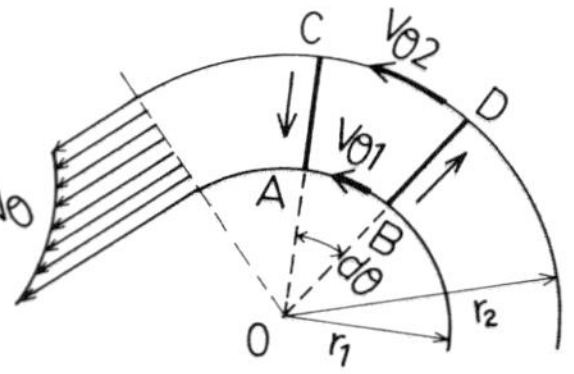

FIGURE 5–4. Schematic illustration of rotational flow and a closed circuit ABCD.

Assuming the fluid flow is the velocity potential Φ, the velocity components can be described as

$$V_x = -\frac{\partial \Phi}{\partial x}, \quad V_y = -\frac{\partial \Phi}{\partial y}, \quad V_z = -\frac{\partial \Phi}{\partial z} \tag{5–14}$$

Then Equation 5–12 becomes

$$\Gamma = \int_C V \cos\theta \, ds = -\int \left(\frac{\partial \Phi}{\partial x} dx + \frac{\partial \Phi}{\partial y} dy + \frac{\partial \Phi}{\partial z} dz \right) = -\int d\Phi \tag{5–15}$$

Therefore Equation 5–15 becomes

$$\Gamma = \Phi_S - \Phi_S \tag{5–16}$$

where the first term of the right-hand side of Equation 5–16 is the value of Φ_A at point A and the second term is the value of Φ at A after completing one revolution along the circuit C. Then when the function Φ is the single valued function, Equation 5–16 becomes zero ($\Gamma = 0$).

As described above, the fluid flow having a velocity potential Φ is the irrotational flow. Hence considering the rotating flow of the incompressible fluid around an arbitrary point and in order to keep the irrotational flow, the tangential velocity V_θ of the flow must be inversely proportional to the radius r; that is to say, this flow must satisfy the following condition:

$$r \cdot V_\theta = K \tag{5–17}$$

Now considering a rotating flow around a point 0 as shown in Figure 5–4, AB, CD are also the routes of flow at the radii r_1 and r_2, respectively, and further these tangential velocities are $V_{\theta 1}$ and $V_{\theta 2}$; so the circulation Γ of this route ABDCA can be calculated as

$$\Gamma = \int \mathbf{V} \cdot ds = \int_{AB} (-V_{\theta 1}) \, d\theta \, r_1 + \int_{BD} \mathbf{V} \cdot ds + \int_{DC} (V_{\theta 2}) \, d\theta \, r_2 + \int_{CA} \mathbf{V} \cdot ds \tag{5–18}$$

The second term and fourth terms of the right-hand side of Equation 5–18 go to zero, since the velocity of flow is perpendicular to the routes BD and CA. However, from Equation 5–17

$$r_1 \cdot V_{\theta 1} = r_2 \cdot V_{\theta 2} = \Gamma \tag{5–19}$$

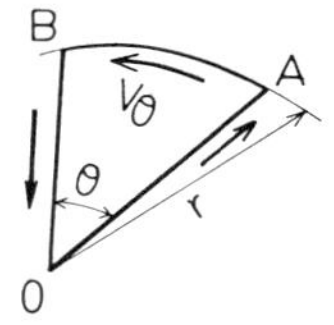

FIGURE 5–5. Circulation along the fan-shaped route.

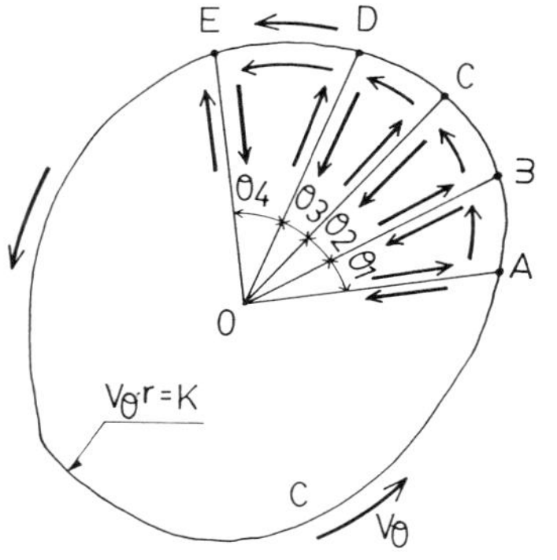

FIGURE 5–6. Circulation of an arbitrary circuit C gathering of the small circuits.

Therefore Equation 5–18 becomes

$$\Gamma = -V_{\theta 1} \cdot r_1 \cdot \theta + V_{\theta 2} \cdot r_2 \cdot \theta = (-V_{\theta 1} \cdot r_1 + V_{\theta 2} \cdot r_2)\,\theta = 0 \qquad (5\text{–}20)$$

However, we try to calculate the circulation along the fan-shaped route of OABO, where the center of the rotating flow is zero as shown in Figure 5–5. The circulation becomes

$$\Gamma = \int_{OA} \mathbf{V} \cdot \mathbf{ds} + \int_{AB} \mathbf{V} \cdot \mathbf{ds} + \int_{BO} \mathbf{V} \cdot \mathbf{ds} = \int_{OA} \mathbf{V} \cdot \mathbf{ds} + V_\theta \cdot r \cdot \theta + \int_{BO} \mathbf{V} \cdot \mathbf{ds} \qquad (5\text{–}21)$$

Then the first and third terms of the right-hand side of Equation 5–21 go to zero, since the tangential velocity V_θ is perpendicular to the routes OA and BO. Therefore in this case the circulation can be written as

$$\Gamma = V_\theta \cdot r \cdot \theta \qquad (5\text{–}22)$$

Further we consider the circulation of an arbitrary circuit C having a center O which can be regarded as the gathering of the small circuits, OABO, OBCO, OCDO, ODEO, . . . as shown in Figure 5–6.

The integrations OA and AO, OB and BO, . . . eliminate each other, since the direction of these routes are opposite each other, OA and AO, OB and BO, . . . Therefore the circulation Γ along the circuit C can be calculated as

$$\Gamma = K(\theta_1 + \theta_2 + \theta_3 + \ldots + \theta_n) \qquad (5\text{–}23)$$

Since this circuit C is traveled on circuit around O

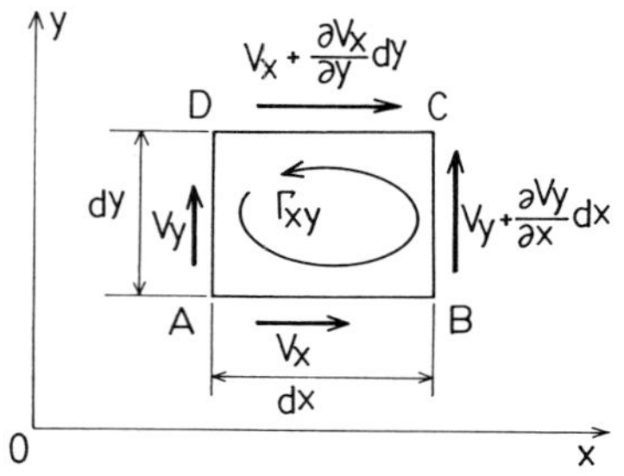

FIGURE 5–7. Circulation along ABCDA of a small rectangle.

$$\theta_1 + \theta_2 + \theta_3 + \theta_n = 2\pi \tag{5–24}$$

Consequently the circulation becomes

$$\Gamma = 2\pi K = 2\pi \cdot V_\theta \cdot r \tag{5–25}$$

To sum up the above described rules, when the center O of this rotating flow is located outside of this circuit C, the circulation is $\Gamma = 0$. When the center O of this rotating flow is located in this circuit C, the circulation becomes $\Gamma = 2\pi K = 2\pi \cdot V_\theta \cdot r$. Thus the circulation is independent of the shapes and the sizes of these circuits. In addition to this, the positive value of this circulation is defined so that the progress of this integration is carried out along the circuit, in which we always see the given surface by the enclosed circuit C to the left-hand side as shown in Figure 5–6.

IV. STOKES THEOREM

Let us consider a small rectangle having two sides dx and dy on the x-y plane as shown in Figure 5–7. Then the fluid velocity on the AB-plane is V_x, so that on the DC-plane it becomes $V_x + (\partial V_x/\partial y)dy$, and on the AD-plane it is V_y, so that on BC-plane it is $V_y + (\partial V_y/\partial x)dx$. Therefore the circulation Γ_{xy} along ABCDA can be represented by

$$\Gamma_{xy} = V_x dx + \left(V_y + \frac{\partial V_y}{\partial y} dx\right) dy - \left(V_x + \frac{\partial V_x}{\partial y} dy\right) dx - V_y dy =$$

$$= \frac{\partial V_y}{\partial x} dxdy - \frac{\partial V_x}{\partial y} dxdy = \left(\frac{\partial V_y}{\partial x} - \frac{\partial V_x}{\partial y}\right) dxdy = \zeta_z\, dxdy \tag{5–26}$$

Hence, in a similar manner we can obtain the circulations Γ_{yz} on the y-z plane and Γ_{zx} on the x-z plane as

$$\Gamma_{yz} = \left(\frac{\partial V_z}{\partial y} - \frac{\partial V_z}{\partial x}\right) dzdx = \zeta_x\, dydz \tag{5–27}$$

$$\Gamma_{zx} = \left(\frac{\partial V_x}{\partial z} - \frac{\partial V_z}{\partial x}\right) dzdx = \zeta_y\, dzdx \tag{5–28}$$

Further we consider the circulation of Figure 5–8. Here, summing up the circulation OABO (x-y plane), OBCO(y-z plane) and OCAO(z-x plane), the integration ($V_x dx$) of OA on the

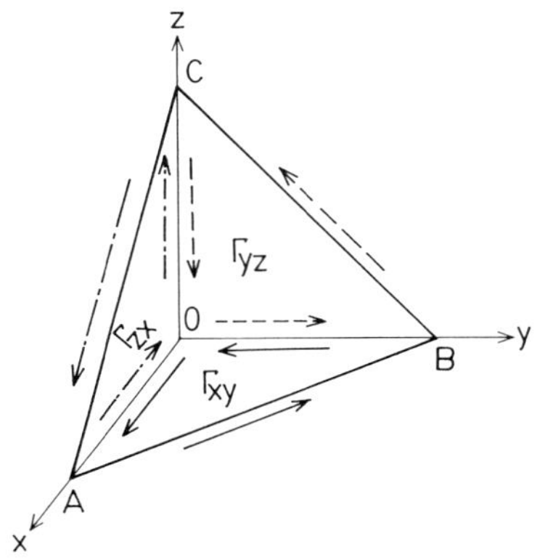

FIGURE 5–8. Schematic illustration of the circulations.

x-y plane and that $(-V_x dx)$ of OA on the z-x plane cancel each other due to the inverse directions of these integrations and, also in a similar manner, OB on y-z plane and BO on x-y plane and also OC on x-z plane and CO on y-z plane, respectively. Hence the summation of the above circulations Γ_{xy}, Γ_{yz}, and Γ_{zx} becomes the circulation of the arbitrary circuit ABCA in the given space. Consequently the circulation $d\Gamma$ of this circuit ABCA can be written as

$$d\Gamma = \Gamma_{xy} + \Gamma_{yz} + \Gamma_{zx} =$$

$$= \left(\frac{\partial V_y}{\partial x} - \frac{\partial V_x}{\partial y}\right) dxdy + \left(\frac{\partial V_z}{\partial y} - \frac{\partial V_y}{\partial z}\right) dydz + \left(\frac{\partial V_x}{\partial z} - \frac{\partial V_z}{\partial x}\right) dzdx \quad (5\text{–}29)$$

Now let us denote the small area of ABC by dS, and the direction cosine of a perpendicular line to this area shows by l, m, and n, so

$$dydz = l\ dS,\ dzdx = m\ dS,\ dxdy = n\ dS \quad (5\text{–}30)$$

Substituting Equation 5–30 into Equation 5–29, the circulation d can be transformed

$$d\Gamma = \left[l\left(\frac{\partial V_x}{\partial y} - \frac{\partial V_y}{\partial z}\right) + m\left(\frac{\partial V_x}{\partial z} - \frac{\partial V_z}{\partial x}\right) + n\left(\frac{\partial V_y}{\partial x} - \frac{\partial V_x}{\partial y}\right)\right] dS$$

$$= (l\ \zeta_x + m\ \zeta_y + n\ \zeta_z)\ dS \quad (5\text{–}31)$$

Therefore the circulation of this small circuit is proportional to the area dS. Therefore as shown in Figure 5–9, the circulation Γ of this circuit C can be written as

$$\Gamma = \int(l\ \zeta_x + m\ \zeta_y + n\ \zeta_z)\ dS \quad (5\text{–}32)$$

Further let us denote that the resultant vorticity of ζ_x, ζ_y, and ζ_z is Ω, and the angle between the direction of Ω and the direction N perpendicular to the small surface dS as shown in Figure 5–10 is O, so the relationship between ζ_x, ζ_y, ζ_z, and Ω can be represented by

$$\Omega \cos\theta = l\ \zeta_x + m\ \zeta_y + n\ \zeta_z \quad (5\text{–}33)$$

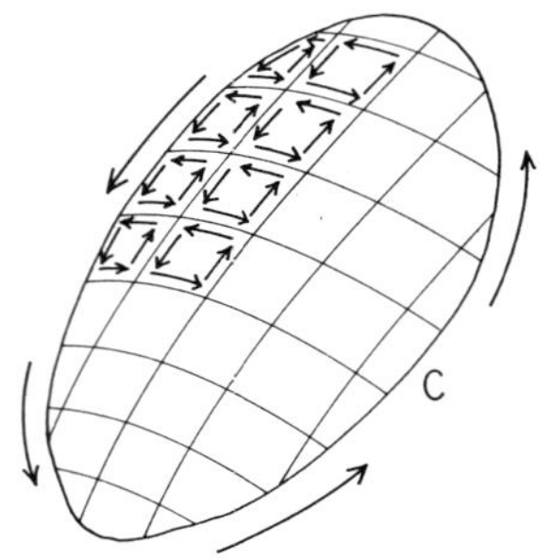

FIGURE 5–9. Circulation along a circuit C.

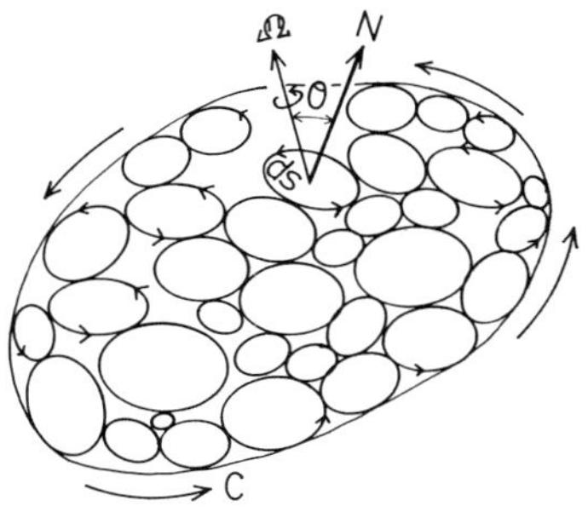

FIGURE 5–10. Vorticity and unit vector.

Hence Equation 5–32 can be transformed as

$$\Gamma = \int \Omega \cos\alpha \, dS \tag{5–34}$$

and Γ could be represented by Equation 5–13 as

$$\Gamma = \int_C \mathbf{V} \cdot \mathbf{ds}$$

Consequently, the relationship between Equation 5–13 and Equation 5–34 can be written by vector notation

$$\Gamma = \int_C \mathbf{V} \cdot \mathbf{ds} = \int_S \text{rot } \mathbf{V} \cdot \mathbf{ds} \tag{5–35}$$

This equation shows that a line integral along the circuit C is equal to the surface integral enclosed by this circuit C. This equation is generally called Stokes theorem.

V. MULTIPLY CONNECTED REGIONS

Let us assume that the given surface is a surface enclosed by the two circuits C_1 and C_2 as shown in Figure 5–11. In this case when we regard ABC BAC A as one continuous circuit by the connected imaginary line AB between C_1 and C_2, the line integral of $\int_{\overrightarrow{AB}}$ (segment $\int_{\overrightarrow{AB}}$) is canceled by that of $\int_{\overrightarrow{BA}}$ (segment $\int_{\overrightarrow{BA}}$); therefore, the circulation Γ of this circuit is the summation of Γ_1 of the circuit C_1 and Γ_2 of the circuit C_2

$$\Gamma = \Gamma_1 + \Gamma_2 \tag{5–36}$$

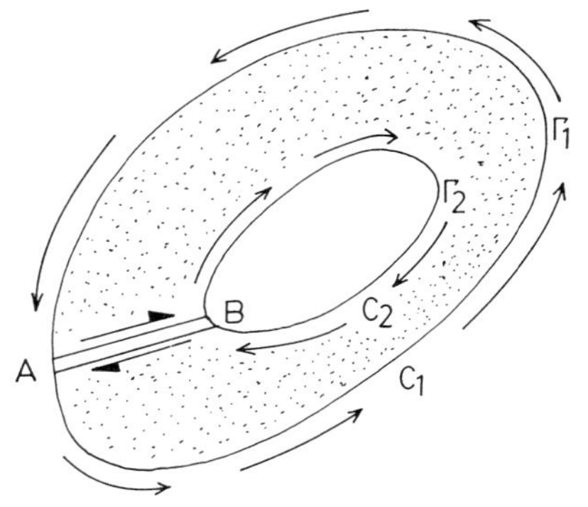

FIGURE 5–11. Schematic illustration of the enclosed surface by the two circuits C_1 and C_2.

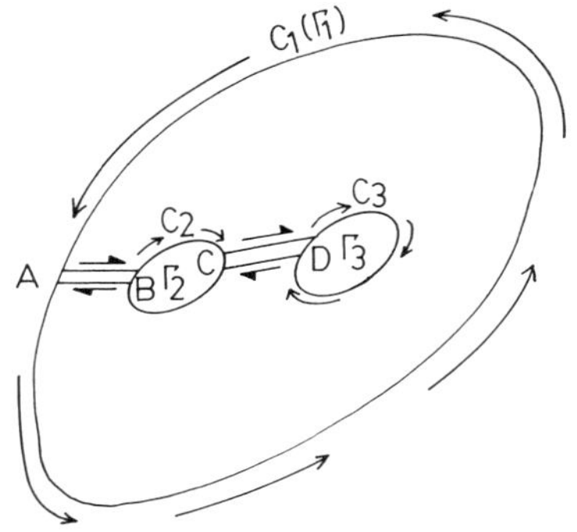

FIGURE 5–12. Circulation along the given surface enclosed by these three circuits C_1, C_2, and C_3.

The positive direction of integration along this circuit must be the direction of the arrows as shown in Figure 5–11; so when we see the given surface enclosed by circuits C_1 and C_2 to the left-hand side like the progressing directions of arrows, we define the positive direction. Similarly as shown in Figure 5–12, when the given surface is enclosed by these circuits C_1, C_2, and C_3, the circulation Γ of the circuit $ABC_2CDC_3DCC_2BAC_1A$ of the circulation Γ_1 of the circuit C_1 and Γ_2 of C_2 and Γ_3 of C_3 can be calculated as

$$\Gamma = \Gamma_1 + \Gamma_2 + \Gamma_3 \tag{5–37}$$

VI. STRENGTH OF VORTEX

The continuous line which is continuously connected in succession to each axis of the resultant vorticity $\mathfrak{N}_i$ is called a vortex line as shown in Figure 5–13. The axis of this vorticity is always tangential to the vortex line. And further a tube which is enclosed by the vortex lines is called a vortex tube as shown in Figure 5–14. Now as shown in Figure 5–15, let us consider the circulation of a part of the vortex tube. Then Γ_1 is the circulation of a circuit C_1 which encloses the sectional area dS_1 of this vortex tube, and Γ_2 is that of C_2 of a circuit C_2 enclosing the sectional area dS_2. Therefore the circulation of this circuit ABC_2BAC_1A can be calculated as

$$\Gamma = \Gamma_1 + \Gamma_2 \tag{5–38}$$

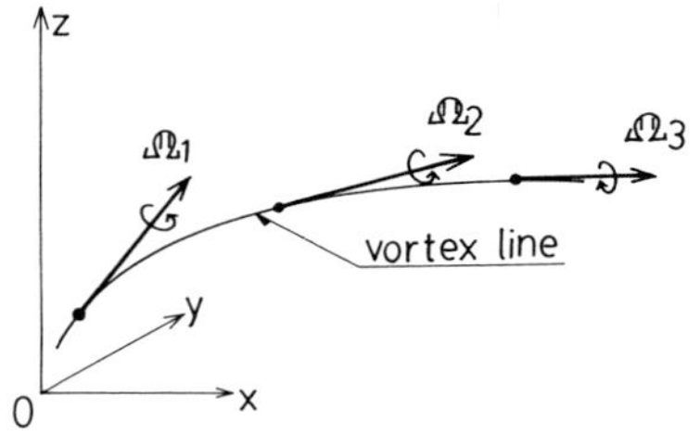

FIGURE 5–13. Illustration of a vortex line.

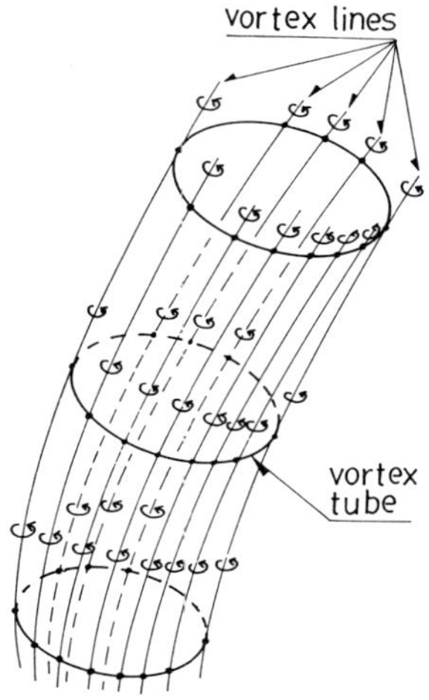

FIGURE 5–14. Illustration of a vortex tube and the vortex lines.

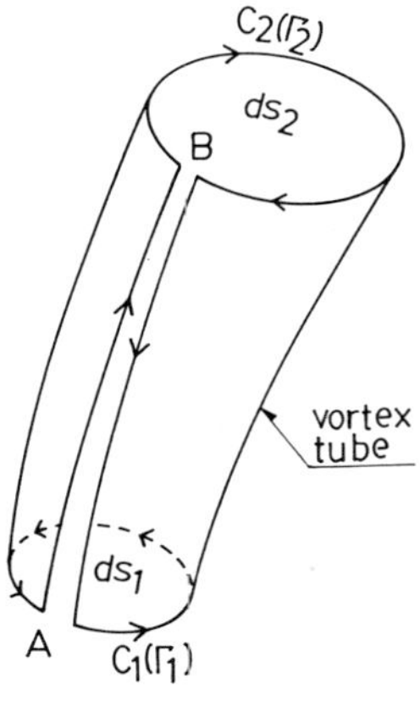

FIGURE 5–15. Schematic illustration of the calculation of the circulation.

since the integral $\int_{\overrightarrow{AB}} \mathbf{V} \cdot \mathbf{ds}$ is canceled by the line integral $\int_{\overrightarrow{BA}} \mathbf{V} \cdot \mathbf{ds}$. Further the axis of this vortex tube is always tangential to its side surface. So, the circulation Γ goes to zero due to the Stokes theorem of Equation 5–35

$$\Gamma = \Gamma_1 + \Gamma_2 = 0 \tag{5–39}$$

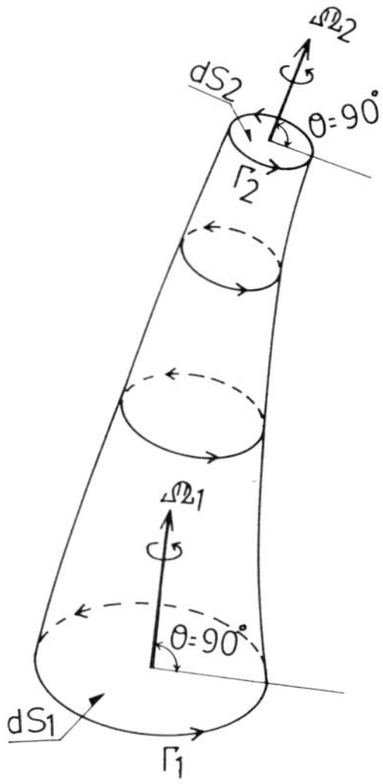

FIGURE 5–16. Illustration of the vorticities and the small cross-sectional areas of the vortex tube.

But the direction of these two circuits C_1 and C_2 are mutually opposite; so, a sign of the circulation Γ_2 becomes negative. Therefore, Equation 5–39 becomes

$$\Gamma_1 - \Gamma_2 = 0 \tag{5–40}$$

Consequently,

$$\Gamma_1 = \Gamma_2 \tag{5–41}$$

On the other hand, as shown in Figure 5–16, the small cross-sectional surface of area dS_1 is perpendicular to the vorticity axis $\boldsymbol{\Omega}_1$, but the direction of the vector of dS_1 is parallel to the vorticity axis $\boldsymbol{\Omega}_1$ and that of dS_2 is also perpendicular to the vorticity axis $\boldsymbol{\Omega}_2$. So by the Stokes theorem

$$\Gamma_1 = \boldsymbol{\Omega}_1 dS_1 \quad , \quad \Gamma_2 = \boldsymbol{\Omega}_2 dS_2 \tag{5–42}$$

respectively. Substituting Equation 5–42 into Equation 5–41, we obtain the following relation

$$\Omega dS_1 = \Omega_2 dS_2 \tag{5–43}$$

Consequently if we denote Ω as the vorticity at the given sectional area dS of the vortex tube, we can obtain the following equation

$$\Omega dS = \text{constant} \tag{5–44}$$

This equation shows that the magnitude of a vorticity $\boldsymbol{\Omega}$ of a given vortex tube is inversely proportional to the sectional area dS. Hence where the sectional area of the vortex tube is small, so the magnitude of the vorticity becomes large due to Equation 5–44. Note that ΩdS = constant is the strength of the vortex. Here as shown in Figure 5–17, assuming that the given surface is a circumferential surface of the opening mouthpiece of the bag like a circuit AB, the circulation Γ is equal to the circulation of the circuit AB. On the other hand,

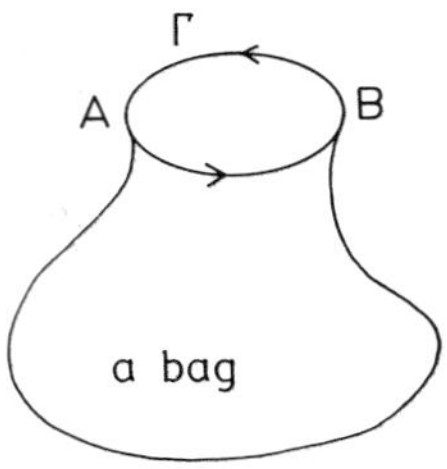

FIGURE 5–17. Opening mouthpiece of the bag like a circuit.

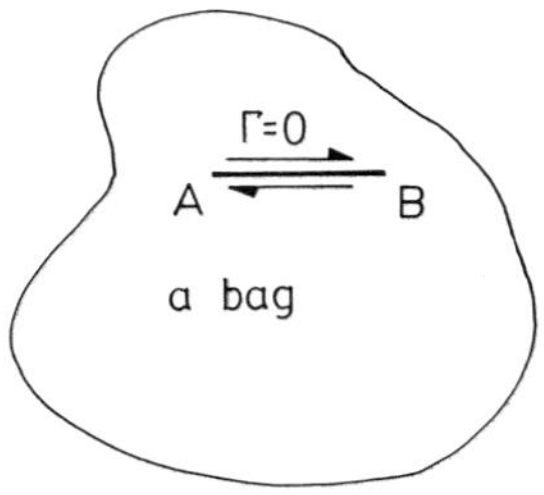

FIGURE 5–18. Completely closed surface of the mouthpiece of the bag.

as shown in Figure 5–18, assuming that the given surface is a completely closed surface of the mouthpiece of the bag like a line AB, the circulation on the closed surface must be zero, since the line integrals AB and BA cancel each other. Therefore in case of the closed surface, the value of the circulation becomes zero due to Stokes theorem of Equation 5–35. Then applying the Gauss theorem of Equation 5–6,

$$\iiint \left(\frac{\partial \zeta_x}{\partial x} + \frac{\partial \zeta_y}{\partial y} + \frac{\partial \zeta_z}{\partial z} \right) dxdydz = \int_S \text{rot}\mathbf{V} \cdot d\mathbf{S} = 0 \tag{5–45}$$

Hence, from Equation 5–45 we can obtain the following relation

$$\frac{\partial \zeta_x}{\partial x} + \frac{\partial \zeta_y}{\partial y} + \frac{\partial \zeta_z}{\partial z} = 0 \tag{5–46}$$

and if we rewrite this equation by a vector notation, Equation 5–46 becomes

$$\text{div rot}\mathbf{V} = 0 \tag{5–47}$$

This equation is the equation of continuity of vorticity.

VII. CONSTANCY OF CIRCULATION (W. THOMSON THEOREM)

As shown in Figure 5–19, assuming that the arbitrary two points A and B in a fluid move with a steady flow of the fluid, then we consider the arbitrary approach to two points P and Q on the connected line of A and B. Here let us denote that the coordinate of a point P is given by (x, y, z) and the fluid velocity at P is given by (Vx, Vy, Vz) and these of Q are given by (x + dx, y + dy, z + dz) and (Vx + dVx, Vy + dVy, Vz + dVz), respectively.

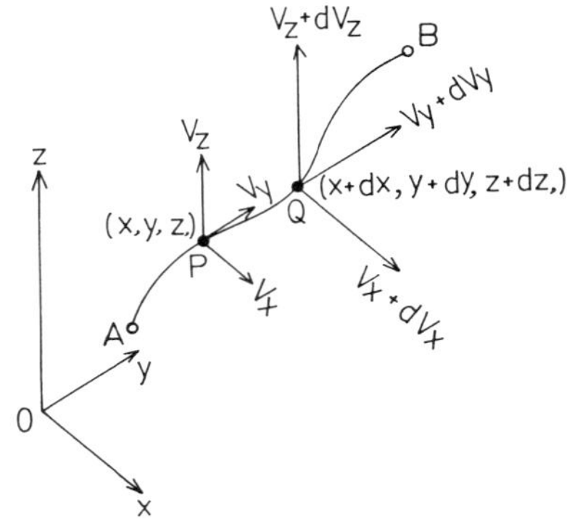

FIGURE 5–19. Schematic illustration of the arbitrary two points in fluid moving with steady flow.

Therefore the rate of change of $V_x dx$ for the x-direction per unit time can be represented as

$$\frac{d}{dt}(V_x dx) = \frac{dV_x}{dt} dx + V_x d\left(\frac{dx}{dt}\right) = \frac{dV_x}{dt} dx + V_x dV_x \tag{5–48}$$

However the equation of motion per unit mass of fluid for x-direction could be written as

$$\frac{dV_x}{dt} = X - \frac{1}{\rho}\frac{\partial p}{\partial x} \tag{5–49}$$

where X is the external force for x-direction on unit mass of fluid. Hence Equation 5–48 can be transformed to

$$\frac{d(V_x dx)}{dt} = \left(X - \frac{1}{\rho}\frac{\partial p}{\partial x}\right) dx + V_x dV_x \tag{5–50}$$

Similarly we can obtain the expression for y- and z-directions as follows

$$\frac{d(V_y dy)}{dt} = \left(Y - \frac{1}{\rho}\frac{\partial p}{\partial y}\right) dy + V_y dV_y \tag{5–51}$$

$$\frac{d(V_z dz)}{dt} = \left(Z - \frac{1}{\rho}\frac{\partial p}{\partial z}\right) dz + V_z dV_z \tag{5–52}$$

where Y and Z are the external forces for y- and z-directions in a unit mass of fluid. Summing up these three equations 5–50, 5–51, and 5–52, we obtain

$$\frac{d}{dt}(V_x dx + V_y dy + V_z dz) = Xdx + Ydy + Zdz - \frac{1}{\rho}$$
$$\left(\frac{\partial p}{\partial x} dx + \frac{\partial p}{\partial y} dy + \frac{\partial p}{\partial z} dz\right) + V_x dV_x + V_y dV_y + V_z dV_z \tag{5–53}$$

In this equation Xdx + Ydy + Zdz show the potential energy of unit mass of fluid; so, in case of the existence of potential Φ, we can represent this equation as

$$Xdx + Ydy + Zdz = -d\Phi \tag{5–54}$$

and further we transform the static pressure terms as

$$\frac{\partial p}{\partial x} dx + \frac{\partial p}{\partial y} dy + \frac{\partial p}{\partial z} dz = dp \tag{5–55}$$

Let us represent the resultant velocity of fluid as

$$V^2 = V_x^2 + V_y^2 + V_z^2 \tag{5–56}$$

So from Equation 5–56 we obtain

$$V_x dV_x + V_d V_y + V_z dV_z = VdV \tag{5–57}$$

Consequently substituting these equations 5–54, 5–55, and 5–56 into Equation 5–53, we obtain

$$\frac{d}{dt}(V_x dV_x + V_y dV_y + V_z dV_z) = -d\Phi - \frac{1}{\rho} dp + VdV \tag{5–58}$$

Then integrating this equation from A to B, we obtain the following relation

$$\frac{d}{dt}\left[\int_A^B (V_x dV_x + V_y dV_y + V_z dV_z)\right] = \left(-\Phi - \int \frac{dp}{\rho} + \frac{V^2}{2}\right)_A^B \tag{5–59}$$

In addition to this, let us assume that the points A and B are the same point on the arbitrary closed circuit C; so, the integration in the parenthesis of the left-hand side is equivalent to the circulation Γ of this circuit C. So the rate of increase of Γ can be represented as

$$\frac{d\Gamma}{dt} = \left(-\Phi - \int \frac{dp}{\rho} + \frac{V^2}{2}\right)_A^{B=A} \tag{5–60}$$

Now the values of $\int(dp/\rho)$ and $V^2/2$ at A and B must be the same value on the same point at A and B = A, but when the potential Φ is assumed to be a single-valued function, which means that the value of Φ is the same value when Φ is returned on to the same point after making a round of the circuit C, the value of the right-hand side must be zero. Accordingly under the assumption of the existence of a single-valued function of potential Φ in steady fluid flow, the left-hand side of Equation 5–60 goes to zero as

$$\frac{d\Gamma}{dt} = 0 \tag{5–61}$$

This equation means that the value of the circulation Γ in the arbitrary circuit C is independent of the lapse of time and also has a constant value under the assumption of the single-valued function of potential Φ as shown in Figure 5–20. This characterization is generally called a Constancy of Circulation from Baron Kelvin's circulation theorem in honor of the first description of this circulation of fluid flow by Sir William Thomson (1824–1907).[1] Moreover this Equation 5–61 means that in steady fluid flow, which has a single-valued function of potential, the irrotational motion continues the irrotational motion to the last; therefore, the

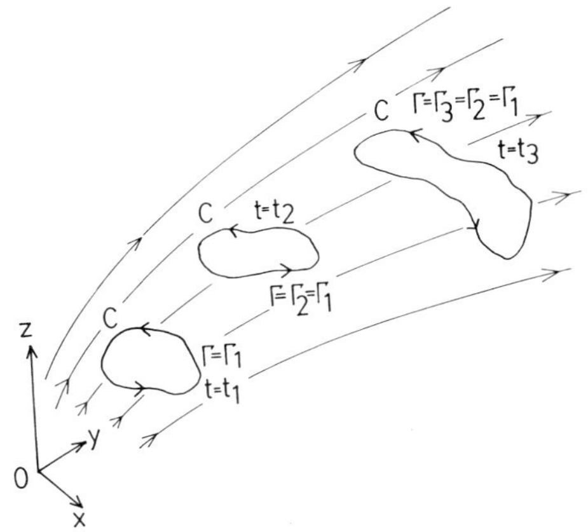

FIGURE 5–20. Constancy of circulation.

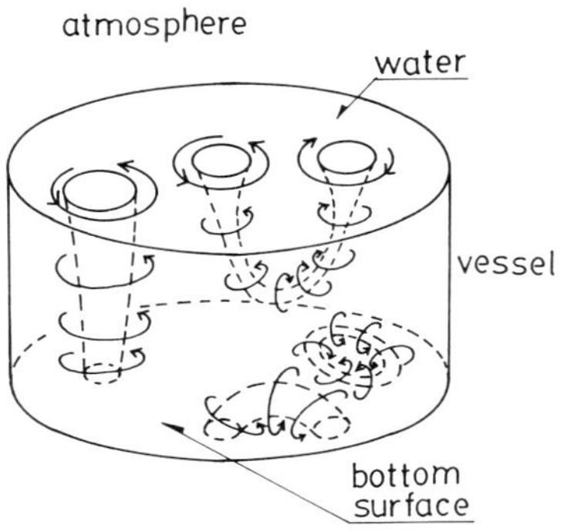

FIGURE 5–21. Schematic illustration of the various kinds of vortex motion in the water vessel.

irrotational motion never changes to rotational motion after the lapse of time. This characterization is called a permanence of irrotational motion.[2]

VIII. HELMHOLTZ THEOREM OF VORTEX

The following theorems can be directly derived by Kelvin's circulation theorem; however, the vortex theorems of the irrotational fluid motion can be arranged as follows:[3]

1. The vortex filament or the vortex line is not created suddenly in the fluid and also does not disappear in the fluid. Hence the vortex filament or the vortex line generated in the form of the closed circuit (vortex ring) is generated on the boundary surface and reaches on another boundary surface as shown in Figure 5–21.
2. The vortex tube which is composed of the vortex lines always consists of the same fluid particle, i.e., the vorticity behaves like freezing in the fluid.
3. The strength of the arbitrary vortex filament always keeps the same value with lapse of time in the inviscid fluid.

These vortex theorems are also well known as Helmholtz vortex theorem (Herman Ludwig Ferdinand von Helmholtz, 1821–1894).

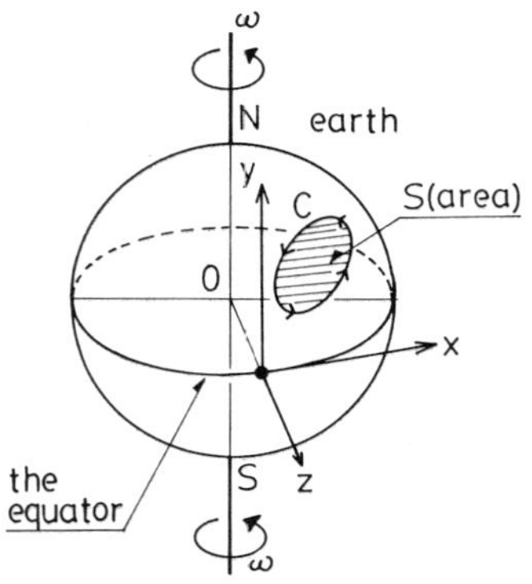

FIGURE 5–22. Rotating coordinates of the earth.

IX. CIRCULATION THEOREM OF V. BJERKNES

As shown in Figure 5–22, considering the atmospheric motion of fluid element with coordinates (x, y, z) rotating with the earth and taking that the coordinates (x, y) are equivalent to the equatorial plane, the equation of motion can be written as

$$\frac{dV_x}{dt} - 2\omega V_y = X - \frac{1}{\rho}\frac{\partial p}{\partial x} \quad (5\text{–}62)$$

$$\frac{dV_y}{dt} + 2\omega V_x = Y - \frac{1}{\rho}\frac{\partial p}{\partial y} \quad (5\text{–}63)$$

$$\frac{dV_z}{dt} = Z - \frac{1}{\rho}\frac{\partial p}{\partial z} \quad (5\text{–}64)$$

where X, Y, and Z are the external forces on the fluid elements per unit mass. Then multiplying dx on Equation 5–62, dy on Equation 5–63, and dz on Equation 5–64 and further summing up these equations, obtain

$$\frac{dV_x}{dt}dx + \frac{dV_y}{dt}dy + \frac{dV_z}{dt}dz = 2\omega(V_y dx - V_x dy) -$$

$$- \frac{1}{\rho}\left(\frac{\partial p}{\partial x}dx + \frac{\partial p}{\partial y}dy + \frac{\partial p}{\partial z}dz\right) + (Xdx + Ydy + Zdz) \quad (5\text{–}65)$$

Hence integrating Equation 5–65 on the closed circuit C moving with fluid, we obtain

$$\frac{d}{dt}\int_C (V_x dx + V_y dy + V_z dz) = 2\omega \int_C (V_y dx - V_x dy) - \int \frac{dp}{\rho} +$$

$$+ \int_C (Xdx + Ydy + Zdz) + \int_C (V_x dV_x + V_y dV_y + V_z dV_z) \quad (5\text{–}66)$$

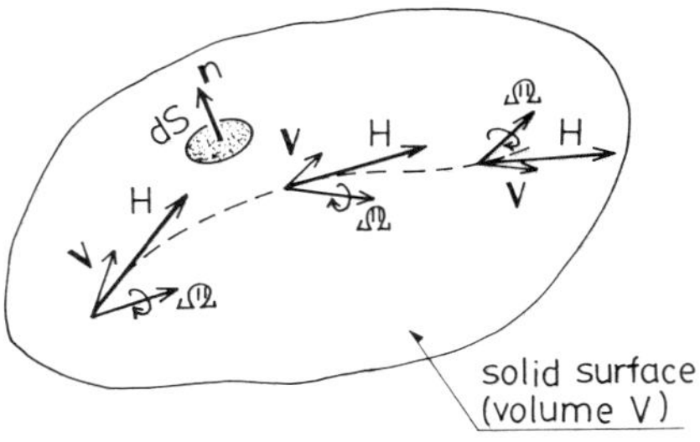

FIGURE 5–23. Schematic illustration of the helicity.

Since the first term of the right-hand side may be transformed to

$$\int_C (V_y dx - V_x dy) = -\iint_S \left(\frac{\partial V_x}{\partial x} + \frac{\partial V_y}{\partial y} \right) dxdy = -\frac{dS'}{dt} \quad (5\text{–}67)$$

where S′ is the projection of the area S enclosed by the circuit C to the equatorial plane, Equation 5–66 becomes

$$\frac{d\Gamma}{dt} = -2\omega \frac{dS'}{dt} - \int_C \frac{dp}{\rho} + \int_C (Xdx + Ydy + Zdz) \quad (5\text{–}68)$$

Now when X, Y, and Z can be derived by the eigenvalued function of potential Φ, Equation 5–68 becomes

$$\frac{d\Gamma}{dt} = -2\omega \frac{dS'}{dt} - \int_C \frac{dp}{\rho} \quad (5\text{–}69)$$

This equation is called the circulation theorem of V. Bjerknes (1898).[4–7]

X. HELICITY

The helicity H which is defined as the scalar product of velocity and vorticity, i.e., $H = \mathbf{V} \cdot \mathbf{\Omega}$, is an interesting scalar quantity of the rotational three-dimensional flows.[8,9] Helicity is a measure of the screw sense at any point of the velocity field as shown in Figure 5–23.

Helicity (scalar quantity) can be written as

$$\begin{aligned} H = \mathbf{V} \cdot \mathbf{\Omega} &= (V_r \mathbf{i} + V_\theta \mathbf{j} + V_z \mathbf{k}) \cdot (\zeta_r \mathbf{i} + \zeta_\theta \mathbf{j} + \zeta_z \mathbf{k}) = \\ &= V_r \zeta_r + V_\theta \zeta_\theta + V_z \zeta_z \end{aligned} \quad (5\text{–}70)$$

Then helicity can be positive or negative depending on the relative orientation of the velocity **V** and vorticity **Ω** vectors. Here we will restrict the following analysis to inviscid incompressible flows subject to conservation force. For this type of flow, the continuity equation and the definition of vorticity can be written as

$$\nabla \cdot \mathbf{V} = 0 \quad (5\text{–}71)$$

$$\nabla \cdot \boldsymbol{\Omega} = 0 \tag{5–72}$$

Then the equation of motion of fluid flow can be represented as

$$\frac{\partial \mathbf{V}}{\partial t} + \mathbf{V} \cdot \nabla \mathbf{V} = -\nabla\left(\frac{p}{\rho} + \Phi\right) \tag{5–73}$$

The vorticity equation can be obtained by taking the curl (rotation) of Equation 5–73 and by applying Equations 5–71 and 5–72

$$\frac{\partial \boldsymbol{\Omega}}{\partial t} + \mathbf{V} \cdot \nabla \boldsymbol{\Omega} = \boldsymbol{\Omega} \cdot \nabla \mathbf{V} \tag{5–74}$$

Now we multiply Equation 5–73 with $\boldsymbol{\Omega}$ and Equation 5–74 with $\mathbf{V}$, respectively, and take the sum of the resulting equations. So, after rearranging terms by the vector analysis we can obtain the helicity equation as

$$\frac{\partial H}{\partial t} + \mathbf{V} \cdot \nabla H = -\boldsymbol{\Omega} \cdot \nabla \left(\frac{p}{\rho} - \frac{\mathbf{V}^2}{2} + \Phi\right) \tag{5–75}$$

It will be recognized that the helicity equation 5–75 is a scalar equation like the equation of kinematic energy. Therefore it is interesting to look for the possible conservation properties of helicity within a closed volume with consideration of the certain boundary conditions. For this purpose we may rewrite the helicity equation in conservation form as

$$\frac{\partial H}{\partial t} + \nabla \cdot (\mathbf{V}H) = -\nabla \cdot \left[\mathbf{V}\left(\frac{p}{\rho} - \frac{\mathbf{V}^2}{2} + \Phi\right)\right] \tag{5–76}$$

Here let us define the total helicity H_T for an arbitrary volume V as

$$H_T = \int_V H \, dv \tag{5–77}$$

so the time rate of change of H_T can be written as

$$\frac{\partial H_T}{\partial t} = -\int_V \nabla \cdot (\mathbf{V}H) \, dv - \int_V \nabla \cdot \left[\mathbf{V}\left(\frac{p}{\rho} - \frac{\mathbf{V}^2}{2} + \Phi\right)\right] dv \tag{5–78}$$

Then applying the relationship (divergence theorem) between the volume integral and the surface integral, Equation 5–72 can be transformed to

$$\frac{\partial H_T}{\partial t} = -\int_S \mathbf{V} \cdot \mathbf{n} H \, dS - \int_S \boldsymbol{\Omega} \cdot \mathbf{n}\left(\frac{p}{\rho} - \frac{\mathbf{V}^2}{2} + \Phi\right) dS \tag{5–79}$$

where S is the surface enclosing this volume V and $\mathbf{n}$ is the unit vector normal to S. Now for the volume V enclosed by the solid boundary, the kinematic boundary condition is

$$\mathbf{V}_n = \mathbf{V} \cdot \mathbf{n} = 0 \tag{5–80}$$

i.e., the velocity component normal to the solid boundary vanishes. Hence the first integral of the right-hand side becomes zero. Then the second integral of that goes to zero, if we put on the surface of S

$$\mathfrak{Q}_n = \mathfrak{Q} \cdot \mathbf{n} = 0 \tag{5–81}$$

This means that if the normal component of vorticity vanishes on the solid boundary of the arbitrary volume V, the total helicity H_T within this volume V enclosed by this surface S is invariant, i.e.,

$$\frac{\partial H_T}{\partial t} = 0 \tag{5–82}$$

Therefore the total helicity H_T of an inviscid incompressible flow is constant with respect to time as

$$\frac{\partial H_t}{\partial t} = \frac{\partial}{\partial t}\int_V \mathbf{V} \cdot \mathfrak{Q}\, dv = 0 \tag{5–83}$$

Hence for these flows the total helicity may be regarded as an invariant like the total kinetic energy.

XI. INDUCED FLOW WITH VORTICITY

A. INCOMPRESSIBLE INVISCID FLOW

Assuming that the fluid is of constant density ρ (incompressible fluid) and also the flow is potential flow, then the velocity field of the fluid flow is given by the velocity potential $\Phi(\mathbf{V} = \nabla\Phi)$. Therefore the equation of continuity (div $\mathbf{V} = 0$) of fluid flow becomes

$$\nabla^2\Phi(\mathbf{r}, t) = 0 \tag{5–84}$$

since $\partial\rho/\partial t = 0$ and $\nabla\rho = 0$. This equation is called a Laplace equation as

$$\nabla^2 \equiv \text{div}\, \nabla = \nabla \cdot \nabla = \frac{\partial^2}{\partial x^2} + \frac{\partial^2}{\partial y^2} + \frac{\partial^2}{\partial z^2} \tag{5–85}$$

Here the potential Φ at each time t can be determined by Equation 5–84, but Equation 5–84 does not contain the differentiation with time. Therefore we can obtain a relationship between Φ and t by Equation 5–84, since the time t is only included as a parameter. Consequently, the variation of the velocity field with lapse of time is created by the variation of the boundary condition, for example, by the movement of the obstacle in the fluid.

Here the equation of Φ = constant = C

$$\Phi(\mathbf{r}) = \Phi(x, y, z) = C \tag{5–86}$$

determines a curved surface which is called an equipotential suface as shown in Figure 5–24. Then as shown in Figure 5–25, taking the two points $E(\mathbf{r})$ and $F(\mathbf{r} + \Delta\mathbf{r})$ on the same equipotential surface Φ, and from the definition of Φ, we obtain

$$\Phi(\mathbf{r}) = C \text{ and } \Phi(\mathbf{r} + \Delta\mathbf{r}) = C$$

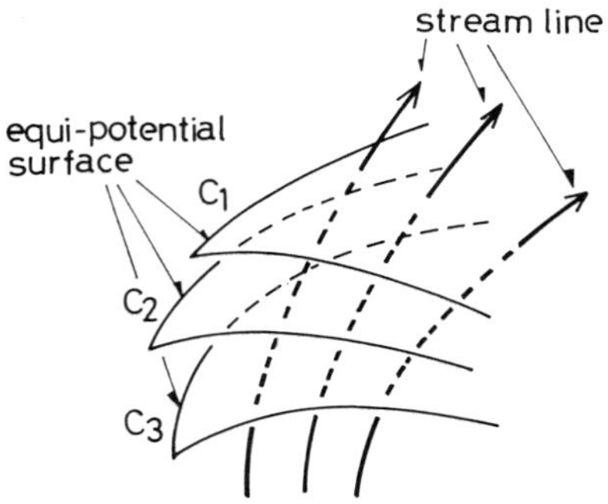

FIGURE 5–24. Relationship between equipotential surface and stream line.

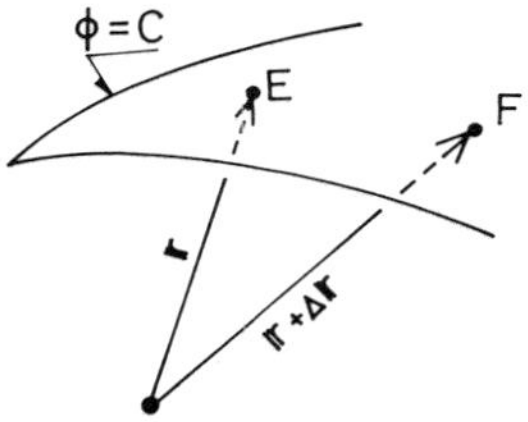

FIGURE 5–25. Two points on the same equipotential surface.

Assuming that $\mathbf{\Delta r}$ is very small,

$$\Phi(\mathbf{r} + \Delta\mathbf{r}) - \Phi(\mathbf{r}) \fallingdotseq \Phi(\mathbf{r}) + \frac{d\Phi}{dr}\Delta\mathbf{r} - \Phi(\mathbf{r}) = 0$$

Then

$$\Delta\mathbf{r} \cdot \nabla\Phi(\mathbf{r}) = 0 \tag{5–87}$$

which means that the velocity $\mathbf{V}(\mathbf{r})$ is perpendicular to the vector $\Delta\mathbf{r}$ of the equipotential surface at the point $\mathbf{r}$.[10]

B. FLOW BY SINK OR SOURCE OF THE INVISCID FLUID

Here taking the origin at the point of source, the potential Φ is a function of radius r of the spherical coordinate. In this case Laplace equation becomes

$$\frac{d^2\Phi}{dr^2} + \frac{2}{r}\frac{d\Phi}{dr} = \frac{1}{r}\frac{d^2}{dr^2}(r\Phi) = 0 \tag{5–88}$$

The solution of this equation except $r = 0$ is

$$\Phi = \frac{B}{r} + A \tag{5–89}$$

Here neglecting the constant A, since A does not depend on the velocity field, the velocity $\mathbf{V} = \nabla\Phi$ has a velocity component for the radial direction as

$$V_r = -\frac{B}{r^2} \qquad (5\text{–}90)$$

Let the flow rate out per unit time be Q. The constant B can be determined as $B = -Q/4\pi$. Consequently Equation 5–89 can be written as

$$\Phi = -\frac{Q}{4\pi r} \qquad (5\text{–}91)$$

This equation shows the velocity potential which gives the flow field from a point source. When the point source is located at $\mathbf{r} = \mathbf{r}_o$, Equation 5–91 becomes

$$\Phi(\mathbf{r}) = -\frac{Q}{4\pi|\mathbf{r} - \mathbf{r}_o|} \qquad (5\text{–}92)$$

Assuming that the source $q(\mathbf{r}')dV'$ is located in the small volume dV' at the point $\mathbf{r}'$, the potential of flow can be given as

$$-\frac{q(\mathbf{r}')dV'}{4\pi}\frac{1}{|\mathbf{r} - \mathbf{r}'|} \qquad (5\text{–}93)$$

Then the potential is obtained by the summation of the total volume element as follows:

$$\Phi(\mathbf{r}) = -\frac{1}{4\pi}\int\frac{q(r)}{|\mathbf{r} - \mathbf{r}'|}\,dV' \qquad (5\text{–}94)$$

This equation becomes

$$\nabla^2\Phi(\mathbf{r}) = q(\mathbf{r}) \qquad (5\text{–}95)$$

C. FLOW WITH VORTICITY

In this section we consider the flow of fluid, in which the fluid is incompressible and the vorticity $\text{rot}\mathbf{V} = \boldsymbol{\Omega}$ is not always zero. Then we must seek the velocity field $\mathbf{V}(\mathbf{r})$ corresponding to the distribution of the vorticity tube of the distribution of the vorticity $\boldsymbol{\Omega}$ $(\mathbf{r})$. Here we introduce a vector $\mathbf{A}(\mathbf{r})$, and a relationship between $\mathbf{V}$ and $\mathbf{A}$ can be written as

$$\mathbf{V} = \text{rot}\,\mathbf{A} \qquad (5\text{–}96)$$

The above relation satisfies the equation of the continuity as $\text{div}\,\mathbf{V} = 0$ $(\text{div rot}\,\mathbf{A} = 0)$. Now applying vector analysis,

$$\text{rot rot}\,\mathbf{A} = -\nabla^2\mathbf{A} + \nabla\text{div}\,\mathbf{A} = \boldsymbol{\Omega} \qquad (5\text{–}97)$$

Here we can exchange the vector $\mathbf{A}$ as follows

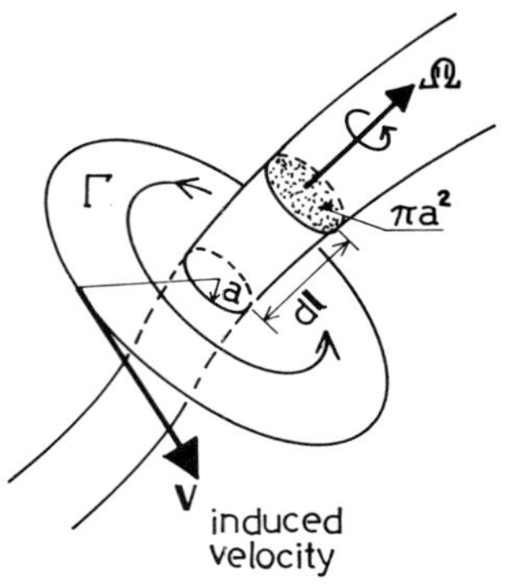

FIGURE 5–26. Core of a vortex filament and induced velocity.

$$\mathbf{A} \rightarrow \mathbf{A} + \nabla\wedge \qquad (5\text{–}98)$$

where $\wedge(\mathbf{r})$ means a scalar field. From the above relationship we can obtain the following equation

$$\text{div } \mathbf{A} = 0 \qquad (5\text{–}99)$$

Therefore the vector **A** obeys Equations 5–98 and 5–99

$$\nabla^2\mathbf{A} = -\boldsymbol{\Omega} \qquad (5\text{–}100)$$

The above equation is the same form of the potential Φ which was described already. Consequently the vector potential **A** for the given $\boldsymbol{\Omega}$ can be written as

$$A(\mathbf{r}) = -\frac{1}{4\pi}\int \frac{\boldsymbol{\Omega}(\mathbf{r}')}{|\mathbf{r} - \mathbf{r}'|} dV' \qquad (5\text{–}101)$$

Then substituting Equation 5–101 into Equation 5–96, we can obtain the following equation of the velocity field as

$$\mathbf{V}(\mathbf{r}) = \frac{1}{4\pi}\int \frac{\boldsymbol{\Omega}(\mathbf{r}) \times (\mathbf{r} - \mathbf{r}')}{|\mathbf{r} - \mathbf{r}'|^3} dV' \qquad (5\text{–}102)$$

This equation corresponds to the Biot-Savart's law which gives a magnetic field generated by electric current.[10–12]

D. INDUCED VELOCITY DUE TO THE VORTEX FILAMENT

Here, assuming that the radius a of a vector tube is infinitesimally small, the magnitude Ω of vorticity has a finite and large value. The magnitude of $\Gamma = \Omega\pi a^2 \cdot K$ is called the strength of the vortex filament. As shown in Figure 5–26, the radius a of this small vortex tube is the core of a vortex filament. The distribution of the vorticity $\boldsymbol{\Omega}$ in this vortex core is not always a constant value.

Now $\boldsymbol{\Omega}(\mathbf{r}')$ has a value only in this small area of the vortex core, and also the direction of this vorticity $\boldsymbol{\Omega}(\mathbf{r}')$ is always a tangential direction of this vortex filament. Then let us take a small volume element dV′ of Equation 5–102 to be a product of the sectional area

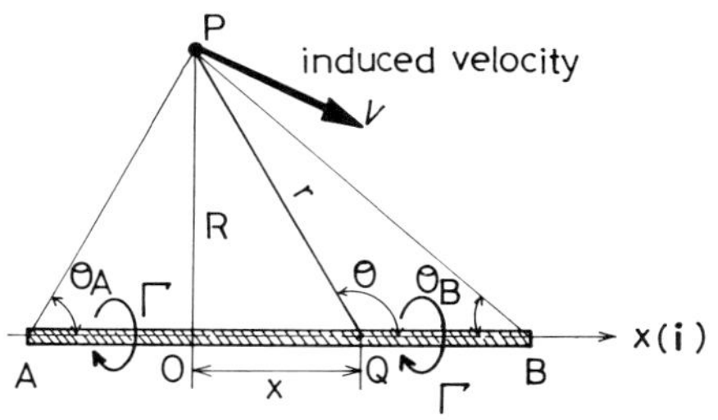

FIGURE 5–27. Induced velocity by a linear vortex filament.

πa^2 of this vortex core and of the small line element dl, so that the following relation can be obtained

$$dV' = \Omega \pi a^2 \, dl = \Gamma \, dl \tag{5–103}$$

Therefore substituting Equation 5–103 into Equation 5–102, Equation 5–102 becomes

$$\mathbf{V}(\mathbf{r}) = \frac{1}{4\pi} \int \frac{\Gamma \, \mathbf{dl} \times (\mathbf{r} - \mathbf{r}')}{|\mathbf{r} - \mathbf{r}'|^3} \tag{5–104}$$

1. Example 1

Determine the induced velocity at a point P by the existence of the linear vortex filament of the length AB as shown in Figure 5–27. As shown in this figure, a distance x = OQ is represented as

$$x = R \tan\left(\theta - \frac{\pi}{2}\right) = -R \cot\theta$$

Therefore,

$$\mathbf{dl} = dx \, \mathbf{i} = R \operatorname{cosec}^2\theta \, d\theta \tag{5–105}$$

where **i** is a unit vector along the x-axis; further

$$r = R \sec\left(\theta - \frac{\pi}{2}\right) = -R \operatorname{cosec}\theta \tag{5–106}$$

and also on the vector product $\mathbf{dl} \times (\mathbf{r} - \mathbf{r}')$ becomes

$$\mathbf{dl} \times (\mathbf{r} - \mathbf{r}') = dx \, r \sin\theta \tag{5–107}$$

Consequently substituting Equations 5–105, 5–106, and 5–107 into Equation 5–104, we obtain

$$\mathbf{V} = \mathbf{K} \frac{1}{4\pi} \int \frac{\Gamma \, dx \, r \sin\theta}{|\mathbf{r}|^3} = \mathbf{K} \frac{1}{4\pi} \int \frac{\Gamma \sin\theta \, R \operatorname{cosec}^2\theta \, d\theta}{R^2 \operatorname{cosec}^2\theta} =$$

$$= \mathbf{k} \frac{\Gamma}{4\pi R} \int_{\theta_A}^{\pi - \theta_B} \sin\theta \, d\theta = \mathbf{k} \frac{\Gamma}{4\pi R} (\cos\theta_A + \cos\theta_B) = V_\theta \tag{5–108}$$

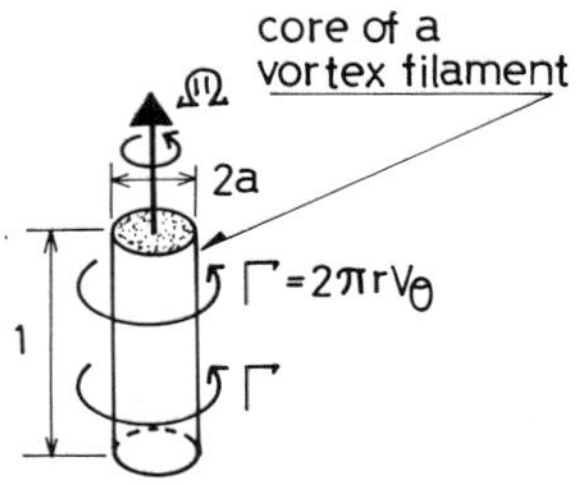

FIGURE 5–28. Vortex filament per unit length.

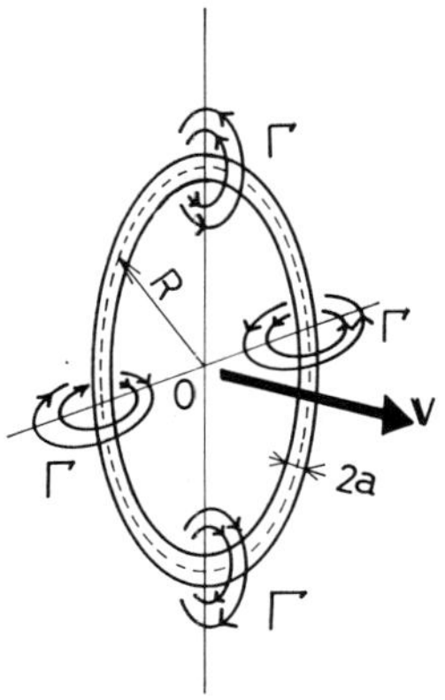

FIGURE 5–29. Ring of a vortex filament and the main symbols.

This induced velocity **V** is the rotational velocity component (V_θ) around this vortex core. Therefore when this straight vortex core filament is infinitely long, angles θ_A and θ_B become zero, so that this induced rotational velocity V_θ can be written as

$$V_\theta = \frac{\Gamma}{2\pi R} \tag{5–109}$$

This result is the well known free vortex in the two-dimensional flow.

2. Example 2

Let us calculate the energy of vortex filament per unit length a shown in Figure 5–28. The kinetic energy of a vortex filament per unit length can be calculated as

$$E = \frac{1}{2}\int_a^R \rho \int_0^{2\pi} V_\theta^2 \, r \, dr \, d\theta = \frac{\rho}{2}\int_a^R 2\pi r \, V_\theta^2 \, dr = \frac{r}{2}\int_a^R 2\pi r\left(\frac{\Gamma}{2\pi r}\right)^2 dr \, d\theta =$$

$$= \frac{\rho}{4\pi}\int_a^R \frac{\Gamma^2}{r} \, dr \, d\theta = \frac{\rho\Gamma^2}{4\pi} \ln \frac{R}{a} \tag{5–110}$$

3. Example 3

When a vortex filament is closed in a fluid, it becomes a ring of a vortex filament of very small cross-area πa^2 as shown in Figure 5–29. For a circular vortex filament of radius

R, the velocity **V**, the kinetic energy E, and the momentum M of this vortex ring can be calculated as follows:

$$|\mathbf{V}| = \frac{\Gamma}{4\pi R}\left(\ln\frac{8R}{a} - \frac{1}{4}\right) \tag{5–111}$$

$$E = \frac{1}{2}\rho\iiint_V (V_\theta^2 + V_r^2 + V_z^2)\,dV = \frac{\rho R\Gamma^2}{2}\left(\ln\frac{8R}{a} - \frac{7}{4}\right) \tag{5–112}$$

$$M = \pi\rho\Gamma R^2 \tag{5–113}$$

Therefore a vortex ring behaves like one of the quantum particles which keep a finite energy E and a finite momentum. Then when the energy E becomes too large, the radius R becomes large also. Consequently the velocity **V** becomes too small.

Further when we apply this model of the vortex ring, it must satisfy a condition of a $<<$ R. In this case a relationship between $|\mathbf{V}|$ and E can be obtained as

$$E \propto |\mathbf{V}^{-1} \tag{5–114}$$

This relation shows quite different characteristics in comparison with an ordinary relation of

$$E \propto m\frac{|\mathbf{V}|^2}{2}$$

where m is a mass of particle.

MAIN SYMBOLS

$H(m/s^2)$;	helicity
m(kg);	mass of the fluid particle
p(Pa);	static pressure
$Q(m^3/s)$;	flow rate
r(m);	radius
ds(m);	small distance
$dS(m^2)$;	small surface area
V(m/s);	velocity vector
V_x(m/s);	velocity component
V_y(m/s);	velocity component
V_z(m/s);	velocity component
V_θ(m/s);	tangential velocity
X(N/kg);	external force per unit mass
Y(N/kg);	external force per unit mass
Z(N/kg);	external force per unit mass
$\Gamma(m^2/s)$;	circulation
ζ_x(1/s);	vorticity component
ζ_y(1/s);	vorticity component
ζ_z(1/s);	vorticity component

θ(rad);	angle
ρ(kg/m^3);	density
Φ(m^2/s);	potential
$\mathfrak{M}$(1/s);	vorticity vector
ω(rad/s);	angular velocity of the earth

REFERENCES

1. **Sir William Thomson (Baron Kelvin),** *Mathematical and Physical Papers,* Vol. IV, 1910.
2. **Poincaré, H.,** Théorie des tourbillons, Georges Carré, Paris, 1893.
3. **Bauer, H. C. G.,** Die Helmholtzsche Wirbeltheorie, München und Berlin, Verlag von Oldenbourg, 1919.
4. **Bjerknes, V.,** *Met. Zeit.,* 97, 1900 and 97, 1902.
5. **Okada, T.,** *Fundamental Meteorology* (in Japanese), Vol. 1, Iwanami-Pub., Tokyo, 1937.
6. **Watanabe, T.,** Function and application to meteorology (in Japanese), Vol. 3, in *Memoirs on Geophysics and Geophysical Education,* Vol. 1, Japan Meteorological Agency, 1960, 518.
7. **Holmboe, J., Forsythe, G. E., and Gustin, W.,** *Dynamic Meteorology,* John Wiley & Sons, New York, 1952.
8. **Etling, D.,** Some aspects on helicity in atmospheric flows, *Contri. Atmos. Phys.,* 58(1), 88, 1985.
9. **Moffatt, H. K.,** The degree of knottedness of tangled vortexlines, *J. Fluid Mech.,* 35(1), 117, 1969.
10. **Tzuneto, T.,** Elastic body and fluid (in Japanese), Iwanami-Pub., Tokyo, 1983.
11. **Biot, J. B.** (1774–1862) was Professor of Collège de France and a physicist, astronomer, and also mathematician.
12. **Savart, F.** (1791–1841) was Professor of Collège de France and a physicist.

Chapter 6

TURBULENT FLOW AND REYNOLDS EQUATION

I. INTRODUCTION

In this chapter the Navier-Stokes equation, the spin-up time which is very important for representing the characteristics in the rotating flow in the boundary layer in the vortex chamber, the equation of motion on turbulent flow, the discontinuous surface and diffusion of vorticity laminar and turbulent flow, the mechanical similarity in a turbulent swirling wake behind the obstacle, and also the physical explanation of the energy dissipation are described in detail.

II. NAVIER-STOKES EQUATION

The Navier-Stokes equation can be represented by a vector form as follows:

$$\rho \frac{D\mathbf{V}}{Dt} = \rho\left[\frac{\partial \mathbf{V}}{\partial t} + (\mathbf{V} \cdot \nabla)\mathbf{V}\right] = -\nabla p + \mathbf{F} + \eta \nabla^2 \mathbf{V} \tag{6–1}$$

where, as shown in Figure 6–1, the following symbols are explained as $\mathbf{F}$, body force per unit volume (N/m^3); p, pressure (Pa); $\mathbf{V}$, velocity vector (m/s); D/Dt, material derivative (1/s); η, viscosity (Pa · s), and also using the vector identity

$$(\mathbf{V} \cdot \nabla)\mathbf{V} = \nabla\left(\frac{|\mathbf{V}|^2}{2}\right) - \mathbf{V} \times (\nabla \times \mathbf{V}) \tag{6–2}$$

Equation 6–1 becomes

$$\rho\left[\frac{\partial \mathbf{V}}{\partial t} + \nabla\left(\frac{\mathbf{V}^2}{2}\right) - \mathbf{V} \times (\nabla \times \mathbf{V})\right] = -\nabla p + \mathbf{F} + \eta \nabla^2 \mathbf{V} \tag{6–3}$$

In this equation the fluid is assumed to be an incompressible fluid, and so $\nabla \cdot \mathbf{V}$ goes to zero. Therefore Equation 6–3 becomes

$$\rho\left[\frac{\partial \mathbf{V}}{\partial t} + \nabla\left(\frac{\mathbf{V}^2}{2}\right) - \mathbf{V} \times (\nabla \times \mathbf{V})\right] = -\nabla p + \mathbf{F} - \eta \nabla \times (\nabla \times \mathbf{V}) \tag{6–4}$$

Further, assuming that the body force $\mathbf{F}$ can be derived from a conservative force field, so $\mathbf{F}$ can be represented by a gradient of a scalar potential Φ as

$$F = -\nabla\Phi$$

Then taking a vector calculation of curl $= (\nabla \times)$ to the above equation, we obtain

$$\frac{\partial \boldsymbol{\Omega}}{\partial t} = \nabla \times (\mathbf{V} \times \boldsymbol{\Omega}) - \nu \nabla^2 \boldsymbol{\Omega} \tag{6–5}$$

This equation is well known as the vorticity transport equation.

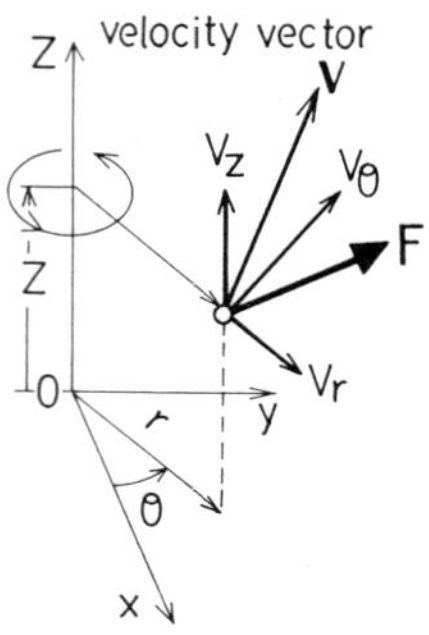

FIGURE 6–1. Velocity vector and the velocity components.

To represent this vorticity equation in dimensionless form, let us introduce the representative quantities of a length R(m), of velocity V_o(m/s), of angular velocity Ω_o (rad/s), and also the dimensionless quantities of length r′, of velocity V′, and of vorticity Ω′. Then the relationship between the dimensionless quantities and the dimensional quantities can be defined as

$$r' = r/R,\ \mathbf{V}' = \mathbf{V}/V_o,\ t' = \Omega_o t,\ \nabla' = R\nabla,\ \nabla'^2 = R^2\nabla \text{ and } \boldsymbol{\Omega}' = \frac{R\boldsymbol{\Omega}}{V_o}$$

Now let us transform each term of the above equations

$$\frac{\partial \boldsymbol{\Omega}}{\partial t} = \frac{\Omega_o V_o}{R}\frac{\partial \boldsymbol{\Omega}'}{\partial t} = \Omega_o^2 \frac{\partial \boldsymbol{\Omega}'}{\partial t}$$

$$\nabla \times (\mathbf{V} \times \boldsymbol{\Omega}) = \frac{V_o^2}{R^2}\nabla' \times (\mathbf{V}' \times \boldsymbol{\Omega}') = \Omega_o^2\, \nabla' \times (\mathbf{V}' \times \boldsymbol{\Omega}')$$

$$\nu\nabla^2\boldsymbol{\Omega} = \frac{\nu\Omega_o}{R^2}\nabla'^2\boldsymbol{\Omega}'$$

We obtain the dimensionless form of Equation 6–5 as

$$\frac{\partial \boldsymbol{\Omega}'}{\partial t'} - \nabla' \times (\mathbf{V}' \times \boldsymbol{\Omega}') = \frac{\nu}{R^2\Omega_o}\nabla'^2\boldsymbol{\Omega}' \tag{6–6}$$

A dimensionless term $\nu/R^2\Omega_o$ of Equation 6–6 can be transformed

$$\frac{R^2\Omega_o}{\nu} = \left(\frac{R}{\sqrt{\nu\Omega_o}}\right)^2 \Omega_o^2 \tag{6–7}$$

In this equation a term $R/\sqrt{\nu\Omega_o}$ can be called a spin-up time T_s in the vortex chamber of the radius R and has a dimension of time defined as

$$T_s = \frac{R}{\sqrt{\nu\Omega_o}} \tag{6–8}$$

While a reciprocal of a term Ω_o may be considered as

$$\frac{1}{\Omega_o} = \frac{R}{V_o} = \frac{R^3}{V_o R^2} = \frac{\text{volume}}{\text{flow rate}} \tag{6–9}$$

hence $(1/\Omega_o)$ may be regarded as the residence time T_r of fluid in the vortex chamber of the radius R defined as

$$T_r = \frac{1}{\Omega_o} \tag{6–10}$$

Consequently as a physical meaning for the vortex flow in the vortex chamber, a dimensionless quantity of Equation 6–7 may be regarded as

$$\frac{R^2\Omega_o}{\nu} = \left(\frac{\text{spin-up time} = T_s}{\text{residence-time} = T_r}\right)^2 \tag{6–11}$$

This dimensionless quantity $T = T_s/T_r$ is very useful for considering the fluid dynamical characteristics of the turbulent rotational flow in the confined vortex chamber.[1–3]

When the vector equation of Navier-Stokes is decomposed into the three directions, the following equations are obtained

$$\rho\left(\frac{\partial V_r}{\partial t} + V_r\frac{\partial V_r}{\partial r} + \frac{V_\theta}{r}\frac{\partial V_r}{\partial \theta} + V_z\frac{\partial V_r}{\partial z} - \frac{V_\theta^2}{r}\right) =$$

$$= F_r - \frac{\partial p}{\partial r} + \eta\left(\frac{\partial^2 V_r}{\partial r^2} + \frac{1}{r}\frac{\partial V_r}{\partial r} + \frac{1}{r^2}\frac{\partial^2 V_r}{\partial \theta^2} + \frac{\partial^2 V_r}{\partial z^2} - \frac{V_r}{r^2} - \frac{2}{r^2}\cdot\frac{\partial V_\theta}{\partial \theta}\right) \tag{6–12}$$

$$\rho\left(\frac{\partial V_\theta}{\partial t} + V_r\frac{\partial V_\theta}{\partial r} + \frac{V_\theta}{r}\frac{\partial V_\theta}{\partial \theta} + V_z\frac{\partial V_\theta}{\partial z} + \frac{V_r V_\theta}{r}\right) =$$

$$= F_\theta - \frac{1}{r}\frac{\partial p}{\partial \theta} + \eta\left(\frac{\partial^2 V_\theta}{\partial r^2} + \frac{1}{r}\frac{\partial V_\theta}{\partial r} + \frac{1}{r^2}\frac{\partial^2 V_\theta}{\partial \theta^2} + \frac{\partial^2 V_\theta}{\partial z^2} + \frac{2}{r^2}\frac{\partial V_r}{\partial \theta} - \frac{V_\theta}{r^2}\right) \tag{6–13}$$

$$\rho\left(\frac{\partial V_z}{\partial t} + V_r\frac{\partial V_z}{\partial r} + \frac{V_\theta}{r}\frac{\partial V_z}{\partial \theta} + V_z\frac{\partial V_z}{\partial z}\right) =$$

$$= F_z - \frac{\partial p}{\partial z} + \eta\left(\frac{\partial^2 V_z}{\partial r^2} + \frac{1}{r}\frac{\partial V_z}{\partial r} + \frac{1}{r^2}\frac{\partial^2 V_z}{\partial \theta^2} + \frac{\partial^2 V_z}{\partial z^2}\right) \tag{6–14}$$

where the rates of the stress-strain relationships can be represented by the cylindrical coordinates as

$$\tau_{r\theta} = \tau_{\theta r} = \eta\left(\frac{\partial V_\theta}{\partial r} - \frac{V_\theta}{r} + \frac{1}{r}\frac{\partial V_r}{\partial \theta}\right) \tag{6–15}$$

$$\tau_{rz} = \tau_{zr} = \eta\left(\frac{\partial V_r}{\partial z} + \frac{\partial V_z}{\partial r}\right) \tag{6–16}$$

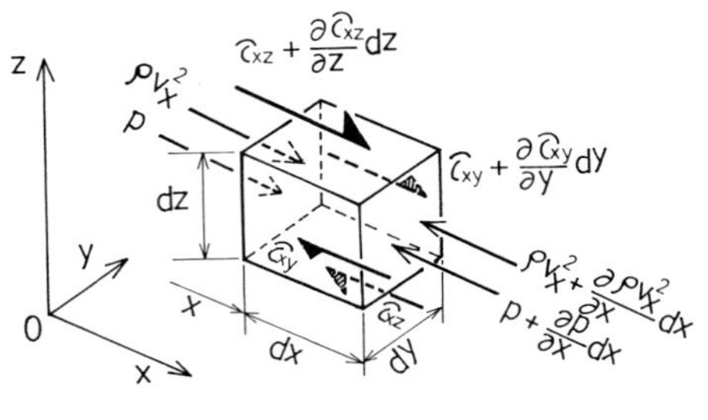

FIGURE 6–2. Turbulent velocity components and shear stresses for the x-direction.

$$\tau_{\theta z} = \partial_{z\theta} = \left(\frac{1}{r}\frac{\partial V_z}{\partial \theta} + \frac{\partial V_\theta}{\partial z}\right) \tag{6–17}$$

Moreover the Reynolds equation for the turbulent rotational flow for the axis-symmetrical motion of an incompressible viscous fluid can be represented by the cylindrical coordinates as

$$\frac{\partial V_r}{\partial t} + V_r\frac{\partial V_r}{\partial r} + V_z\frac{\partial V_r}{\partial z} - \frac{V_\theta^2}{r} = -\frac{1}{\rho}\frac{\partial p}{\partial r} + \nu\left(\frac{\partial^2 V_r}{\partial r^2} + \frac{1}{r}\frac{\partial V_r}{\partial r} - \frac{V_r}{r^2} + \frac{\partial^2 V_r}{\partial z^2}\right) -$$

$$-\frac{\partial}{\partial r}(\overline{v_r^2}) - \frac{\partial}{\partial z}(\overline{v_r v_z}) - \frac{\overline{v_r^2}}{r} + \frac{\overline{v_\theta^2}}{r} \tag{6–18}$$

$$\frac{\partial V_\theta}{\partial t} + V_r\frac{\partial V_\theta}{\partial r} + V_z\frac{\partial V_\theta}{\partial z} + \frac{V_r V_\theta}{r} = \nu\left(\frac{\partial^2 V_\theta}{\partial r^2} + \frac{1}{r}\frac{\partial V_\theta}{\partial r} - \frac{V_\theta}{r^2} + \frac{\partial^2 V_\theta}{\partial z^2}\right) -$$

$$-\frac{\partial}{\partial r}(\overline{v_r v_\theta}) - \frac{\partial}{\partial z}(\overline{v_\theta v_z}) - 2\frac{\overline{v_r v_\theta}}{r} \tag{6–19}$$

$$\frac{\partial V_z}{\partial t} + V_r\frac{\partial V_z}{\partial r} + V_z\frac{V_z}{\partial z} = -\frac{1}{\rho}\frac{\partial p}{\partial z} + \nu\left(\frac{\partial^2 V_z}{\partial r^2} + \frac{1}{r}\frac{\partial V_z}{\partial r} + \frac{\partial^2 V_z}{\partial z^2}\right)$$

$$-\frac{\partial}{\partial r}(\overline{v_r v_z}) - \frac{\partial}{\partial z}(\overline{v_z^2}) - \frac{1}{r}(\overline{v_r v_z}) \tag{6–20}$$

III. EQUATION OF MOTION ON THE TURBULENT FLOW

In order to derive the equation of motion for high turbulent flow, we consider a small volume element dxdydz of fluid particle as shown in Figure 6–2, where v_x, v_y, and v_z are the turbulent velocities for the x-, y-, and z-components.

First we consider the going in and out of the momentum through the small area dydz by the turbulent velocity v_x for the x-direction. When an eddy particle of mass m with the velocity $v_x > 0$ runs away from the surface dydz, an impulse mv_x as a reaction pushes the fluid in the small volume dxdydz. When the eddy particle of mass m with the velocity $v_x < 0$ pushes in from the area dydz to the small volume dxdydz, the impulse corresponding to the momentum $|mv_x|$ pushes the fluid in this small volume. Therefore, such a kind of the turbulent velocity v_x pushes the fluid in this small volume through the area dydz by an apparent pressure. Then, the mean value of this apparent pressure is equivalent to the mean value of the sum total of the impulses of the eddy particles

$$\Sigma|mv_x| \tag{6–21}$$

which are coming and going to the small volume per unit time and per unit area. We may regard the mass m of the eddy particle as

$$m = \rho v_x \tag{6–22}$$

so the apparent pressure can be expressed as

$$\rho\overline{v_x^2}$$

Consequently the difference of the apparent pressures acting on the two opposite surfaces dydz at x and at x + dx becomes

$$-\rho\left(\frac{\partial\overline{v_x^2}}{\partial x}\right)dxdydz \tag{6–23}$$

Next we consider the acting shear stress τ_{xy} on the surface area dxdz to the x-direction and the acting shear stress τ_{xz} on the surface area dxdy to the x-direction. These shear stresses can be represented as

$$\tau_{xy} + \tau_{xz} = (\tau_{xy} + \tau_{xz})_{laminar} + (\tau_{xy} + \tau_{xz})_{turbulent} \tag{6–24}$$

In this equation the viscous shear stress on this small volume to the x-direction can be written by Navier-Stokes equation as

$$\eta\left(\frac{\partial^2 V_x}{\partial x^2} + \frac{\partial^2 V_x}{\partial y^2} + \frac{\partial V_x}{\partial z^2}\right) dxdydz = \eta\nabla^2 V_x\ dxdydz \tag{6–25}$$

In addition to this, the Reynolds stresses due to the turbulent velocities can be represented as

$$(\tau_{xy})_t = -\rho\overline{v_x v_y} \quad , \quad (\tau_{xz})_t = -\rho\overline{v_x v_z} \tag{6–26}$$

Therefore the sum total of the forces acting on this small hexahedron by the Reynolds stresses to the x-direction can be written as

$$\frac{\partial\tau_{xy}}{\partial y}\,dxdydz + \frac{\partial\tau_{xz}}{\partial z}\,dxdydz = -\rho\left(\frac{\partial\overline{v_x v_y}}{\partial y} + \frac{\partial\overline{v_x v_z}}{\partial z}\right) dxdydz \tag{6–27}$$

Accordingly the sum total of the forces per unit volume of this small fluid particle by the turbulent velocities to the x-direction can be represented as

$$-\rho\left(\frac{\partial\overline{v_x^2}}{\partial x} + \frac{\partial\overline{v_x v_y}}{\partial y} + \frac{\partial\overline{v_x v_z}}{\partial z}\right) \tag{6–28}$$

In conclusion of the above stated results, the following equations of motion on the turbulent flow for the x-, y-, and z-directions are obtained

$$\frac{\partial \overline{V}_x}{\partial t} + \overline{V}_x \frac{\partial \overline{V}_x}{\partial x} + \overline{V}_y \frac{\partial \overline{V}_x}{\partial y} + \overline{V}_z \frac{\partial \overline{V}_x}{\partial z} = X - \frac{1}{\rho}\frac{\partial p}{\partial x} + \nu \nabla^2 \overline{V}_x -$$

$$- \left(\frac{\partial \overline{v_x^2}}{\partial x} + \frac{\partial \overline{v_x v_y}}{\partial y} + \frac{\partial \overline{v_x v_z}}{\partial z} \right), \qquad (6\text{–}29)$$

$$\frac{\partial \overline{V}_y}{\partial t} + \overline{V}_x \frac{\partial \overline{V}_y}{\partial x} + \overline{V}_y \frac{\partial \overline{V}_y}{\partial y} + \overline{V}_z \frac{\partial \overline{V}_y}{\partial z} = Y - \frac{1}{\rho}\frac{\partial p}{\partial y} + \nu \nabla^2 \overline{V}_y -$$

$$- \left(\frac{\partial \overline{v_x v_y}}{\partial x} + \frac{\partial \overline{v_y^2}}{\partial y} + \frac{\partial \overline{v_y v_z}}{\partial z} \right), \qquad (6\text{–}30)$$

$$\frac{\partial \overline{V}_z}{\partial t} + \overline{V}_x \frac{\partial V_z}{\partial x} + \overline{V}_y \frac{\partial V_z}{\partial y} + V_z \frac{\partial V_z}{\partial z} = Z - \frac{1}{\rho}\frac{\partial p}{\partial z} + \nu \nabla^2 \overline{V}_z -$$

$$- \left(\frac{\partial \overline{v_x v_z}}{\partial x} + \frac{\partial \overline{v_y v_z}}{\partial y} + \frac{\partial \overline{v_z^2}}{\partial z} \right), \qquad (6\text{–}31)$$

where X, Y, Z are body forces per unit mass, respectively.

The equations of the turbulent energy in cylindrical coordinates, after simplification by the assumption of the axial symmetry, are reproduced from a paper of Stephen C. Traugott[4] as follows:

radial direction

$$\frac{1}{2}\left(V_r \frac{\partial \overline{v_r^2}}{\partial r} + V_z \frac{\partial \overline{v_r^2}}{\partial z} \right) + \overline{v_r v_\theta} \frac{2V_\theta}{r} + \overline{v_r^2} \frac{\partial V_r}{\partial r} + \overline{v_{rz}} \frac{\partial V_r}{\partial z} =$$

$$= -\frac{1}{\rho} \overline{v_r \frac{\partial p}{\partial r}} - \frac{1}{2}\left(\frac{1}{r}\frac{\partial \overline{r v_r^3}}{\partial r} + \frac{\partial \overline{v_z v_r^2}}{\partial z} - 2\frac{\overline{v_r v_\theta^2}}{r} \right) - \qquad (6\text{–}32)$$

$$- \nu \left[\overline{\left(\frac{\partial V_r}{\partial r} \right)^2} + \overline{\left(\frac{\partial V_r}{\partial z} \right)^2} + \overline{\left(\frac{\partial V_r}{r \partial \theta} \right)^2} + \frac{\overline{V_r^2}}{r^2} + 2\overline{\frac{V_r}{r}\frac{\partial V_\theta}{r \partial \theta}} \right]$$

tangential direction

$$\frac{1}{2}\left(V_r \frac{\partial \overline{v_\theta^2}}{\partial r} + V_z \frac{\partial \overline{v_\theta^2}}{\partial z} \right) - \overline{v_r v_\theta} \frac{1}{r}\frac{\partial r V_\theta}{\partial r} + \overline{v_r v_\theta} \frac{\partial V_\theta}{\partial z} + \overline{v_\theta^2} \frac{V_r}{r} =$$

$$= -\frac{1}{\rho} \overline{v_\theta \frac{\partial p}{r \partial \theta}} - \frac{1}{2}\left(\frac{1}{r}\frac{\partial r\, \overline{v_r v_\theta^2}}{\partial r} + \frac{\partial \overline{v_z v_\theta^2}}{\partial z} + 2\frac{\overline{v_r v_\theta^2}}{r} \right) -$$

$$- \nu \left[\overline{\left(\frac{\partial V_\theta}{\partial r} \right)^2} + \overline{\left(\frac{\partial V_\theta}{\partial z} \right)^2} + \overline{\left(\frac{\partial V_\theta}{r \partial \theta} \right)^2} + \frac{\overline{V_\theta^2}}{r^2} + 2\overline{\frac{V_\theta}{r}\frac{\partial V_r}{r \partial \theta}} \right] \qquad (6\text{–}33)$$

axial direction

$$\frac{1}{2}\left(V_r\frac{\partial\overline{v_z^2}}{\partial r} + V_z\frac{\partial\overline{v_z^2}}{\partial z}\right) + \overline{v_zv_r}\,\frac{\partial V_z}{\partial r} + \overline{v_z^2}\,\frac{\partial V_z}{\partial z} =$$

$$= -\frac{1}{\rho}\overline{v_z\frac{\partial p}{\partial z}} - \frac{1}{2}\left(\frac{1}{r}\frac{\partial r\,\overline{v_rv_z}}{\partial r} + \frac{\partial\overline{v_z^3}}{\partial z}\right) -$$

$$-\nu\left[\overline{\left(\frac{\partial V_z}{\partial r}\right)^2} + \overline{\left(\frac{\partial V_z}{\partial z}\right)^2} + \overline{\left(\frac{\partial V_z}{r\partial\theta}\right)^2}\right] \tag{6–34}$$

Therefore, putting

$$q^2 = \overline{v_r^2} + \overline{v_\theta^2} + \overline{v_z^2}$$

and also adding these three equations, we obtain an equation of turbulent energy as follows:

$$\frac{1}{2}\left(V_r\frac{\partial\overline{q^2}}{\partial r} + V_z\frac{\partial\overline{q^2}}{\partial z}\right) + \overline{v_\theta^2}\,\frac{V_r}{r} + \overline{v_r^2}\,\frac{\partial V_r}{\partial r} + \overline{v_z^2}\,\frac{\partial V_z}{\partial z} + \overline{v_zv_\theta}\,\frac{\partial V_\theta}{\partial z} +$$

$$+ \overline{v_rv_z}\left(\frac{\partial V_z}{\partial r} + \frac{\partial V_r}{\partial z}\right) + \overline{v_rv_\theta}\,r\frac{\partial}{\partial r}\left(\frac{V_\theta}{r}\right) =$$

$$= -\frac{1}{r}\frac{\partial}{\partial r}\,r\,\overline{v_r\left(\frac{q^2}{2} + \frac{p}{\rho}\right)} + \frac{\partial}{\partial z}\,\overline{v_z\left(\frac{q^2}{2} + \frac{p}{\rho}\right)}$$

$$-\nu\left[\overline{\left(\frac{\partial V_r}{\partial r}\right)^2} + \overline{\left(\frac{\partial v_r}{\partial z}\right)^2} + \overline{\left(\frac{\partial v_r}{r\partial\theta}\right)^2} + \overline{\left(\frac{\partial v_\theta}{\partial r}\right)^2} + \overline{\left(\frac{\partial v_\theta}{\partial z}\right)^2} + \overline{\left(\frac{\partial v_\theta}{r\partial\theta}\right)^2} +\right.$$

$$+ \overline{\left(\frac{\partial v_z}{\partial r}\right)^2} + \overline{\left(\frac{\partial v_z}{\partial z}\right)^2} + \overline{\left(\frac{\partial v_z}{r\partial\theta}\right)^2} + \frac{\overline{v_\theta^2} + \overline{v_r^2}}{r^2} -$$

$$\left. - 2\overline{\left(\frac{v_\theta}{r}\cdot\frac{\partial v_r}{r\partial\theta} - \frac{v_r}{r}\cdot\frac{\partial v_\theta}{r\partial\theta}\right)}\right] \tag{6–35}$$

where a symbol p is the instantaneous pressure fluctuation.

IV. DISCONTINUOUS SURFACE AND DIFFUSION OF VORTICITY

A. HOW TO CHANGE THE DISCONTINUOUS SURFACE DUE TO THE VELOCITY DIFFERENCES

Let us consider the diffusion phenomenon of the discontinuous surface of the velocities with lapse of time as shown in Figure 6–3. Here let us define the vorticity component ζ_z as

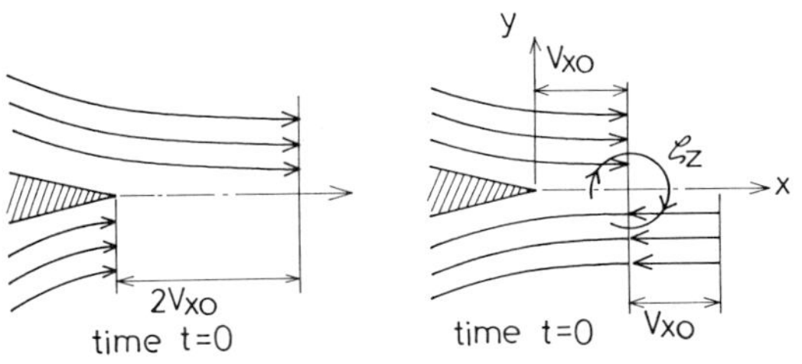

FIGURE 6–3. Illustration of the discontinuous surface.

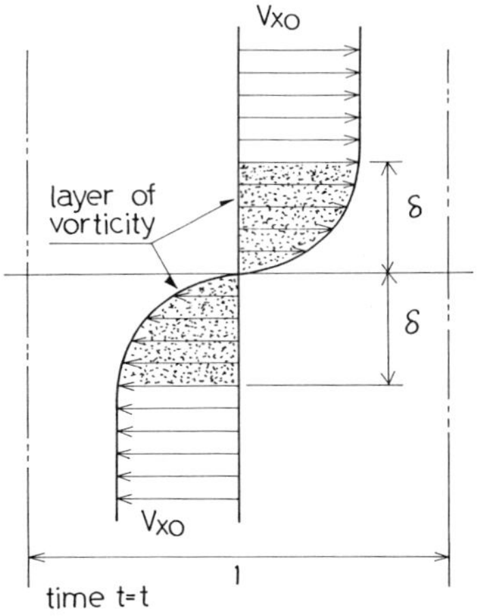

FIGURE 6–4. Diffusion of vorticity at the lapse of time.

$$\zeta_z = \frac{\partial V_y}{\partial x} - \frac{\partial V_x}{\partial y} \tag{6–36}$$

As shown in Figure 6–4, we may assume that the width of the vorticity layer is zero at $t = 0$ and then 2δ at $t = t$, and so the equation of velocity distribution which satisfies the boundary conditions at $y = \pm\delta$ is

$$\frac{V_x}{V_{xo}} = \frac{3}{2}\frac{y}{\delta} - \frac{1}{2}\left(\frac{y}{\delta}\right)^3 \tag{6–37}$$

Therefore the vorticity component ζ_z can be represented as

$$\zeta_z = \frac{\partial V_y}{\partial x} - \frac{\partial V_x}{\partial y} = -\frac{3}{2}\frac{V_{xo}}{\delta} + \frac{3}{2}\frac{V_{xo}}{\delta}\left(\frac{y}{\delta}\right)^2 \tag{6–38}$$

for the dimensionless form. This becomes

$$\frac{\zeta_z}{V_{xo}/\delta} = \frac{3}{2}\left[\left(\frac{y}{\delta}\right)^2 - 1\right] \tag{6–39}$$

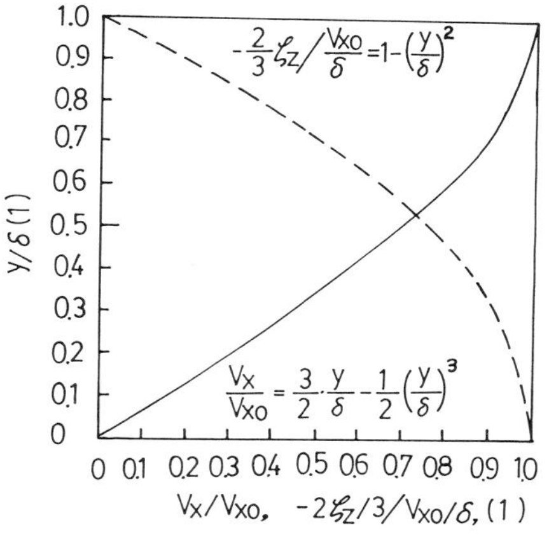

FIGURE 6–5. Numerical example of the velocity and vorticity distributions.

The velocity and vorticity distributions are shown in Figure 6–5. Here we consider the two kinds of flow conditions of laminar and turbulent flows.[5]

B. LAMINAR FLOW

The difference of the momentum between t = 0 and t = t is

$$\rho V_{xo}\delta - \rho\int_0^\delta V_x\,dy = \frac{3}{8}\rho\delta V_{xo} \tag{6–40}$$

The change of the momentum during the time interval dt can be represented as

$$\frac{3}{8}\rho V_{xo}\frac{d\delta}{dt}dt \tag{6–41}$$

Therefore the change of momentum per unit time can be written as

$$\frac{3}{8}\rho V_{xo}\frac{d\delta}{dt} \tag{6–42}$$

This quantity is equivalent to the shear force acting on the surface y = 0 of unit span as shown in Figure 6–4, while the shear stress on the surface y = 0 can be represented as

$$\tau_o = \eta\left(\frac{dV_x}{dy}\right)_{y=0} = \frac{3}{2}\cdot\frac{\eta V_{xo}}{\delta} \tag{6–43}$$

From Equations 6–42 and 6–43, we can obtain the following differential equation

$$\frac{3}{8}\rho V_{xo}\frac{d\delta}{dt} = \frac{3}{2}\frac{\eta V_{xo}}{\delta} \tag{6–44}$$

The solution of Equation 6–44 can be obtained for the boundary condition δ = 0 at t = 0 as follows:

$$2\delta(\text{width of the vorticity layer}) = 2\sqrt{8\nu t} \tag{6–45}$$

so that the width of the vorticity layer is proportional to $\sqrt{\nu t}$.

C. TURBULENT FLOW

In case of the turbulent flow the shear stress τ_o on the surface $y = 0$ should be applied using the theory of the mixing length. The shear stress can be represented as

$$\tau_o = \rho l^2\left(\frac{dV_x}{dy}\right)^2_{y=0} = \frac{9\rho}{4}\left(\frac{l}{\delta}\right)^2 \cdot V_{xo}^2 \tag{6–46}$$

where a symbol l is the mixing length. Here let us assume that the mixing length l is proportional to the mixing layer δ; so, Equation 6–46 can be transformed to

$$\tau_o = \frac{9}{4}\rho C^2 V_{xo}^2 \tag{6–47}$$

where C is a proportional constant of $l = C\delta$. Then the change of the momentum per unit time of Equation 6–42 is equivalent to Equation 6–47. So we can obtain the following differential equation

$$\frac{3}{8}\rho V_{xo}\frac{d\delta}{dt} = \frac{9}{4}\rho C^2 V_{xo}^2 \tag{6–48}$$

Therefore the width (2δ) of the vorticity layer for the boundary condition $\delta = 0$ at $t = 0$ is

$$2\delta = 12C^2 V_{xo} \cdot t \tag{6–49}$$

If in this equation we assume $C \fallingdotseq 0.14$ in the turbulent flow, then Equation 6–49 becomes

$$2\delta = 0.235V_{xo} \cdot t \tag{6–50}$$

So the width of the vorticity layer is proportional to t. Consequently the width of the vorticity layer in turbulent flow becomes diffuse faster than that in the laminar flow. A numerical example of these laminar and turbulent cases is shown in Figure 6–6.

V. MECHANICAL SIMILARITY IN TURBULENT SWIRLING WAKE

Here we consider the mechanical similarity in the turbulent swirling wake of the incompressible fluid on the downstream region passing over a stationary obstacle as shown in Figure 6–7. For the axial momentum representation, we can obtain the following simplified equation:[6]

$$V_r\frac{\partial V_z}{\partial r} + V_z\frac{\partial V_z}{\partial z} = V_z\frac{\partial V_z}{\partial z} + \frac{\partial}{\partial r}(V_r V_z) - V_z\frac{\partial V_r}{\partial r} = 0 \tag{6–51}$$

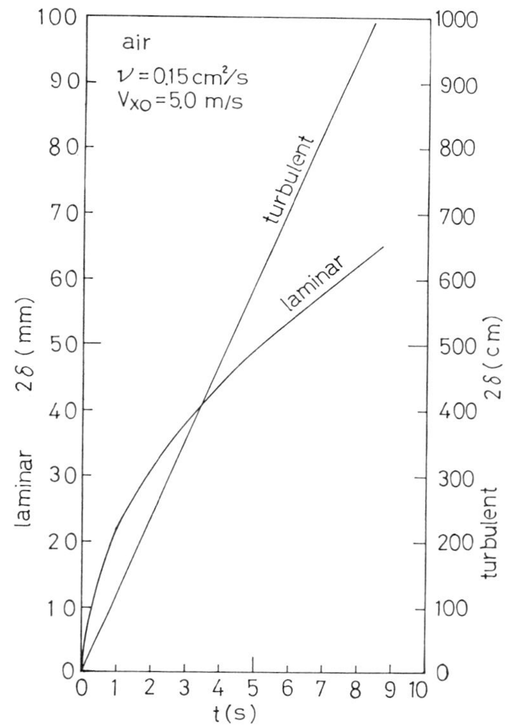

FIGURE 6–6. Numerical example of the thickness of the vorticity layer for the laminar and turbulent flows.

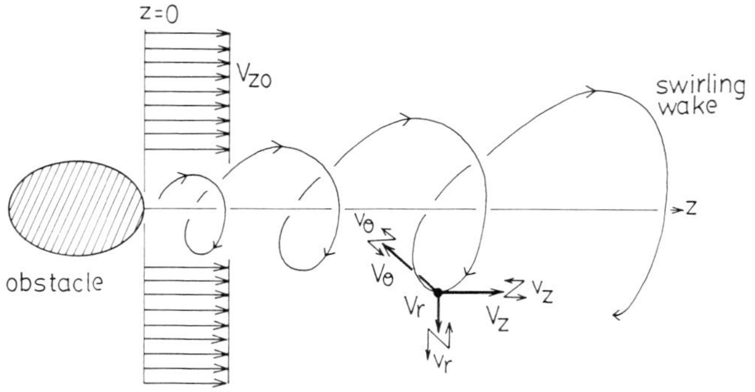

FIGURE 6–7. Schematic illustration of the velocity components of the swirling wake behind an obstacle.

continuity equation

$$\frac{\partial V_r}{\partial r} + \frac{V_r}{r} + \frac{\partial V_z}{\partial z} = 0 \quad , \quad \frac{\partial V_r}{\partial r} = -\frac{V_r}{r} \tag{6–52}$$

Therefore,

$$V_z \frac{\partial V_z}{\partial z} + \frac{1}{r}\frac{\partial}{\partial r}(rV_rV_z) = 0 \tag{6–53}$$

Then in order to simplify the assumed flow properties, let us designate V_{zo} as a homogeneous steady axial velocity. So the above equation can be transformed to

$$V_{zo}\frac{\partial V_z}{\partial z} + \frac{1}{r}\frac{\partial}{\partial r}(r\,\overline{v_r v_z}) = 0 \tag{6–54}$$

Further, for the tangential momentum representation we obtain the following simplified equation

$$V_r\frac{\partial V_\theta}{\partial r} + \frac{V_r V_\theta}{r} + V_z\frac{\partial V_\theta}{\partial z} = 0 \tag{6–55}$$

$$V_z\frac{\partial V_\theta}{\partial z} + \frac{\partial}{\partial r}(V_r V_\theta) + \frac{V_r V_\theta}{r} - V_\theta\frac{\partial V_r}{\partial r} = 0 \tag{6–56}$$

Substituting the continuity equation, we transform the above equation to

$$V_z\frac{\partial V_\theta}{\partial z} + \frac{\partial}{\partial r}(V_r V_\theta) + \frac{2V_r V_\theta}{r} = 0 \tag{6–57}$$

Therefore, the simplified tangential momentum equation is

$$V_{zo}\frac{\partial V_\theta}{\partial z} + \frac{1}{r^2}\frac{\partial}{\partial r}(r^2\overline{v_r v_\theta}) = 0 \tag{6–58}$$

If we assume that flow properties do not change on the downstream region far from the obstacle, we can define the fluxes of the axial and angular momenta as follows:

$$V_{zo}\int_0^\infty (V_{zo} - V_z)\cdot r\cdot dr = G_z(m^4/s^2) \text{ for axial momentum flux} \tag{6–59}$$

$$V_{zo}\int_0^\infty V_\theta\cdot r^2\cdot dr = G_\theta(m^5/s^2) \text{ for angular momentum flux} \tag{6–60}$$

Here, we consider the solutions which can be determined only by G_z, G_θ, and V_{zo}. These solutions are equivalent to the solutions which control the swirling wake motion on the downstream far from the obstacle. In that place we introduce a symbol L as a wake width, so L can be represented by the following terms

$$L = L(G_z, G_\theta, V_{zo}, Z) \tag{6–61}$$

The dimensional analysis can be written as

$$M^\xi = \left(\frac{M^4}{s^2}\right)^a \left(\frac{M^5}{s^2}\right)^b \left(\frac{M}{s}\right)^c M^1 \tag{6–62}$$

where the dimension of the length is

$$\xi = 4a + 5b + c + 1 \tag{6–63}$$

and also the dimension of the second (time) is

$$0 = 2a + 2b + c \tag{6–64}$$

Therefore, we may consider the two kinds of flow condition. When the axial momentum flux G_z is sufficiently larger than the angular momentum flux G_θ, Equations 6–63 and 6–64 can be represented as

$$\xi = 4a + c + 1 \tag{6–65}$$

$$0 = 2a + c \tag{6–66}$$

In that case, the undetermined values a, c and ξ are

$$a = 1, c = -2, \xi = 3 \tag{6–67}$$

The length L can be represented as a function

$$\frac{G_z \cdot z}{V_{zo}^2 \cdot L^3} = F\left(\frac{G_z \cdot L}{G_\theta}\right) \tag{6–68}$$

When the angular momentum flux G_θ is sufficiently larger than the axial momentum flux G_z, Equations 6–63 and 6–64 can be represented as

$$\xi = 5b + c + 1 \quad , \quad 0 = 2b + c \tag{6–69}$$

In that case the undetermined values b, c, and ξ are

$$b = 1 \quad , \quad c = -2 \quad \text{and} \quad \xi = 4 \tag{6–70}$$

The length L can be represented by

$$\frac{G_\theta \cdot Z}{V_{zo}^2 \cdot L^4} = G\left(G_z \cdot \frac{L}{G_\theta}\right) \tag{6–71}$$

In case of $G_zL/G_\theta >> 1$, let us assume that the development of the turbulent swirling wake is independent of G_θ. So we may estimate from Equation 6–68 that

$$L \sim \left(\frac{G_z \cdot Z}{V_{zo}^2}\right)^{1/3} \sim Z^{1/3} \tag{6–72}$$

and the scale of velocity defect is

$$\Delta V_z = V_{zo} - V_z \propto \frac{L}{Z/V_{zo}} \sim V_{zo}\frac{L}{Z} \sim Z^{-2/3} \tag{6–73}$$

In case of $G_zL/G_\theta << 1$, let us assume that the development of the turbulent swirling wake is independent of G_z. So we may estimate from Equation 6–71

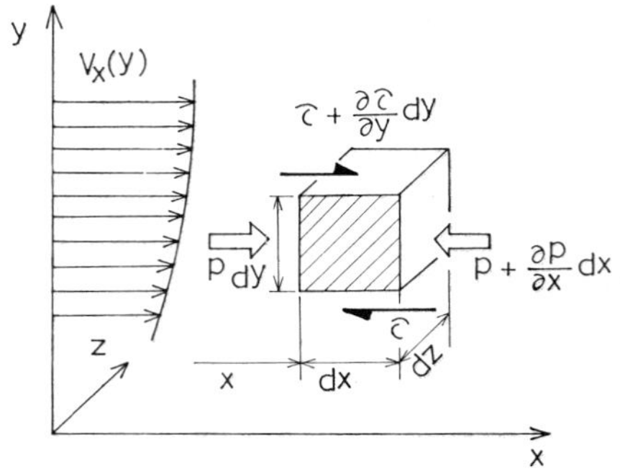

FIGURE 6–8. Schematic illustration of the energy dissipation processes on the two-dimensional turbulent flow.

$$L \sim \left(\frac{G_\theta \cdot Z}{V_{zo}}\right)^{1/4} \sim Z^{1/4} \tag{6–74}$$

and the scale of the rotational velocity is

$$V_\theta \sim Z^{-3/4} \tag{6–75}$$

VI. PHYSICAL EXPLANATION OF THE ENERGY DISSIPATION IN TWO-DIMENSIONAL TURBULENT FLOW

Here let us assume a small fluid particle of the volume dxdydz in the two-dimensional turbulent flow as shown in Figure 6–8, and also the mean velocity $V_x(y)$ to the x-direction depends only on a coordinate y-axis.

Then the work done of

$$V_x \cdot \tau \cdot dxdz(N \cdot m/s) \tag{6–76}$$

acts through the small area dxdz of the under surface of this fluid particle to the exterior, while the work done of

$$\left[V_x \cdot \tau + \frac{\partial(V_x \cdot \tau)}{\partial y} dy\right] dxdz\ (N \cdot m/s) \tag{6–77}$$

acts through the small area dxdz of the upper surface of this fluid particle into this small fluid particle. Therefore subtracting Equation 6–77 from Equation 6–76, the accumulative energy in this small volume per unit time is

$$\frac{\partial(V_x \cdot \tau)}{\partial y} dxdydz = \left(V_x \frac{\partial \tau}{\partial y} + \tau \frac{\partial V_x}{\partial y}\right) dxdydz \tag{6–78}$$

The fluid flux entering from the surface dydz at x into the small fluid volume gives the work done by pressure on this fluid volume as

$$V_x \cdot p \cdot dydz\ (N \cdot m/s) \tag{6–79}$$

and when the fluid flux flows out from the surface dydz at x + dx, the work done by pressure

$$V_x \cdot \left(p + \frac{\partial p}{\partial x} dx\right) dydz \ (N \cdot m/s) \tag{6–80}$$

is carried out from this small volume. Therefore the accumulative energy in this small volume per unit time is

$$-V_x \cdot \frac{\partial p}{\partial x} \cdot dxdydz \ (N \cdot m/s) \tag{6–81}$$

Consequently the total accumulative energy in this small volume per unit time is

$$\left(V_x\frac{\partial \tau}{\partial y} + \tau\frac{\partial V_x}{\partial y} - V_x\frac{\partial p}{\partial x}\right) dxdydz = \tau \cdot \frac{\partial V_x}{\partial y} dxdydz + V_x\left(\frac{\partial \tau}{\partial y} - \frac{\partial p}{\partial x}\right) dxdydz \tag{6–82}$$

Since the sum total of the forces by the shear stress and by the pressure contribute to the accelerative force for this small volume as follows:

$$\left(\frac{\partial \tau}{\partial y} - \frac{\partial p}{\partial x}\right) dxdydz = \rho V_x \cdot \frac{\partial V_x}{\partial x} dxdydz = \rho\frac{\partial}{\partial x}\left(\frac{V_x^2}{2}\right) dxdydz \tag{6–83}$$

so the second term of the right-hand side of Equation 6–82 can be transformed with the equation of motion

$$V_x\left(\frac{\partial \tau}{\partial y} - \frac{\partial p}{\partial x}\right) dxdydz = \underbrace{\rho V_x \, dydz}_{\text{mass flux}} \underbrace{\frac{\partial}{\partial x}\left(\frac{V_x^2}{2}\right) dx}_{\text{kinetic energy}} \tag{6–84}$$

This equation means that the energy is given as a kinetic energy per unit time to the penetrating fluid through the small fluid volume. Then subtracting the energy transforming the kinetic energy from the total accumulative energy of Equation 6–82, we can obtain the equation of the energy dissipation as

$$\tau \frac{\partial V_x}{\partial y} dxdydz \tag{6–85}$$

Therefore the energy dissipation $Ee(N/m^2s)$ per unit time and per unit volume is

$$Ee = \eta\frac{\partial V_x}{\partial y} \tag{6–86}$$

The above equation can be applied to the laminar flow. In case of the laminar flow the shear stress τ can be represented as

$$\tau = \eta\frac{\partial V_x}{\partial y}$$

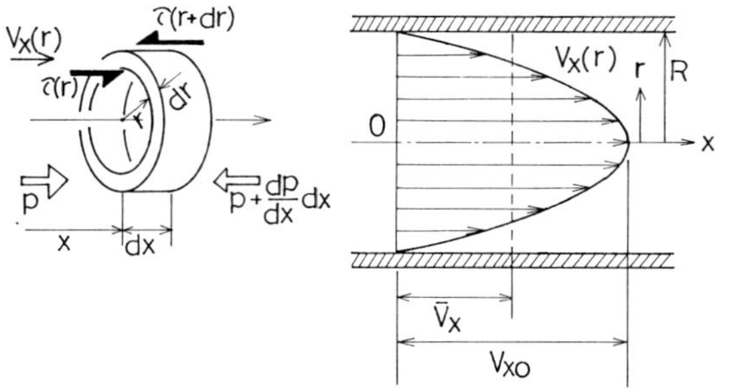

FIGURE 6–9. Schematic illustration of the velocity distribution of laminar flow in a circular pipe.

So the energy dissipation Ee can be calculated as

$$Ee = \eta\left(\frac{\partial V_x}{\partial y}\right)^2 \tag{6–87}$$

VII. ENERGY DISTRIBUTIONS IN LAMINAR FLOW IN A CIRCULAR PIPE

In order to derive the energy distributions in laminar flow in a cylindrical pipe, we consider a small ring of distribution of the laminar velocity in a circular pipe of a radius R as shown in Figure 6–9. Then the work done by the pressure p with the velocity V_x to the small area $2\pi \cdot r \cdot dr$ at a position x can be written as

$$V_x(r) \cdot p(x) \cdot 2\pi r \cdot dr \tag{6–88}$$

while the fluid flowing away from the small area $2\pi r \cdot dr$ at the position $x + dx$ performs work to the exterior as

$$V_x(r) \cdot p(x + dx) \cdot 2\pi r \cdot dr \tag{6–89}$$

Therefore the accumulative energy per unit time in the small volume $2\pi r \cdot dr \cdot dx$ of this ring is

$$-V_x(r) \cdot \frac{dp}{dr} \cdot 2\pi r \cdot dr \cdot dx \tag{6–90}$$

So the transported energy per unit time and per unit volume of this small ring through the pressure is

$$E_p = -V_x(r) \cdot \frac{dp}{dx} = 2\bar{V}_x\left(-\frac{\partial p}{\partial x}\right)\left(1 - \frac{r^2}{R^2}\right) \tag{6–91}$$

where an equation of the distribution of the velocity can be expressed as

$$V_x(r) = V_{xo}\left(1 - \frac{r^2}{R^2}\right) = 2\bar{V}_x\left(1 - \frac{r^2}{R^2}\right) \tag{6–92}$$

Next we consider the energy transfer by the shear stress. Since the inside surface area $2\pi r \cdot dx$ of the small ring is trailed by the velocity $V_x(r)$ with the shear stress τ, the given energy to this small ring through this inside surface area per unit time can be expressed as

$$V_x(r) \cdot \tau(r) \cdot 2\pi r \cdot dx \tag{6–93}$$

While the outside surface area $2\pi(r + dr) \cdot dx$ of this small ring coming in contact with the exterior fluid trails by the velocity $V_x(r + dr)$ with the shear stress $\tau(r + dr)$, so this small ring element consumes energy per unit time of

$$V_x(r + dr) \cdot \tau(r + dr) \cdot 2\pi(r + dr) \cdot dx \tag{6–94}$$

Therefore, the accumulative energy in this small ring volume is

$$-2\pi \cdot \frac{d(r \cdot V_x \cdot \tau)}{dr} \cdot drdx \tag{6–95}$$

so that the transported energy E_τ per unit time and per unit volume of this small ring through the shear stress is

$$E_\tau = -\frac{1}{r} \cdot \frac{d(r \cdot V_x \cdot \tau)}{dr} = -\frac{\eta}{r} \cdot \frac{d}{dr}\left(r \cdot V_x \cdot \frac{dV_x}{dr}\right) \tag{6–96}$$

Substituting Equation 6–92 into Equation 6–96, Equation 6–96 can be transformed as

$$E_\tau = 2\overline{V}_x \frac{dp}{dx}\left(1 - 2\frac{r^2}{R^2}\right) \tag{6–97}$$

Accordingly this small ring element receives energy through the pressure and through the shear stress

$$\begin{aligned} E_p + E_\tau &= 2\overline{V}_x\left(-\frac{dp}{dx}\right)\left(1 - \frac{r^2}{R^2}\right) + 2\overline{V}_x\left(\frac{dp}{dx}\right)\left(1 - 2\frac{r^2}{R^2}\right) \\ &= 2\overline{V}_x\left(-\frac{dp}{dx}\right) \cdot \frac{r^2}{R^2} \end{aligned} \tag{6–98}$$

These energies E_p and E_τ do not exchange to the available heat like the kinetic energy, but change to heat.

On the other hand the energy dissipation Ee can be expressed by Equation 6–87. So Ee can be calculated for the laminar flow in the circular pipe as follows:

$$Ee = \eta\left(\frac{dV_x}{dy}\right)^2 = \eta\left(\frac{dV_x}{dr}\right)^2 = 2\overline{V}_x\left(-\frac{dp}{dx}\right) \cdot \frac{r^2}{R^2} \tag{6–99}$$

Therefore Ee of Equation 6–99 is equivalent to $E_p + E_\tau$ of Equation 6–98. These energy distributions in dimensionless forms are

$$E'_\tau = \frac{E_\tau}{2\overline{V}_x(-dp/dx)} = 2\frac{r^2}{R^2} - 1$$

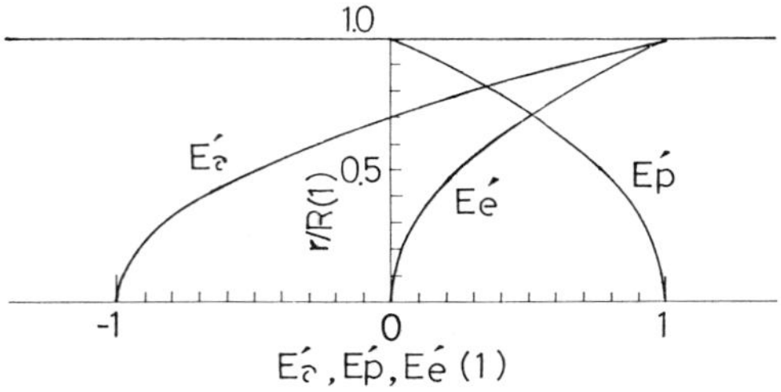

FIGURE 6–10. Distribution of the energy dissipation on the laminar flow in the circular pipe.

$$E_p' = \frac{E_p}{2\overline{V}_x(-dp/dx)} = 1 - \frac{r^2}{R^2}$$

$$E_e' = \frac{E_e}{2\overline{V}_x(-dp/dx)} = \frac{r^2}{R^2}$$

These distributions are shown in Figure 6–10.

MAIN SYMBOLS

$Ee(N/m^2s)$;	energy dissipation per unit time per unit volume
$\mathbf{F}(N/m^3)$;	body force per unit volume
$G_z(m^4/s^2)$;	axial momentum flux
$G_\theta(m^5/s^2)$;	angular momentum flux
$L(m)$;	wake width
$p(Pa)$;	static pressure
$R(m)$;	length
$T_s(s)$;	spin-up time
$T_r(s)$;	residence time
$\mathbf{V}(m/s)$;	velocity vector
$V_o(m/s)$;	velocity
$V_x(m/s)$;	velocity component
$V_y(m/s)$;	velocity component
$V_z(m/s)$;	velocity component
$\delta(m)$;	thickness of the mixing layer or vorticity layer
$\nu(m^2/s)$;	kinematic viscosity
$\zeta_z(1/s)$;	vorticity component
$\rho(kg/m^3)$;	density
$\tau_{xy}(Pa)$;	shear stress, Reynolds stress
$\tau_{xz}(Pa)$;	shear stress, Reynolds stress
$\tau_{yz}(Pa)$;	shear stress, Reynolds stress
$\mathbf{\Omega}(1/s)$;	vorticity vector
$\Omega_o(rad/s)$;	angular velocity

REFERENCES

1. **Textor, R. E.,** A numerical investigation of a confined vortex problems, Union Carbide Corporation, Nuclear Division, Tennessee, January 1968.
2. **Ogawa, A., Seito, O., and Nagabayashi, H.,** On the spin-up time and the residue time of the turbulent rotational air flow in the special forms of the cyclone dust collectors, *Bull. JSME,* 29(252), 1698, 1986.
3. **Ogawa, A., Fujita, Y., and Nagabayashi, H.,** Development and collapse of Taylor vortex on the rotating inner cylinder in the coaxial rest outer cylinder, *Chem. Eng. Commun.,* 37, 209, 1985.
4. **Traugott, S. C.,** Influence of solid-body rotation on screen-produced turbulence, *NACA-TN* 4135, August 1958.
5. **Harada, S.,** *Fluid Mechanics* (in Japanese), Maki-Pub. Co., Tokyo, 1958.
6. **Reynolds, A. J.,** Similarity in swirling wakes and jets, *J. Fluid Mech.,* 14, 241, 1962.

Chapter 7

MIXING LENGTH AND SCALE OF TURBULENCE

I. INTRODUCTION

In this chapter the physical explanation of the Reynolds stress and also the scale of turbulence in the turbulent flow are described in detail. It is very important to recognize the characteristics of turbulence and the scales of turbulence which supply the inferable criteria on the mechanical behaviors of the fine solid particles in turbulent flow.[1,2] Now before advancing the explanation of the scale of turbulence, the empirical equation of the pressure drop along the special forms of pipes by Schiller's experiments[3] are described. The pressure drop Δp along the pipe length L for the mean fluid velocity V_o(m/s) in the pipe was written by Schiller:

$$\Delta p = \lambda' \frac{L}{(D/2)} \frac{\rho V_o^2}{2} = \lambda' \frac{L}{m'} \cdot \frac{\rho V_o^2}{2} \tag{7–1}$$

where $\lambda'(1)$ is the frictional coefficient and D(m) is the diameter of a pipe. He used the hydraulic mean radius m'(m) defined as

$$m' = \frac{2A}{P} \tag{7–2}$$

where A(m^2) is the cross-sectional area and P(m) is the length of a periphery of a pipe. In case of the diameter D of the cylindrical pipe, m' becomes $m' = D/2$. Then the Reynolds number R_e' was defined as

$$R_e' = \frac{V_o \cdot D}{2\nu} = \frac{V_o \cdot m'}{\nu} \tag{7–3}$$

Therefore if we change R_e' to Re ($= V_oD/\nu$), the relationship between R_e' and R_o may be written as

$$R_e' = R_e/2 \tag{7–4}$$

Figure 7–1 shows the experimental results of Schiller for the pipe forms of triangle, square, rectangle, and special shapes. In addition to this, Figure 7–2 shows the experimental results of the dimensionless apparent eddy viscosity ν_T/ν vs. the Reynolds number $R_e(V_oD/\nu)$ in the circular pipe for a smooth surface ($\delta = 0$) and for the roughness δ of the pipe surface by Albring.[4] The value of ν_T/ν is proportional to Re for the smooth pipe.

For turbulent rotational flow, the experimental result of the dimensionless apparent eddy viscosity ν_T/ν vs. the Reynolds number R_{em} is

$$R_{em} = \frac{V_{\theta M} \cdot a}{\nu} \tag{7–5}$$

This is shown in Figure 7–3, where a(m) is the radius corresponding to the maximum tangential velocity $V_{\theta M}$(m/s) by Albring. In this figure the experimental result of the diameter

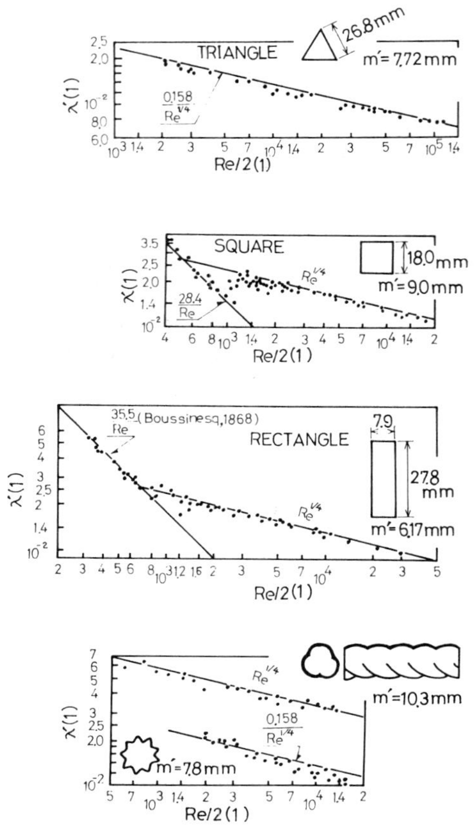

FIGURE 7–1. Experimental results of the frictional coefficient for the special forms of the pipes by Schiller.

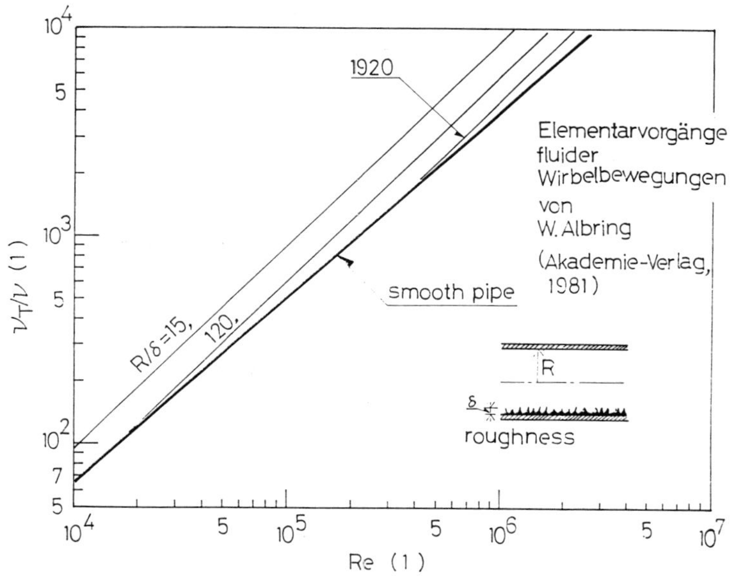

FIGURE 7–2. Relationship between the apparent eddy viscosity and the Reynolds number in the circular pipe.

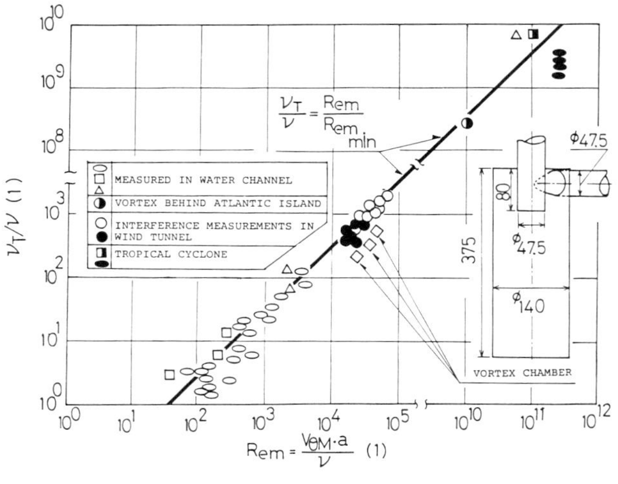

FIGURE 7–3. Relationship between the apparent eddy viscosity of the turbulent rotational flow and the Reynolds number defined by Equation 7–5.

$D_1 = 140$ mm by the author is included.[5] In this representation, a relationship between the Reynolds stress $\tau_{r\theta}$ and the eddy viscosity ν_T can be defined as

$$\tau_{r\theta} = -\rho\overline{v_r v_\theta} = \rho\nu_T\left(\frac{dV_\theta}{dr} - \frac{V_\theta}{r}\right) \tag{7–6}$$

In this vortex chamber the empirical equation of the Kolmogorov micro-scale η_* and of the Prandtl mixing length l in the quasi-free vortex region are:

$$\frac{\eta_*}{R_1} = \frac{7}{R_{ec}}\frac{r}{R_1} \tag{7–7}$$

$$\frac{l}{R_1} = 6 \times 10^{-6} \cdot R_{ec}\frac{r}{R_1} \tag{7–8}$$

where R_1(m) is the radius of the vortex chamber and $R_{ec}(1) = Q_o/H_i\nu$ is the Reynolds number, and where $Q_o(m^3/s)$ is the flow rate into the vortex chamber, H_i(m) is the imaginary cylindrical length, and also $\nu(m^2/s)$ is the kinematic viscosity. Here the Kolmogorov micro-scale η_* and the Prandtl mixing length l can be defined as

$$\eta_* = (\nu^3/\epsilon)^{0.25} = \left[\frac{\nu^3}{-\overline{v_r v_\theta} \cdot r \cdot \frac{d}{dr}\left(\frac{V_\theta}{r}\right)}\right]^{0.25} \tag{7–9}$$

$$l = \sqrt{\frac{-\overline{v_r v_\theta}}{\left(\frac{dV_\theta}{dr} - \frac{V_\theta}{r}\right)^2}} \tag{7–10}$$

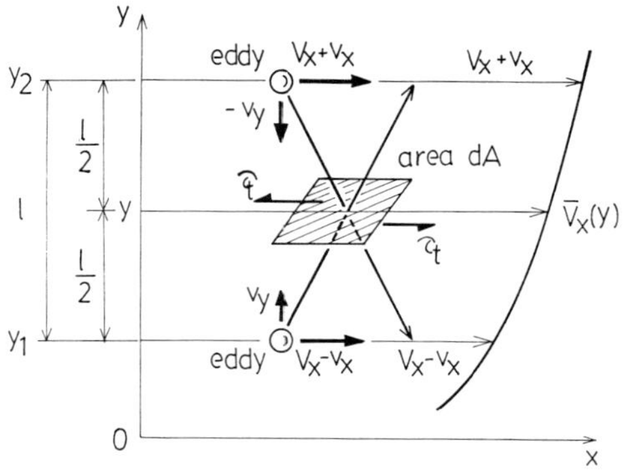

FIGURE 7–4. Physical illustration of the relationship between the fluctuating velocities and Reynolds stress in turbulent flow.

where $\epsilon(m^2/s^3)$ is the energy dissipation. From the above experimental results, the scale of the Prandtl mixing length is at least ten times longer than that of the Kolmogorov micro-scale in the turbulent rotational flow in the vortex chamber.

II. FLUCTUATING VELOCITIES AND REYNOLDS STRESS IN TURBULENT FLOW

The shear stress τ in the turbulent flow is composed of the laminar shear stress τ_ℓ which is based upon the momentum exchange of the molecules and of the turbulent shear stress τ_t which is based upon the momentum exchange of the turbulent eddies. Therefore

$$\tau = \tau_\ell + \tau_t \tag{7–11}$$

where τ_ℓ can be expressed as

$$\tau_\ell = \eta \cdot \frac{d\overline{V}_x}{dy} \tag{7–12}$$

The turbulent shear stress τ_t can be obtained as follows. As shown in Figure 7–4, the momentum of the lump of eddy to the x-direction due to passing through the small area dA can be derived as follows. The eddy mass now flowing through an area dA from the lower position y_1 to the upper position y_2 during the time interval dt is estimated as

$$\rho \cdot v_y \cdot dA \cdot dt \tag{7–13}$$

This mass flux has the velocity $V_x - v_x$ in the x-direction at positive y_1. The upper position y_2 can be obtained from the momentum flux

$$\rho v_y(V_x - v_x)dA\ dt \tag{7–14}$$

by the movement of this mass flux. Therefore the time averaged momentum flux per unit area and per unit time

$$\rho\overline{v_y(V_x - v_x)} = -\rho\overline{v_x v_y} \tag{7–15}$$

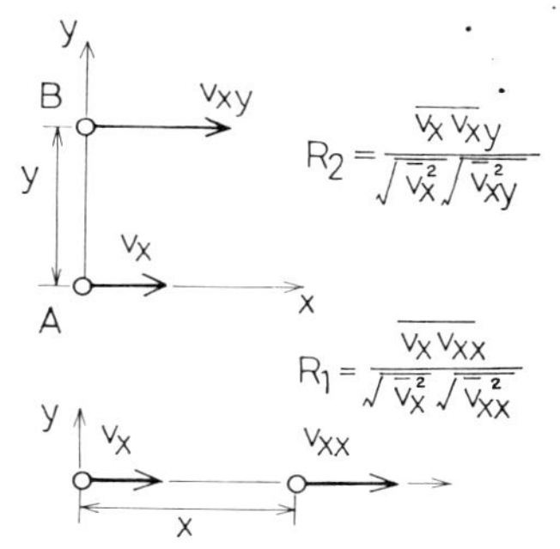

FIGURE 7–5. Illustration of the two kinds of correlations.

will be transported from y_1 to y_2. Here when the shear stress on the upper surface is acted in the decelerating direction, the shear stress is assigned a positive value. The time averaged momentum flux per unit time and per unit area which is transported from the upper position y_2 to the lower position y_1 is

$$\overline{\rho(-v_y)(V_x + v_x)} = -\rho\overline{v_x v_y} \tag{7–16}$$

This quantity $-\rho\overline{v_x y_y}$ is the Reynolds stress in turbulent flow and is defined

$$\tau_t = -\rho\overline{v_x v_y} \tag{7–17}$$

In terms of the idea of the mixing length l, the fluctuating velocities v_x and v_y are

$$v_x \fallingdotseq l\frac{dV_x}{dy} \quad \text{and} \quad v_y \fallingdotseq l\frac{dV_x}{dy} \tag{7–18}$$

Therefore the shear stress τ_t is

$$\tau_t = l^2\left(\frac{dV_x}{dy}\right)^2 \tag{7–19}$$

III. INTEGRATED SCALE OF TURBULENCE (EULER CORRELATION FUNCTION)

An instantaneous correlation between the fluid velocities at two arbitrary points in the flow is called an Euler correlation.[6] When the distance between the two points is smaller than the diameter of an eddy, a high correlation is obtained. But when the distance between the two points is larger than the diameter of this eddy, the correlation is lower. Here let us take the two points A and B of the section of the distance y parallel to the y-axis, and also let us denote v_x and v_{xy} as the disturbance velocity for the x-direction at the two points as shown in Figure 7–5. So the correlation R_2 between v_x and v_{xy} is

$$R_2 = \frac{\overline{v_x v_{xy}}}{\sqrt{\overline{v_x^2}}\,\sqrt{\overline{v_{xy}^2}}} \tag{7–20}$$

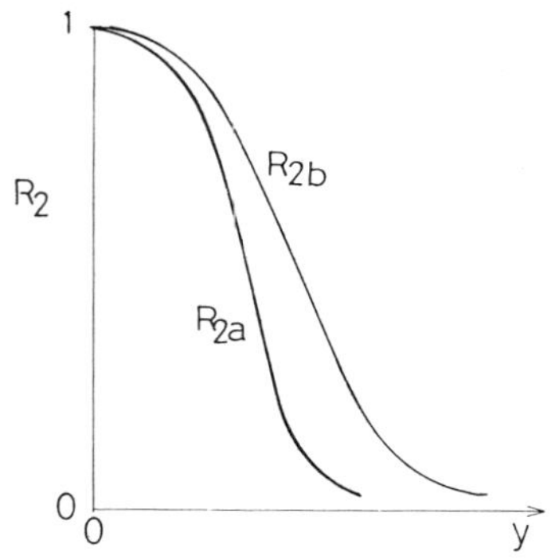

FIGURE 7–6. Physical meaning of the correlation R_2.

Taking the two points A and B which are distance x apart parallel to the x-axis, and denoting v_x and v_{xx} as the disturbance velocity for the x-axis as shown in Figure 7–5, the correlation R_1 between v_x and v_{xx} is

$$R_1 = \frac{\overline{v_x v_{xx}}}{\sqrt{\overline{v_x^2}}\sqrt{\overline{v_{xx}^2}}} \tag{7–21}$$

When the turbulent flow is assumed to be homogeneous, so

$$\overline{v_x^2} = \overline{v_{xx}^2} = \overline{v_{xy}^2} = \text{constant} \tag{7–22}$$

and Equations 7–20 and 7–21 become

$$R_2 = \frac{\overline{v_x v_{xy}}}{\overline{v_x^2}} \tag{7–23}$$

$$R_1 = \frac{\overline{v_x v_{xx}}}{\overline{v_x^2}} \tag{7–24}$$

Then in order to explain the physical meaning of the correlation R_2, let us show the illustration of the curves R_{2a} and R_{2b} as a function of the distance y in Figure 7–6. Now when the distances between the two points A and B approach each other, so that the same volume of fluid easily passes through the two points at the same time, the correlation R_2 therefore approaches one. When the distance between the two points becomes longer, the correlation R_2 gradually approaches zero. In addition to this, comparing a curve R_{2a} with a curve R_{2b}, the curve R_{2a} shows the small mixing eddy, and the curve R_{2b} shows the larger mixing eddy.

Consequently the scale of turbulence as an integrated scale of turbulence can be defined as

$$L_2 = \int_0^y R_2 dy \tag{7–25}$$

This length gives an idea of the averaged size of the turbulent eddy or of the volume of turbulence. Figure 7–7 shows one example of the correlation function R_2 for the longitudinal

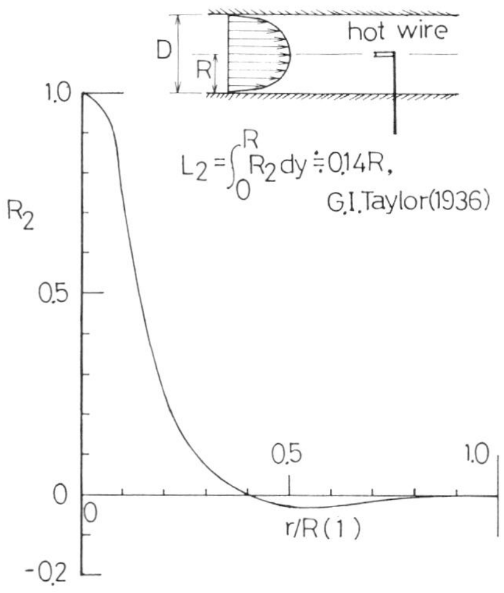

FIGURE 7–7. Correlation function for the longitudinal component of velocity in the circular pipe by G. I. Taylor.

fluctuations which were measured in the center of a circular pipe of a diameter D(= 2R) due to Simmons reported by G. I. Taylor.[7] The averaged size L_2 of the turbulent eddies is about $L_2 \fallingdotseq 0.14(D/2) = 0.07D = 0.14R$. Further when we expand the fluctuating velocity V_x as the Taylor series and assuming that the turbulence is homogeneous, the correlation function R_2 can be transformed to

$$R_2(y) = 1 - \frac{1}{2!}\frac{y^2}{\overline{v_x^2}} \cdot \overline{\left(\frac{\partial v_x}{\partial y}\right)^2} + \frac{1}{4}\frac{y^4}{\overline{v_x^2}}\overline{\left(\frac{\partial v_x}{\partial y}\right)^2} \tag{7–26}$$

When the distance y is very small, the above equation becomes

$$R_2(y) = 1 - \frac{y^2}{2\overline{v_x^2}} \cdot \overline{\left(\frac{\partial v_x}{\partial y}\right)^2} \tag{7–27}$$

Transforming this equation

$$\begin{aligned}\overline{\left(\frac{\partial v_x}{\partial y}\right)^2} &= \lim_{y\to 0}\left\{\frac{2\overline{v_x^2}[1 - R_2(y)]}{y^2}\right\} = 2\overline{v_x^2}\lim_{y\to 0}\left[\frac{1 - R_2(y)}{y^2}\right]\frac{\overline{(\partial \overline{v}_x/\partial y)^2}}{2\overline{v_x^2}} \\ &= \lim_{y\to 0}\frac{1 - R_2}{y^2} = \frac{1}{\lambda^2}\end{aligned} \tag{7–28}$$

Here a length λ corresponding to a parabolic curve coming in contact with R_2 shows the distance y which is intersected at the y-axis as shown in Figure 7–8. Therefore when y becomes very small, Equation 7–28 reduces to

$$\lambda^2 = \frac{y^2}{1 - R_2} \tag{7–29}$$

We can obtain a relationship between $R_1(x)$ and $R_2(y)$

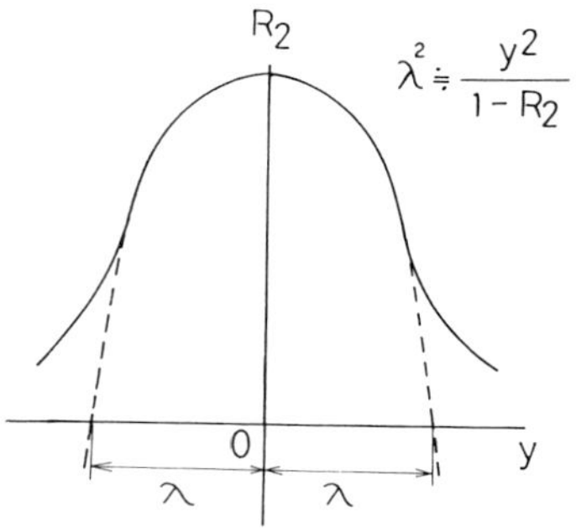

FIGURE 7–8. Illustration of the micro-scale of turbulence.

$$R_2(r) = R_1(r) + \frac{1}{2} r \frac{dR_1(r)}{dr} \tag{7–30}$$

And then the relationship between $(\partial v_x/\partial y)$, $(\partial v_x/\partial x)$, and λ is

$$\frac{1}{\lambda^2} = \frac{\overline{(\partial v_x/\partial y)^2}}{2\overline{v_x^2}} = \frac{\overline{(\partial v_x/\partial x)^2}}{\overline{v_x^2}} \tag{7–31}$$

This length λ may be called a micro-scale turbulence. The detailed treatment of the derivation of the above-stated correlations can be found in many books on the fluid mechanics of turbulence.

IV. TURBULENT SHEAR STRESS (PRANDTL MIXING LENGTH HYPOTHESIS)

The mechanism of the generation of the turbulent shear stress in the turbulent flow could be explained by L. Prandtl, who had introduced a quantity corresponding to the mean free path of the behaviors of the molecules of gases.

Let us explain the physical meaning of a Prandtl hypothesis.[8] Assume that, when an eddy fluid particle moves a distance l as an average, this particle migrates together with another eddy fluid particle by collision, and they merge at that collision place. This distance l is called the mixing length. As shown in Figure 7–9, we consider the motion of one eddy particle of a symbol a which is passing a plane $y = y_o$ to the upward y-direction. An eddy particle passing through the y_o-plane and ascending to the upward direction begins to depart at the $y = y_o - l$ plane on the average. When the eddy particle a keeping a velocity $(\overline{V}_{xo} - l \cdot d\overline{V}_{xo}/dy)$ of the x-component at a plane $y = y_o - l$ reaches to the y_o-plane, the fluctuating velocity v_x of the x-component at the plane $y = y_o$ becomes

$$v_x = \overline{V}_{xo}(y_o - l) - V_{xo}(y_o) = -l \cdot \frac{d\overline{V}_{xo}}{dy} \tag{7–32}$$

While both the fluctuating velocity v_x which comes from the $y_o - l$ plane to the y_o-plane and also comes from the $y_o + l$ plane to the y_o-plane occur at the same time, so the mean value of the absolute value $|v_x|$ may be regarded as

$$|v_x| \doteqdot l \cdot \left|\frac{d\overline{V}_{xo}}{dy}\right| \tag{7–33}$$

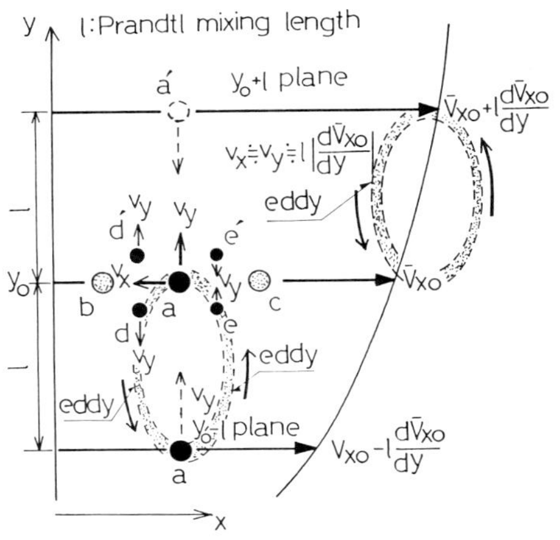

FIGURE 7–9. Physical illustration of the Prandtl mixing length hypothesis.

Here the eddy particle passing through the $y_o - l$ plane and ascending to the y_o-plane has the fluctuating velocity of $v_x < 0$ at the y_o plane. Therefore an eddy particle b being on the y_o plane approaches the eddy particle a with a relative velocity $|v_x|$. The eddy particles d and d′ between the eddy particles a and b push forward with the fluctuating velocity v_y to the y-direction. However the distance between the eddy particles c and a reaches in the distance with the relative velocity $|v_x|$ and at the same time the eddy particles e and e′ existing between the eddy particles a and c are swallowed up with the velocity $|v_y|$. Therefore the absolute value of the fluctuating velocity $|v_y|$ for the y-direction which is generated by rushing an eddy particle a into the y_o-plane may be regarded as nearly the same order of $|v_x|$ as

$$|v_y| \doteqdot |v_x| \doteqdot l \cdot \left|\frac{d\bar{V}_{xo}}{dy}\right| \tag{7–34}$$

And also the order of $|v_y|$ of an eddy particle a may be regarded as nearly the same order of $|v_y|$ at the $y_o - l$ plane and at the y_o-plane. In addition, when we consider the signs of v_x and v_y, it will be found that, when $d\bar{V}_x/dy$ has a positive value, v_x becomes the negative corresponding to the positive value of v_y. Consequently we may estimate

$$\overline{v_x v_y} \propto |v_x| \cdot |v_y| \doteqdot -l^2 \cdot \left(\frac{d\bar{V}_x}{dy}\right)^2 \tag{7–35}$$

Here we introduce the mixing length l as a mean free path of the eddy particle, but we cannot determine directly this mixing length l by the measurements. Therefore we must revise the definition of the mixing length l which includes the proportional constant as follows:

$$-\overline{v_x v_y} = l^2 \cdot \left(\frac{d\bar{V}_x}{dy}\right)^2 \tag{7–36}$$

Then the turbulent shear stress τ_t called Reynolds stress can be written as

$$\tau_t = -\rho\overline{v_x v_y} = \rho l^2 \left|\frac{d\bar{V}_x}{dy}\right| \left(\frac{d\bar{V}_x}{dy}\right) \tag{7–37}$$

Now in order to give the positive shear stress for the positive velocity gradient and to give the negative shear stress for the negative velocity gradient, we define the equation of the turbulent shear stress by the form of Equation 7–37. This equation is called the Prandtl mixing hypothesis and is applied to the estimation of turbulent shear stress.[9]

V. VELOCITY DISTRIBUTION BY PRANDTL THEORY

We can derive the velocity distribution in the turbulent flow by the Prandtl mixing length hypothesis. Consider the turbulent flow along a smooth flat wall and denote the distance perpendicular to the wall by a symbol y. Further, let us assume that the mixing length l is proportional to the distance y near the wall surface

$$l = ky \tag{7–38}$$

where k denotes a dimensionless constant value which can be deduced by experiment. Therefore the turbulent shear stress τ_t can be described as follows

$$\tau_t = \rho k^2 y^2 \left(\frac{dV_x}{dy}\right)^2 \tag{7–39}$$

We introduce a reasonable assumption that the shear stress near the wall surface is to be constant

$$\tau_o = \tau \tag{7–40}$$

where τ_o means the shear stress at the wall. And moreover we define the frictional velocity as

$$v_* = \sqrt{\frac{\tau_o}{\rho}} \tag{7–41}$$

Substituting Equation 7–41 into Equation 7–39, we obtain

$$\frac{dV_x}{dy} = \frac{v_*}{ky} \tag{7–42}$$

Integrating Equation 7–42, we obtain

$$V_x(y) = \frac{V_*}{k} \ln y + C \tag{7–43}$$

In order to make an equation describing the whole region of the turbulent flow, we may extend to the maximum velocity V_{xM} at $y = h$. So we can determine the equation as

$$\frac{V_{xM} - V_x}{v_*} = \frac{1}{k}(\ln h - \ln y) \tag{7–44}$$

where, generally speaking, a constant value k can be given to be $k = 0.4$ by the experimental results. A schematic illustration is shown in Figure 7–10.

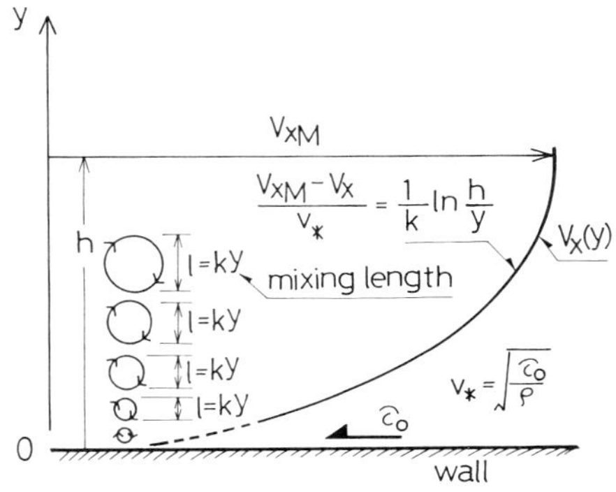

FIGURE 7–10. Schematic illustrations of the velocity profile and the mixing length.

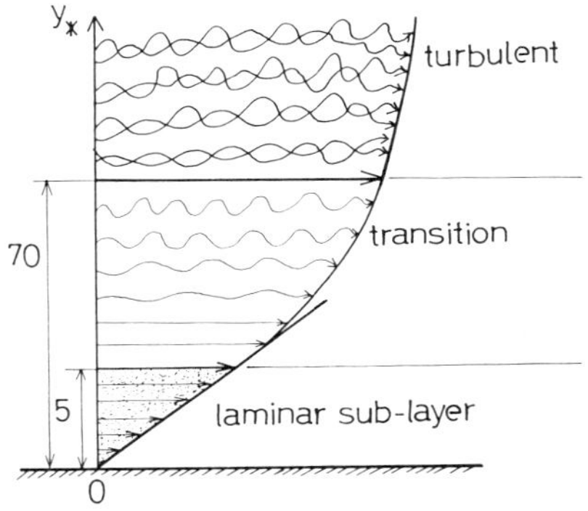

FIGURE 7–11. Schematic illustration of the velocity profile divided into the three regions.

VI. PRANDTL MIXING LENGTH

In the case of fully turbulent flow in the smooth pipe, the velocity distribution can be divided into the three regions as shown in Figure 7–11. Here in order to make clear the characteristics of the turbulent flow, we must introduce a dimensionless distance y_* which is defined as

$$y_* = \frac{v_* y}{\nu} \tag{7–45}$$

In the laminar sublayer ($y_* = v_* y/\nu < 5$), the dimensionless velocity V_x/v_* can be written as

$$\frac{V_x}{v} = y_* = \frac{v_* y}{\nu} \tag{7–46}$$

and in the fully turbulent flow ($y_* > 70$), the dimensionless velocity V_x/v_* can be written as

$$\frac{V_x}{v_*} = 2.5\ln y_* + 5.5 \tag{7–47}$$

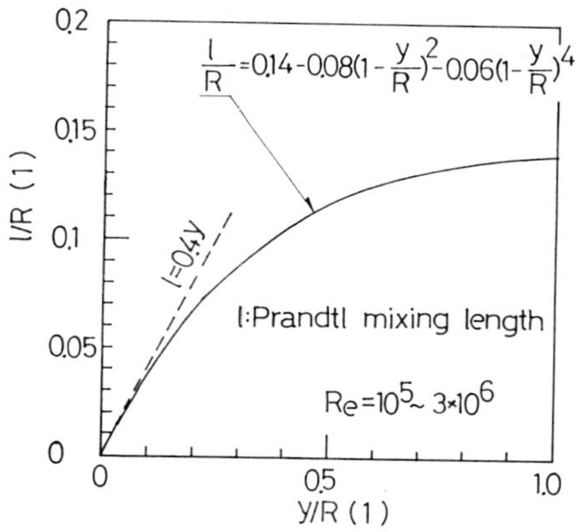

FIGURE 7–12. Mixing length in the smooth pipe.

J. Nikuradse[10] measured the velocity distributions in the pipe. The distribution of the shear stress can be represented by

$$\tau = \tau_o\left(1 - \frac{y}{R}\right) \tag{7–48}$$

Combining Equation 7–48 with Equation 7–37, an empirical equation of the Prandtl mixing length is obtained

$$\frac{l}{R} = 0.14 - 0.08\left(1 - \frac{y}{R}\right)^2 - 0.06\left(1 - \frac{y}{R}\right)^4 \tag{7–49}$$

and also in the near region of the pipe wall this equation can be simplified to

$$\frac{l}{R} = 0.4\frac{y}{R} - 0.44\left(\frac{y}{R}\right)^2 \tag{7–50}$$

This equation shows that the Prandtl hypothesis $l = ky$ can be applied near the wall. Figure 7–12 shows the curves of Equation 7–50 and $l = ky$ of the mixing lengths for the smooth pipes which were estimated by Nikurade experiments.[11] This estimated equation of the mixing length can be applied to the rough pipe wall ($R/k = 507$, 126, and 30.6 for $R_e \doteqdot 10^6$; k means the mean height of the wall roughness).

Dr. X. Yan has measured intensity of turbulence in turbulent flow in a roughened pipe.[12]

VII. KOLMOGOROV MICRO-SCALE

When we measure the velocity V(t) at a point P in the turbulent air flow with a hot-wire anemometer, we can obtain the instantaneous velocity disturbance v(t) as shown in Figure 7–13. Then the velocity disturbance v(t) can be resolved into the complex Fourier expansion as

$$v(t) = \Sigma A_j \exp(i \cdot k_j \cdot t) \tag{7–51}$$

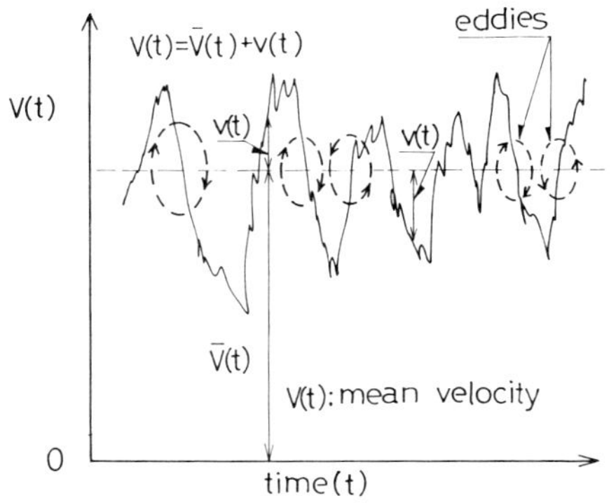

FIGURE 7–13. Schematic illustrations of the mean velocity and of the fluctuating velocities of the turbulent flow.

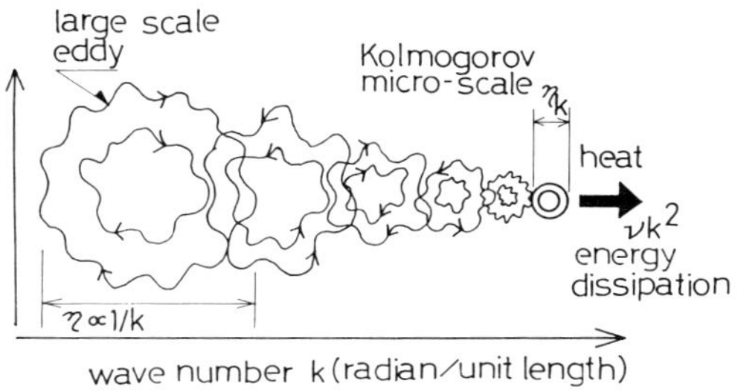

FIGURE 7–14. Schematic illustration between the scales of turbulence and the wave number.

where k_j is the wave number (radian/unit length), and also this wave number may be recognized as a reciprocal of the diameter of the turbulent eddy. Therefore the diameter of the turbulent eddy may be distributed from zero to infinity. Here let us denote the kinetic energy of this eddy with the wave number k as E(k), which is generally called an energy spectrum of turbulence.

The term of the energy dissipation by the viscosity can be represented by the transformation of Navier-Stokes equation as

$$\nu k^2 \tag{7–52}$$

This equation means that the large scale eddy becomes unstable and decomposes step by step into the smaller scale eddies, while the kinetic energy maintaining in the large eddy transfers little by little into the smaller eddy and finally dissipates to heat by viscosity. This physical explanation can be shown in Figure 7–14. Now in order to explain more clearly the mechanism of the energy transfer in the turbulent flow, transforming the Navier-Stokes equation with the energy spectrum of turbulence, we obtain

$$\frac{\partial E(k,t)}{\partial t} = \frac{\partial S(k,t)}{\partial k} - 2\nu k^2 E(k,t) \tag{7–53}$$

where $S(m^2/s^3)$ is the energy transfer per unit time from the large eddy into the smaller eddy, and also a unit of a symbol E(k) is $E(m^3/s^2)$. Then the inertia subrange is the range which

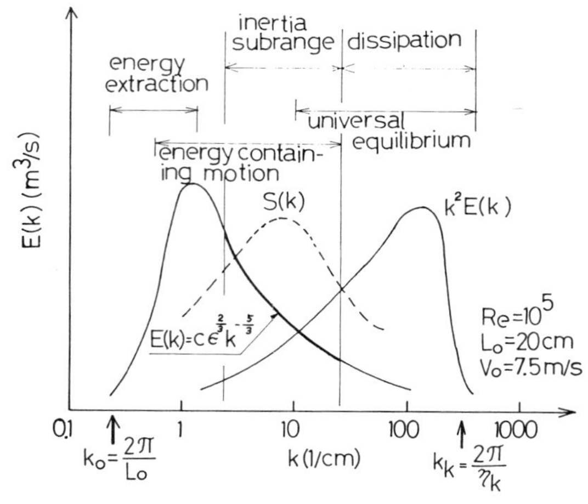

FIGURE 7–15. Schematic illustration of the characteristic range of a turbulent motion of $R_e = 10^5$ by A. J. Reynolds.[13]

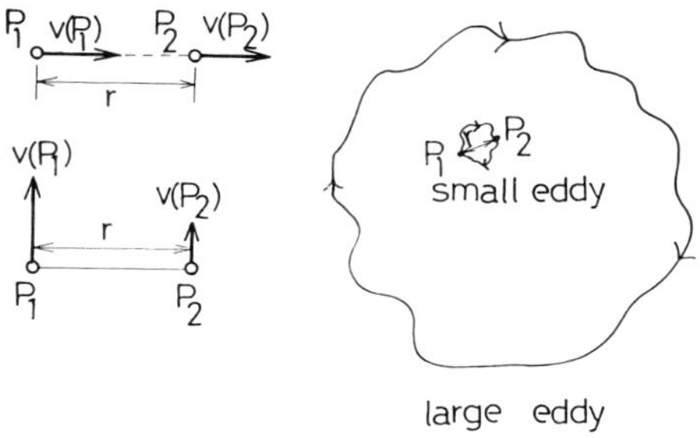

FIGURE 7–16. Scale of the small eddy in the large eddy.

is not influenced extremely by viscosity. Figure 7–15 shows an example of the characteristic range of a turbulent motion for $R_e = 10^5$, L_o (scale of mean motion) = 20 cm, and V_o (mean velocity) = 7.5 m/s by A. J. Reynolds.[13]

Further, according to the theory of Kolmogorov, the behavior of the small-scale eddy produced by the large eddy is statistically independent of that of the large-scale eddy, and also the physical information in the small-scale eddy only depends on the kinetic energy which is transferred by the large-scale eddy. Consequently the statistical quantities of the small-scale eddy can be determined by the energy dissipation $\epsilon(m^2/s^3)$ per unit volume. Here, as shown in Figure 7–16, the relative velocity v(r) on the small distance r between a point P_1 and a point P_2 is

$$\overline{v(r)^2} = \overline{[v(P_1) - v(P_2)]^2} \tag{7–54}$$

So $\overline{v(r)^2}$ can be determined by the small-scale eddy which has no influence upon the larger scale eddy than the distance r. Thereupon when the scale r is smaller than the scale L of the main flow, so $\overline{v(r)^2}$ can be estimated by the energy dissipation per unit volume and by the viscosity of fluid, and further $\overline{v(r)^2}$ is independent of the direction of the distance r and may be called a local isotropy. In addition to this, Kolmogorov assumed that, when the distance r is smaller than the scale L of the main flow but is larger than the eddy scale η_k which dissipates the energy, $\overline{v(r)^2}$ is independent of the kinematic viscosity $\nu(m^2/s)$ of fluid and also depends on a function of the energy dissipation ϵ. So he derived an equation by dimensionless analysis

$$\overline{v(r)^2} = C_1 \epsilon^{2/3} r^{2/3} \quad , \quad (L >> r >> \eta_k) \tag{7–55}$$

When the distance r is smaller than the scale η_k,

$$\overline{v(r)^2} = C_2 \frac{\epsilon r^2}{\nu} \qquad (r << \eta_k) \tag{7–56}$$

where the scale of η_k can be given as

$$\eta_k = (\nu^3/\epsilon)^{1/4} \tag{7–57}$$

and the velocity scale v_k can be given by

$$v_k = (\nu\epsilon)^{1/4} \tag{7–58}$$

Therefore η_k has Reynolds number one and also has a characteristic of the maximum energy dissipation.[14–21]

VIII. SCALES OF EDDIES IN CIRCULAR PIPE FLOW

In the case of the isotropic turbulence in a circular pipe of the diameter D, the scales of the integrated largest size L, of the Prandtl mixing length λ, and of the Kolmogorov micro scale length η_* in the central region of the pipe have been stated by J. T. Davies[22] as follows:

$$L/D \fallingdotseq 0.5 \tag{7–59}$$

$$\lambda/D \fallingdotseq 0.08 \tag{7–60}$$

$$\eta_*/D \fallingdotseq 4R_e^{-0.78} \tag{7–61}$$

where R_e is the Reynolds number defined as $R_e = \overline{V}_o D/\nu$. Here $\overline{V}_o$ is the mean velocity of fluid in the pipe. Therefore when water is flowing in a pipe of the diameter D = 50 mm with the mean velocity $\overline{V}_o$ = 1.8 m/s and $R_e = 10^5$, each scale is estimated as follows: L = 25 mm, λ = 4 mm, η_* = 25 μm.

IX. TURBULENT FLOW IN A CURVED CHANNEL

S. Eskinazi and H. Yeh[23] measured the intensity of turbulence, the shear stress, and the power-spectrum by the hot-wire anemometer with a tungsten wire 7.62 μm in diameter of fully developed turbulent flow in a curved channel between the two concentric circular walls. The width B and the height of this channel were B = 76.2 mm and H = 1181.1 mm, and also the radii of this circular channel were R_1 (outer radius) = 762 mm and R_2 (inner radius) = 685.8 mm. The Reynolds number R_e defined by

$$R_e = \frac{Q_o}{\nu H} \tag{7–62}$$

was R_e = 148,400.

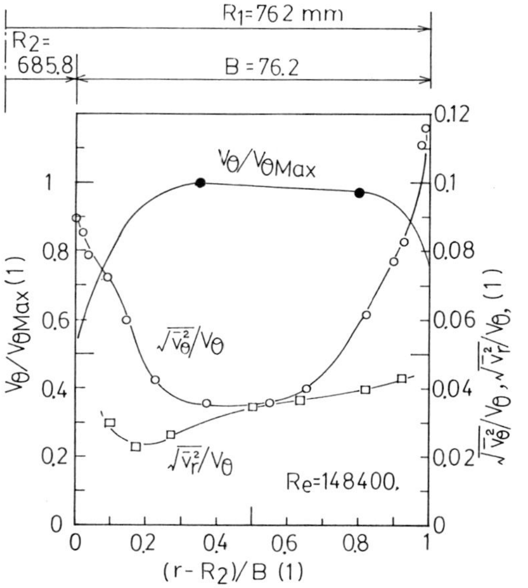

FIGURE 7–17. Distributions of the mean tangential velocity and of the intensity of turbulence in the fully developed curved section in the channel.

Figure 7–17 shows the distributions of the mean tangential velocity V_θ, the intensity of turbulence $\sqrt{\overline{v_\theta^2}}/V_\theta$, and $\sqrt{\overline{v_r^2}}/V_\theta$ in the fully developed curved section in the channel, where $V_{\theta Max}$ means the maximum tangential velocity. From these experimental results, $\sqrt{\overline{v_\theta^2}}/V_\theta$ shows the high values near the outer and inner wall surfaces. Then Figure 7–18 and 7–19 show the micro-scales (λ_θ and λ_r) and the integral scales (L_θ and L_r) of turbulence in the fully developed curved section for $R_e = 148{,}400$, where λ_θ and λ_r are the micro-scales for the tangential and radial directions. They had obtained experimentally the micro-scales by the one-dimensional energy spectrum and by the following equation defined as

$$\lambda_\theta^2 = \overline{v_\theta^2}/\overline{(\partial v_\theta/r\partial\theta)^2} \tag{7–63}$$

$$\lambda_r^2 = 2\overline{v_r^2}/\overline{(\partial v_r/r\partial\theta)^2} \tag{7–64}$$

The integral scales L_θ and L_r in the fully developed curved channel are defined

$$L_\theta = \int_0^\infty R_\theta(v_\theta) \cdot r \cdot d\theta \tag{7–65}$$

$$L_r = \int_0^\infty R_\theta(v_r) \cdot r \cdot d\theta \tag{7–66}$$

where a symbol R_θ is the space-correlation function which can be calculated by the Fourier transform of the one-dimensional energy spectrum. The integral scale L_θ for the tangential direction is longer than the scale L_r for the radial direction and is nearly equivalent to half of the width B of the channel.

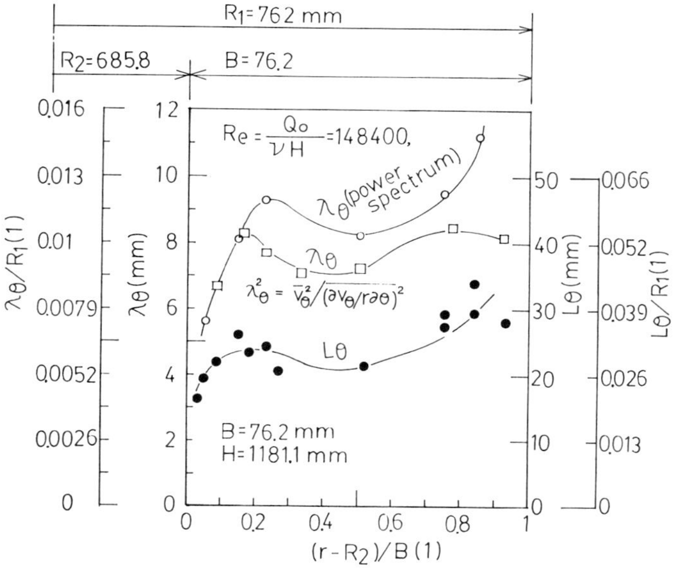

FIGURE 7–18. Distributions of the micro-scales and of the integral scale in the fully developed curved section.

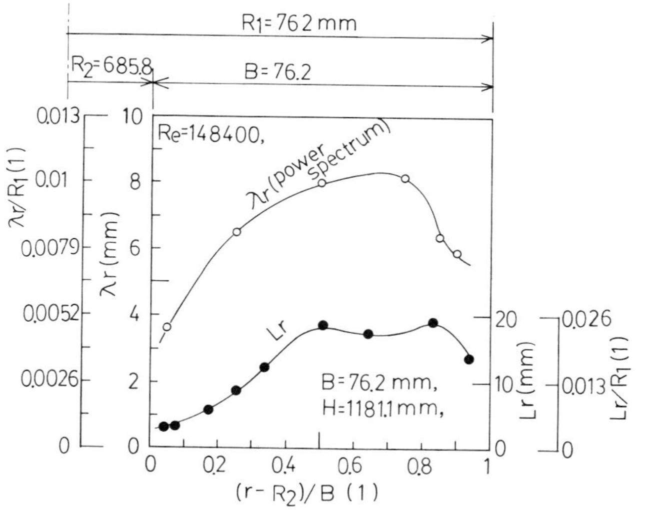

FIGURE 7–19. Distributions of the micro-scales and of the integral scale in the fully developed curved section.

MAIN SYMBOLS

a(m);	radius of the maximum tangential velocity
D(m);	diameter of a pipe
H_i(m);	imaginary cylindrical length
k(1/m);	wave number
L_2(m);	integrated scale of turbulence

l(m);	mixing length
m′(m);	hydraulic mean radius
$Q_o(m^3/s)$;	flow rate
$R_1(m)$;	radius of the vortex chamber
$R_1(1)$;	correlation between the two points
$R_2(1)$;	correlation between the two points
$v_o(m/s)$;	mean velocity in the pipe
$v_\theta(m/s)$;	tangential velocity
$v_*(m/s)$;	frictional velocity
$\eta_*(m)$;	Kolmogorov micro-scale
$\lambda(m)$;	micro-scale of turbulence
$\lambda'(1)$;	frictional coefficient on the pipe flow
$\nu(m^2/s)$;	kinematic viscosity
$\nu_T(m^2/s)$;	kinematic eddy viscosity
$\tau_t(Pa)$;	Reynolds stress

REFERENCES

1. **Hinze, J. O.,** Turbulent fluid and particle interaction, *Prog. Heat Mass Transfer,* 6, 433.
2. **Ogawa, A.,** Axial and radial diffusions of the fine solid particles in the turbulent rotational air flow in the returned flow type of the vortex chamber, *Nihon Univ. J. Coll. Eng.,* 27-A, 23, 1986.
3. **Schiller, L.,** Ü ber den Strömungswiderstand von Rohren verschiedenen Querschnitts und Rauhigkeitsgrades, *Z. A. M. M.,* 3(1), 2, 1923.
4. **Albring, W.,** *Elementarvorgänge fluider Wirbelbewegungen,* Akademie-Verlag, Berlin, 1981.
5. **Ogawa, A.,** *Separation of Particles from Air and Gases,* Vol. II, CRC Press, Boca Raton, FL, 1984.
6. **Harada, S.,** *Mechanics of Fluid* (in Japanese), Maki-Pub. Co., Tokyo, 1959.
7. **Taylor, G. I.,** Correlation measurements in a turbulent flow through a pipe, *Proc. R. Soc. London,* A-157, 537, 1936.
8. **Shirakura, M. and Ohashi, H.,** *Fluid Mechanics* (in Japanese), Corona-Pub. Co., Tokyo, 1969.
9. **Rothfus, R. R., Archer, D. H., and Sikchi, K. G.,** Distribution of eddy viscosity and mixing length in smooth tubes, *AIChE J.,* 4(1), 27, 1958.
10. **Nikuradse, J.,** Gesetzmkäßigkeit der turbulenten Strömung in glatten Rohren, *Forsch. Arb. Ing.-Wes.,* 356, 132.
11. **Schlichting, H.,** *Boundary Layer Theory,* McGraw-Hill, New York, 1968.
12. **Yan, X.,** Experimental study on the turbulent flow in a forcefully roughened pipe, *Acta Aerodynamica Sinica,* 4, 57, 1983.
13. **Reynolds, A. J.,** *Turbulent Flows in Engineering,* John Wiley & Sons, New York, 1974.
14. **Endo, K. and Nishikawa, A.,** *Chem. Eng.* (in Japanese), 28, 495, 1964.
15. **Hinze, J. O.,** *Turbulence,* McGraw-Hill, New York,
16. **Taniyama, I.,** Behaviours of the droplets in the mixer (in Japanese), in *Development of Chemical Engineering,* Nikan Kogyo Schinbun-sha, 1969.
17. **Batchelor, G. K.,** *The Theory of Homogeneous Turbulence,* Cambridge, 1953.
18. **Frisch, U., Sulem, P. I., and Nelkin, M.,** A simple dynamical model of intermittent fully developed turbulence, *J. Fluid Mech.,* 87, 719, 1978.
19. **Batchelor, G. K.,** Kolmogoroff's theory of locally isotropic turbulence, *Proc. Cambridge Philos. Soc.,* 43, 553, 1947.
20. **Doshi, M. R. and Gill, W. N.,** A note on the mixing length theory of turbulent flow, *AIChE J.,* 16, 885, 1970.
21. **Rothfus, R. R., Archer, D. H., and Sikchi, K. G.,** Distribution of eddy viscosity and mixing length in smooth tubes, *AIChE J.,* 4(1), 27, 1958.
22. **Davies, J. T.,** *Turbulence Phenomena,* Academic Press, New York, 1972.
23. **Eskinazi, S. and Yeh, H.,** An investigation on fully developed turbulent flows in a curved channel, *J. Aeronaut. Sci.,* 23, 23, 1956.

Chapter 8

VORTEX FLOW IN THE CONFINED VORTEX CHAMBER

I. INTRODUCTION

In this chapter, the fluid dynamical characteristics in the returned and axial flow types of the vortex chambers which can be applied to the various kinds of chemical and mechanical or power-handling engineering are described.

Then in order to recognize the rotational flow properties, the experimental results of the distributions of the static pressure, the vorticity, the angular momentum flux, and the axial momentum flux are illustrated in detail. The idea of the spin-up time is introduced.

II. THE FUNDAMENTAL EQUATION OF THE PRESSURE DROP

Let us assume that the viscous fluid particle of the small volume dv of the fixed mass (ρdv) in Figure 8–1 has kinetic energy

$$\frac{1}{2}(\rho dv)\mathbf{V}^2 \tag{8–1}$$

where **V** is a vector of velocity of a lump of fluid. So the rate of gain of kinetic energy at time t can be expressed as

$$\frac{d}{dt}\left[\frac{1}{2}(\rho dv)\mathbf{V}^2\right] = \rho dv\, \mathbf{V} \cdot \frac{d\mathbf{V}}{dt} \tag{8–2}$$

Then the total rate of gain of kinetic energy for the incompressible fluid of volume V is

$$\int_V \rho \mathbf{V} \cdot \frac{d\mathbf{V}}{dt}\, dv = \rho \int_V \mathbf{V} \cdot \frac{d\mathbf{V}}{dt}\, dv \tag{8–3}$$

Here substituting the Navier-Stokes equation of incompressible fluid into Equation 8–3, Equation 8–3 becomes

$$\begin{aligned} \rho \int_V \mathbf{V} \cdot \frac{d\mathbf{V}}{dt}\, dv &= \rho \int_V \mathbf{V} \cdot \left(K - \frac{1}{\rho}\nabla p + \nu \nabla^2 \mathbf{V}\right) dv = \\ &= \rho \int_V \mathbf{V} \cdot K\, dv - \int_V \mathbf{V} \cdot \nabla p\, dv - \eta \int_V \mathbf{V} \cdot [\nabla \times (\nabla \times \mathbf{V})] dv, \end{aligned} \tag{8–4}$$

where an external force **K** can be neglected. While from the result of the vector analysis the second and third terms of the right-hand side of Equation 8–4 can be transformed by Gauss and Stokes theorem to

$$\int_V \mathbf{V} \cdot \nabla p\, dv = \int_V \nabla \cdot (p\mathbf{V})\, dv = \int_S p\mathbf{V} \cdot d\mathbf{S} \tag{8–5}$$

$$\eta \int_V \mathbf{V} \cdot [\nabla \times (\nabla \times \mathbf{V})]\, dv = \eta \int_V \{(\nabla \times \mathbf{V})^2 - \nabla \cdot [\mathbf{V} \times (\nabla \times \mathbf{V})]\}\, dv \tag{8–6}$$

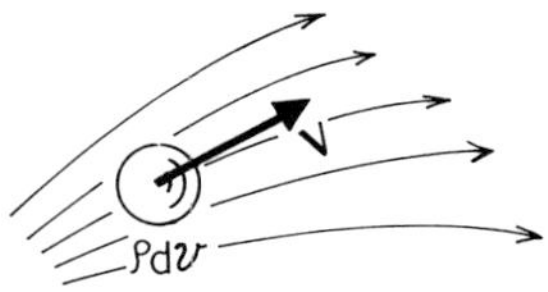

FIGURE 8–1. Flow and fluid particle.

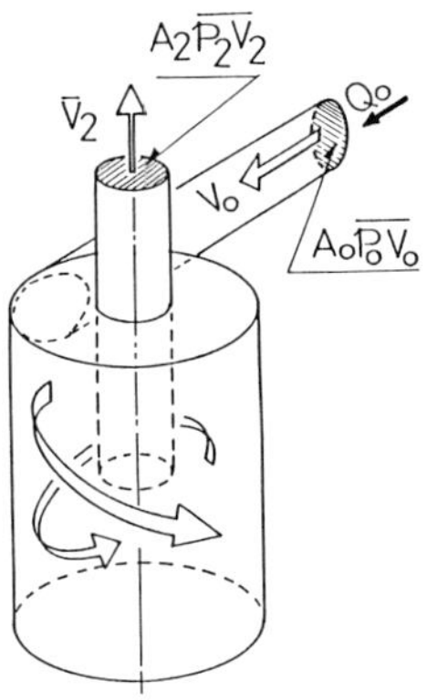

FIGURE 8–2. Schematic illustration of the vortex chamber.

S is the total surface area enclosing the volume V. Therefore, the rate of energy dissipation W(N · m/s = W) due to viscosity is

$$\begin{aligned} W &= \eta\int_V \mathbf{V}\cdot[\nabla\times(\nabla\times\mathbf{V})]\,dv = \eta\int_V (v\times\mathbf{V})^2\,dv - \eta\int_V \nabla\cdot[\mathbf{V}\times(\nabla\times\mathbf{V})]\,dv = \\ &= \eta\int_V (\nabla\times\mathbf{V})^2\,dv - \eta\int_S \mathbf{n}\cdot[\mathbf{V}\times(\nabla\times\mathbf{V})]\,dS \end{aligned} \tag{8–7}$$

Further let us assume that the fluid particle is at rest on the fixed boundary surface S in the confined vortex chamber; consequently, there is no slip between fluid and boundary surface in the vortex chamber. So Equation 8–7 reduces to

$$W = \eta\int_V (\nabla\times\mathbf{V})^2\,dv = \eta\int_V \boldsymbol{\Omega}^2\,dv \tag{8–8}$$

where a symbol $\boldsymbol{\Omega}$ is generally called a vorticity and is defined as

$$\boldsymbol{\Omega} = \nabla\times\mathbf{V} \tag{8–9}$$

In addition to this, the second term of the right-hand side of Equation 8–4 can be transformed as shown in Figure 8–2,

$$\int_V \mathbf{V}\cdot\nabla p\,dv = \int_S p\mathbf{V}\cdot d\mathbf{S} = A_2\bar{p}_2\bar{V}_2 - A_o\bar{p}_o\bar{V}_o \tag{8–10}$$

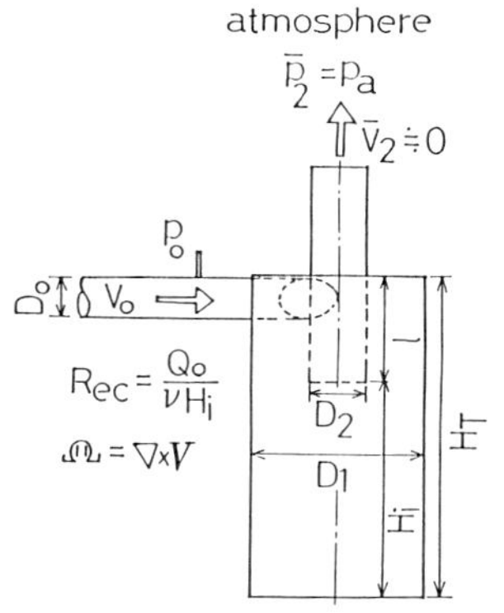

FIGURE 8–3. Schematic illustration of the vortex chamber and the main symbols.

In this equation, the symbols $\overline{V}_o$ and $\overline{V}_2$ are the mean axial velocity perpendicular to the inlet and exit pipes, and also the symbols $\overline{p}_o$ and $\overline{p}_2$ are the averaged static pressure in the inlet and exit pipes, respectively. Therefore, Equation 8–4 can be reduced to

$$\rho\int_V \mathbf{V}\cdot\frac{d\mathbf{V}}{dt}\,dv = -(A_2\overline{p}_2\overline{V}_2 - A_o\overline{p}_o\overline{V}_o) - \eta\int \Omega^2 dv \tag{8–11}$$

Here let us assume that fluid flow is to be steady state, so the rate of gain of kinetic energy of Equation 8–2 goes to zero. Accordingly, Equation 8–11 can be transformed to

$$A_o\overline{p}_o\overline{V}_o - A_2\overline{p}_2\overline{V}_2 = Q_o(\overline{p}_o - \overline{p}_2) = \eta\int_V \Omega^2 dv \tag{8–12}$$

where $A_o\overline{V}_o$ and $A_2\overline{V}_2$ are equal to the flow rate Q_o. Finally, an equation of the pressure drop Δp_v due to the vorticity can be expressed as

$$\Delta p_v = \overline{p}_o - \overline{p}_2 = \frac{\eta}{Q_o}\int_V \Omega^2\, dv \tag{8–13}$$

Here the equation of the pressure drop Δp_c of the vortex chamber can be derived. As shown in Figure 8–3, the total pressure at the inlet pipe and at the exit pipe is

$$P_o = p_o + \rho\frac{\overline{V}_o^2}{2} \tag{8–14}$$

$$P_2 = p_2 + \rho\frac{V_2^2}{2} \tag{8–15}$$

Since the pressure drop Δp_c of the vortex chamber is equal to the difference of the total pressures at the inlet and exit pipes, Δp_c can be obtained as

$$\Delta p_c = \overline{p}_o - \overline{p}_2 + \rho\frac{V_o}{2}\left[1 - \left(\frac{\overline{V}_2}{\overline{V}_o}\right)^2\right] \tag{8–16}$$

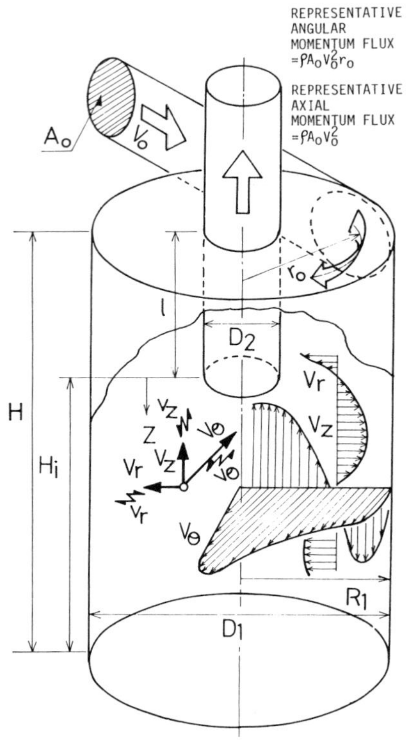

FIGURE 8–4. Schematic illustration of the velocity distributions and of the main sizes of the vortex chamber.

Here when the exit pipe is diffused to atmosphere, the static pressure p_2 is equal to $\bar{p}_2 = p_a$ (atmospheric pressure), and also the mean exit velocity $\bar{V}_2$ becomes zero in the infinite space, so that Equation 8–16 becomes

$$\Delta p_c = \bar{p}_o - p_a + \rho \frac{V_o^2}{2} \tag{8–17}$$

Therefore, as a result of the above considerations, substituting Equation 8–13 into Equation 8–17, a fundamental equation of the pressure drop Δp_c in the vortex chamber can be derived as

$$\Delta p_c = \frac{\eta}{Q_o} \int_V \Phi^2 \, dv + \rho \frac{V_o}{2} \tag{8–18}$$

III. PRESSURE DROP DUE TO VISCOSITY

As shown in Figure 8–4, let us assume that the rotational flow in the vortex chamber is axial-symmetrical, so the energy dissipation of the turbulent rotational air flow due to the eddy viscosity η_T in the cylindrical vortex chamber of the radius R_1 and height H_i can be expressed as

$$D = H_i \int_0^{R_1} \eta_T \, \zeta^2 \, 2\pi r \, dr \tag{8–19}$$

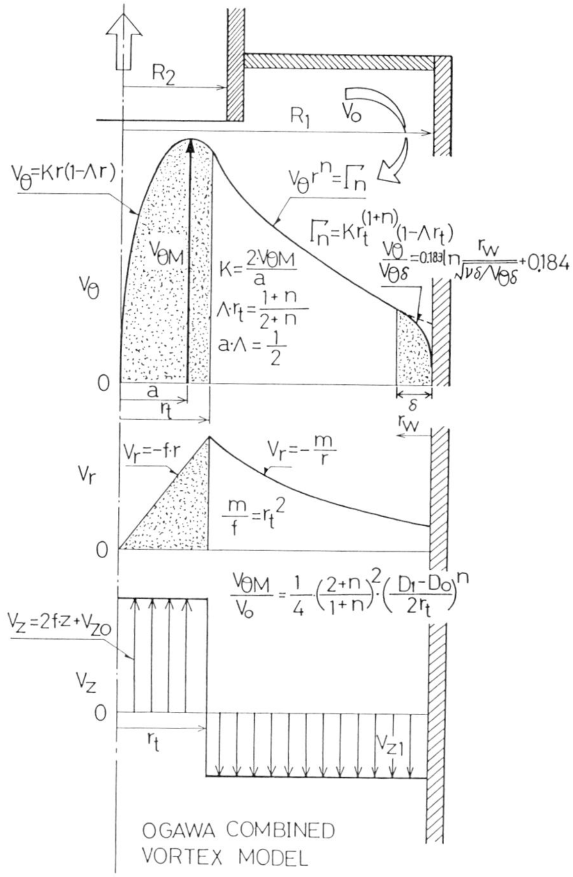

FIGURE 8–5. Schematic illustration of the Ogawa combined vortex model.

where a symbol η_T(Pa · s) is an eddy viscosity of the turbulent air flow and a symbol ζ(1/s) is the vorticity component of $\boldsymbol{\Omega}$ defined as

$$\zeta = \frac{\partial V_\theta}{\partial r} + \frac{V_\theta}{r} \tag{8–20}$$

Then in order to simplify Equation 8–19, the domain of the vortex flow should be divided into the quasi-forced vortex flow region ($0 \leq r \leq r_t$) and the quasi-free vortex region ($r_t \leq r \leq R_1$). These are called the Ogawa combined vortex model as shown in Figures 8–5, 8–6, and 8–7, in which the physical characteristics of the tangential, radial, and axial velocity components and also the distributions of the static pressure and the distributions of the vorticity and the circulation are illustrated in detail with figures.[1] Therefore, the eddy viscosity η_T in the quasi-forced vortex may be expressed as $\eta_T = \eta_{TF}$, and in the quasi-free vortex the eddy viscosity may be expressed as $\eta_T = \eta_{Tf}$, respectively. Consequently, Equation 8–19 can be transformed to

$$D = H_i \int_0^{r_t} \eta_{TF} \zeta^2 \, 2\pi r \cdot dr + H_i \int_{r_t}^{R_1} \eta_{Tf} \, \zeta^2 \, 2\pi r \cdot dr \tag{8–21}$$

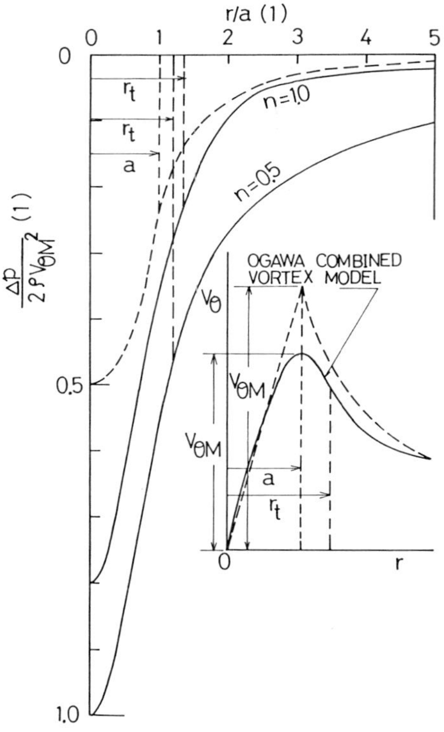

FIGURE 8–6. Distributions of the static pressures of the Ogawa combined vortex.

Now in the quasi-forced vortex flow region of the tangential velocity $V_\theta = Kr(1 - \wedge r)$, ζ can be represented as

$$\zeta = K(2 - 3\wedge r) \qquad (8\text{–}22)$$

Therefore the energy dissipation D_F in the quasi-forced vortex region can be calculated as

$$D_F = 64\pi\eta_{TF}H_iV_{\theta M}^2\left(\frac{1 + n}{2 + n}\right)^2\left[-\frac{n}{2 + n} + \frac{9}{8}\left(\frac{1 + n}{2 + n}\right)^2\right] \qquad (8\text{–}23)$$

In the quasi-free vortex region of the tangential velocity $V_\theta r^n = \Gamma_n$, ζ can be written as

$$\zeta = \Gamma_n(1 - n) \cdot r^{(1+n)} \qquad (8\text{–}24)$$

Therefore the energy dissipation D_f in the quasi-free vortex region is

$$D_f = 16\pi\eta_{Tf}H_iV_{\theta M}^2 \cdot \frac{1}{n}\left(\frac{1 + n}{2 + n}\right)^2 \cdot \left(\frac{1 - n}{2 + n}\right)^2\left[1 - \left(\frac{r_t}{R_1}\right)^{2n}\right] \qquad (8\text{–}25)$$

Finally the energy dissipation D of Equation 8–21 can be represented by the summation of Equation 8–23 and 8–25

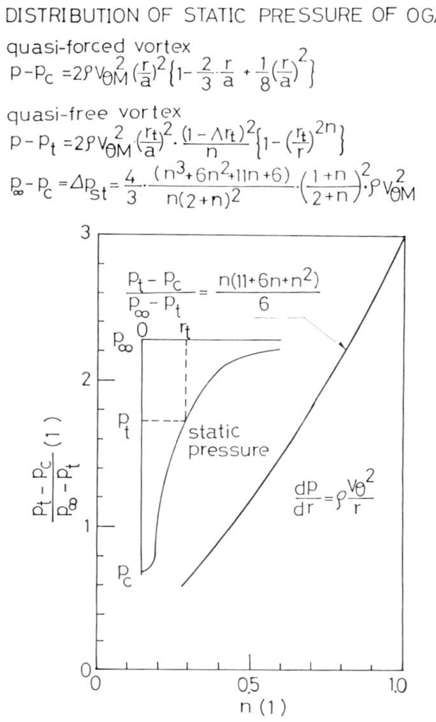

FIGURE 8–7. Characteristics of the depressions of the static pressures of the Ogawa combined vortex.

$$D = 64\pi H_i V_{\theta M}^2\,\eta_{TF}\left(\frac{1+n}{2+n}\right)^2\left\{-\frac{n}{2+n}+\frac{9}{8}\left(\frac{1+n}{2+n}\right)^2 + \frac{\eta_{Tf}}{\eta_{TF}}\frac{1}{4n}\left(\frac{1-n}{2+n}\right)^2\left[1-\left(\frac{r_t}{R_1}\right)^{2n}\right]\right\} \qquad (8\text{–}26)$$

Then transforming Equation 8–26, we obtain

$$\frac{D}{64\pi H_i V_{\theta M}^2\,\eta_{TF}} = \frac{9}{8}\left(\frac{1+n}{2+n}\right)^4\cdot\left\{\underbrace{1-\frac{8}{9}\frac{n(2+n)}{(1+n)^2}}_{\text{quasi-forced vortex}}+\underbrace{\frac{2}{9}\frac{\eta_{Tf}}{\eta_{TF}}\frac{1}{n}\left(\frac{1-n}{1+n}\right)^2\left[1-\left(\frac{r_t}{R_1}\right)^{2n}\right]}_{\text{quasi-free vortex}}\right\} \qquad (8\text{–}27)$$

Fujita and Ogawa had measured the Reynolds stresses $-\rho\overline{v_r v_\theta}$ and $-\rho\overline{v_v v_\theta}$ in the returned flow type of the vortex chamber of $D_1 = 140$ mm with the hot-wire of the constant temperature system. From these experimental results, they estimated the eddy kinematic viscosity ν_T, Prandtl mixing length, and Kolmogorov micro-scale. As a result of the estimated value of ν_T, neglecting the second term (quasi-free vortex) in the parentheses of the right-hand side of Equation 8–27 in comparison with the first term (quasi-forced vortex) of that and also putting the right-hand side as

$$\Phi(n) = \frac{9}{8}\cdot\left(\frac{1+n}{2+n}\right)^4\left[1-\frac{8}{9}\cdot\frac{n(2+n)}{(1+n)^2}\right] \qquad (8\text{–}28)$$

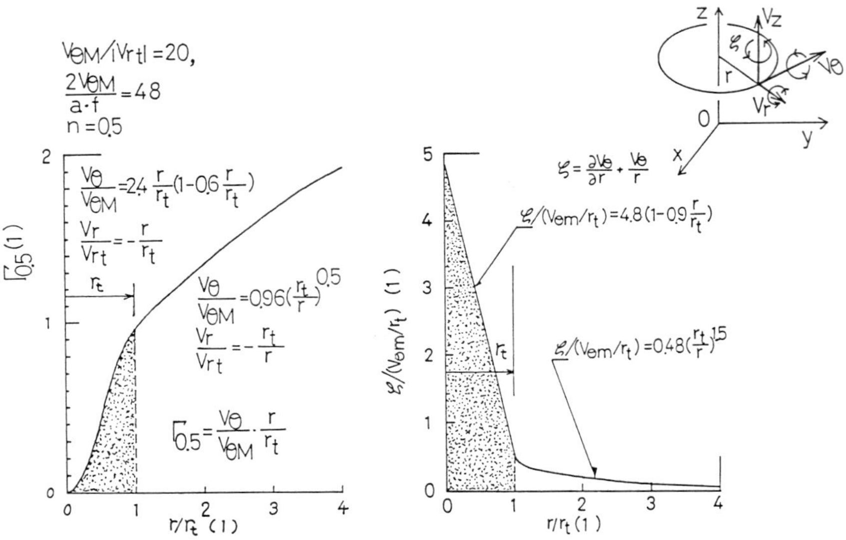

FIGURE 8–8. Distributions of Γ and ζ for n = 0.5 of the Ogawa combined vortex.

the value of the above equation approaches to $\Phi(n) \fallingdotseq 0.074$, when the exponent n of the quasi-free vortex is put between $n \fallingdotseq 0.4$ and $n \fallingdotseq 1.0$, in which 1.0 corresponds to the free vortex. Accordingly the pressure drop Δp_v due to the viscosity is

$$\Delta p_v = \frac{D}{Q_o} = \frac{64\pi H_i V_{\theta M}^2 \, \eta_{TF} \, \Phi(n)}{Q_o} = 64\rho \frac{\eta_{TF}}{\eta} \cdot \frac{V_{\theta M}^2 \, \Phi(n)}{R_{ec}} \tag{8–29}$$

where Reynolds number R_{ec} is defined as

$$R_{ec} = Q_o/H_i\nu$$

Figure 8–8 shows the examples of the distributions of Γ and ζ for n = 0.5 of the Ogawa combined vortex model.

IV. PRESSURE DROP OF THE VORTEX CHAMBER

Substitute Equation 8–29 into Equation 8–18 so the pressure drop Δp_c of the vortex chamber or the cyclone dust collectors can be represented as

$$\Delta p_c = 64\pi\rho \cdot \frac{\eta_{TF}}{\eta} \cdot \frac{\Phi(n) \cdot V_{\theta M}^2}{R_{ec}} + \rho\frac{V_o^2}{2} \tag{8–30}$$

The coefficient ζ_c of the pressure drop Δp_c can be defined that the pressure drop Δp_c is equivalent to a multiple of the mean dynamic pressure $\rho V_o^2/2$ in the inlet pipe

$$\Delta p_c = \zeta_C \, \rho\frac{V_o^2}{2} \tag{8–31}$$

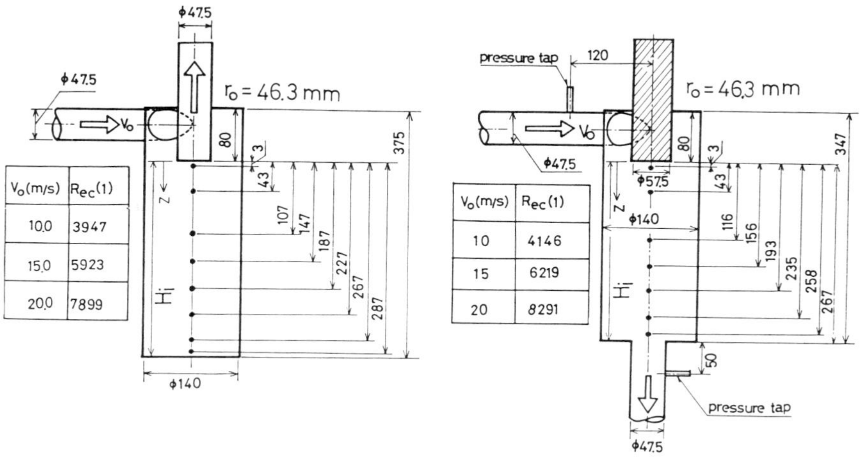

FIGURE 8–9. Main sizes of the two kinds of vortex chambers. (A) Main sizes of the returned flow type of the vortex chamber of D_1 = 140 mm. (B) Axial flow type of the vortex chamber of D_1 = 140 mm.

Therefore, the coefficient of the pressure drop can be represented as

$$\zeta_C = 1 + 128\pi \frac{\eta_{TF}}{\eta} \cdot \frac{\Phi(n)}{R_{ec}} \cdot \left(\frac{V_{\theta M}}{V_o}\right)^2 \tag{8–32}$$

This equation indicates the relationship between ζ_C and η_{TF}/η.From the equation one may estimate that ζ_C has a constant value which is dependent on the construction of the vortex chamber or of the cyclone dust collector, since the eddy viscosity η_{TF} may be supposed to be proportional to Reynolds number R_{ec}, and also $(V_{\theta M}/V_o)$ is a function of the Reynolds number R_{ec} and the construction of the vortex chamber or the cyclone dust collector as

$$\eta_{TF}/\eta = k_T \cdot R_{ec} \tag{8–33}$$

$$\frac{V_{\theta M}}{V_o} = k_c \cdot \left(\frac{D_1 - D_o}{D_2}\right)^n \tag{8–34}$$

Consequently, Equation 8–32 can be transformed to

$$\zeta_C = 1 + 128\pi\Phi(n) \cdot k_T k_c^2 \cdot \left(\frac{D_1 - D_o}{D_2}\right)^{2n} \tag{8–35}$$

V. EXPERIMENTAL RESULTS

A. CONSTRUCTION OF THE VORTEX CHAMBERS

Figure 8–9 shows the two kinds of vortex chamber which were made of transparent acrylic resin. The body diameter of the vortex chambers is D_1 = 140 mm, the diameter of the tangential inlet pipe is D_o = 47.5 mm, and the diameter of the exit pipe is D_2 = 47.5 mm, respectively.

Figure 8–9(a) is called a returned flow type of the tangential inlet vortex chamber which has the imaginary cylindrical length H_i = 295 mm. Figure 8–9(b) is named as an axial

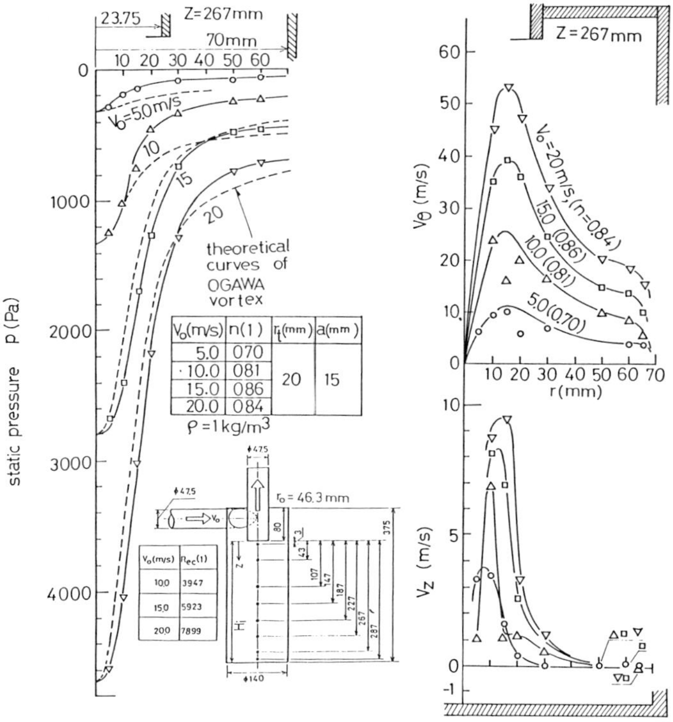

V_o(m/s)	n(1)	r_t(mm)	a(mm)
5.0	070	20	15
10.0	081		
15.0	086		
20.0	084		

V_o(m/s)	R_{ec}(1)
100	3947
15.0	5923
200	7899

FIGURE 8–10. Distributions of the velocities and the static pressure in the returned flow type of the vortex chamber.

flow type of the tangential inlet vortex chamber which has the imaginary cylindrical length $H_i = 267$ mm.

B. DISTRIBUTIONS OF THE TANGENTIAL VELOCITIES AND OF THE STATIC PRESSURES

Figure 8–10 shows the experimental results of the distributions of the tangential and axial velocities and of static pressures which were measured by a cylindrical Pitot tube at the distance $Z = 267$ mm beneath the edge of the exit pipe in the returned flow type of the vortex chamber. The mean inlet velocity V_o is 5.0, 10.0, 15.0, and 20.0 m/s, and the exponent n of the quasi-free vortex is $n = 0.7$ for $V_o = 5.0$, $n = 0.8$ for $V_o = 10.0$ m/s, $n = 0.86$ for $V_o = 15.0$ m/s, and $n = 0.84$ for $V_o = 20.0$ m/s, respectively. The dotted lines of the static pressure are the calculated results by the OGAWA combined vortex model. And in this figure the boundary radius r_t and the radius a corresponding to the maximum tangential velocity are shown.

Figure 8–11 shows the experimental results of the distributions of the tangential and axial velocities and of the static pressures at $Z = 193$ mm in the axial flow type of vortex chamber. The exponent n of the quasi-free vortex is $n = 0.84$ for $V_o = 10.0$ m/s, $n = 0.84$ for $V_o = 15.0$ m/s, and $n = 0.90$ for $V_o = 20.0$ m/s, respectively. The dotted lines of the static pressure were the calculated results by the Ogawa combined vortex model.

C. EMPIRICAL EQUATIONS OF THE VARIOUS CHARACTERISTICS FOR THE VORTEX FLOW

Figure 8–12 shows the relationship between the spin-up time T_2 which is defined as $T_2 = R_2/\sqrt{\nu\Omega_o}$ and the residue time F which is defined as $F = V_{CB}/Q_o$, where $V_{CB}(m^3)$

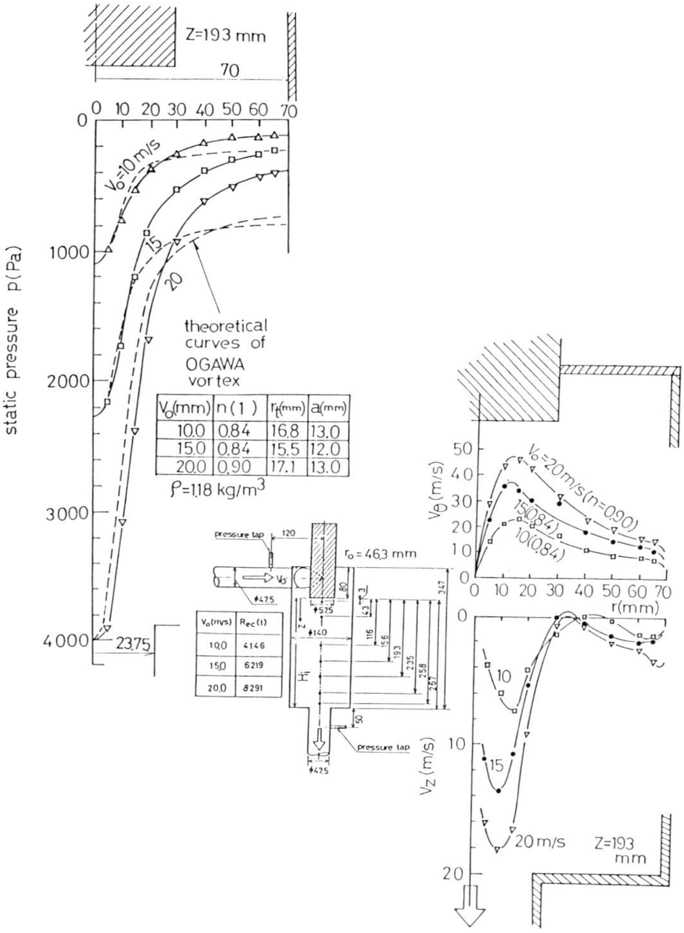

FIGURE 8–11. Distributions of the velocities and the static pressure in the axial flow type of the vortex chamber.

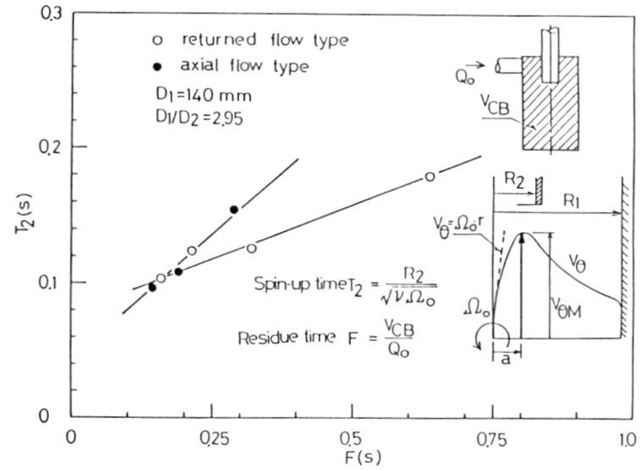

FIGURE 8–12. Relationship between spin-up time and the residue time.

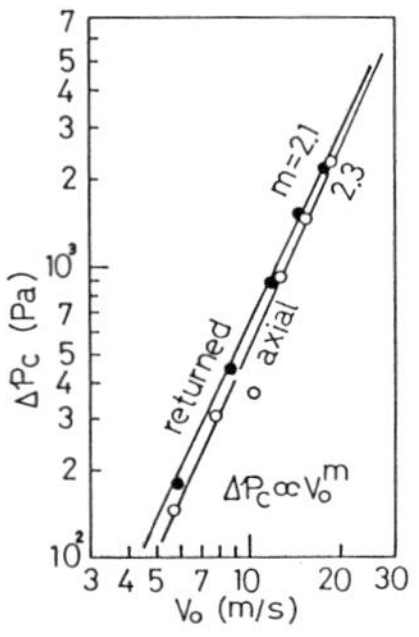

FIGURE 8–13. Relationship between the pressure drop and the mean inlet velocity for both types of the vortex chambers.

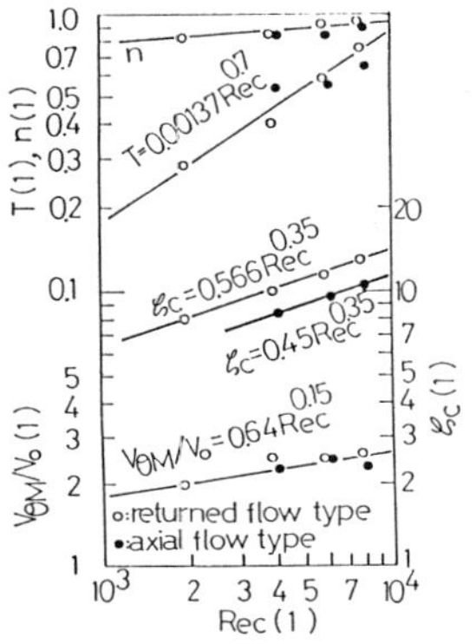

FIGURE 8–14. Relationship between T, n, $V_{\theta M}/V_o$, ζ_C, and R_{ec} for both types of the vortex chambers.

means the volume of the vortex chamber excluded the volume of the exit pipe in the vortex chamber and also Ω_o (rad/s) is the angular velocity of the tangential velocity at the vortex core. From this figure T_2 is increased with increasing F; however, T_2 is dependent on the construction of the vortex chambers and of the cyclone dust collectors.

Figure 8–13 shows the relationship between the pressure drops of the two kinds of the vortex chambers and the mean inlet velocity. The pressure drop of the returned flow type of the vortex chamber is a little larger than that of the axial flow type.

Figure 8–14 shows the experimental results of the velocity exponent n of the quasi-free vortex, of the dimensionless time $T = T_2/F$, of the maximum tangential velocity $V_{\theta M}/V_o$ in the quasi-forced vortex, and of the coefficient of the pressure drop ζ_C for the Reynolds number R_{ec} on the above-stated two kinds of vortex chambers. The relationship between them can be described empirically

$$T = 0.00137 \cdot R_{ec}^{0.7} \quad \text{(for the both types)} \tag{8–36}$$

$$V_{\theta M}/V_o = 0.64 \cdot R_{ec}^{0.15} \quad \text{(for the both types)} \tag{8–37}$$

$$\zeta_C = 0.566 \cdot R_{ec}^{0.35} \quad \text{(for the returned flow type)} \tag{8–38}$$

$$\zeta_C = 0.45 \cdot R_{ec}^{0.35} \quad \text{(for the axial flow type)} \tag{8–39}$$

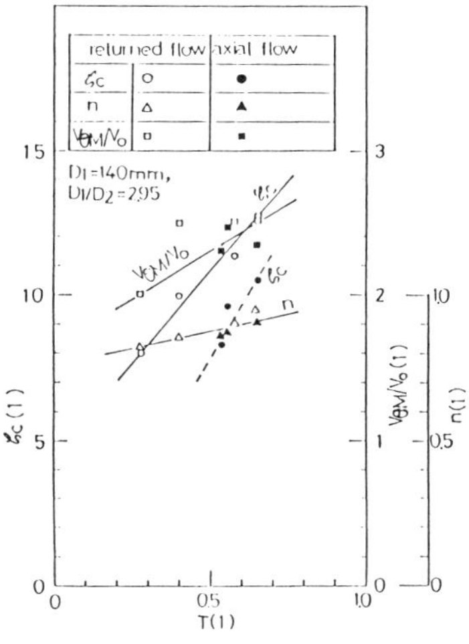

FIGURE 8–15. Relationship between ζ_C, n, $V_{\theta M}/V_o$, and T for both types of the vortex chambers.

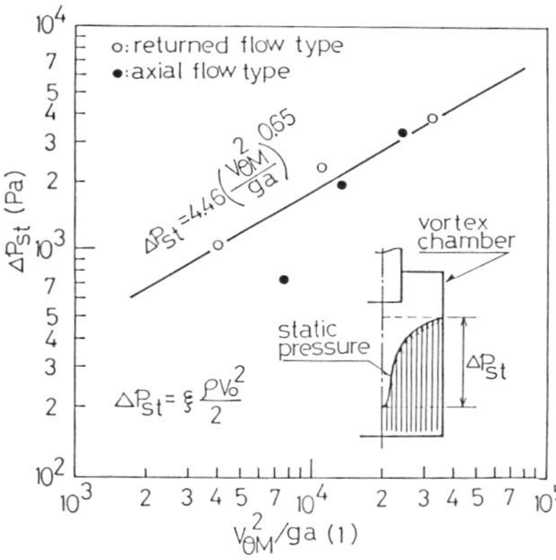

FIGURE 8–16. Relationship between Δp_{st} and the centrifugal effect.

Concerning the characteristics of $V_{\theta M}/V_o$, Bolchkobana and Smulbskii showed that the value of $V_{\theta M}/V_o$ is strongly related to the size ratio D_1/D_2 in their experiments in the returned flow type of the vortex chambers.[2] In the tangential inlet type of $D_1 = 160$ mm, $H = 200$ mm, and $R_{ec} \geqslant 1300$, when D_1/D_2 becomes $D_1/D_2 \fallingdotseq 20$, the value of $V_{\theta M}/V_o$ approaches about 45. Further when D_1/D_2 becomes larger than 20, $V_{\theta M}/V_o$ decreases suddenly.

Figure 8–15 shows the experimental results of ζ_C, $V_{\theta M}/V_o$, and n for $T = T_2/F$ for both types of vortex chambers.

Figure 8–16 shows the relationship between ΔP_{st} (the difference of the static pressures at the center [r = 0] and at the outer wall of the cylinder [$r = R_1$]) and the centrifugal effect represented by the empirical equation

$$\Delta p_{st} = 4.46 \left(\frac{V_{\theta M}^2}{ga}\right)^{0.65} \tag{8–40}$$

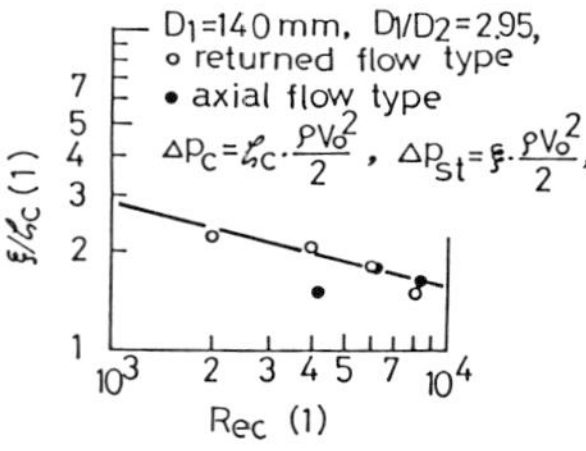

FIGURE 8–17. Relationship between ξ/ζ_C and R_{ec} for both types of the vortex chambers.

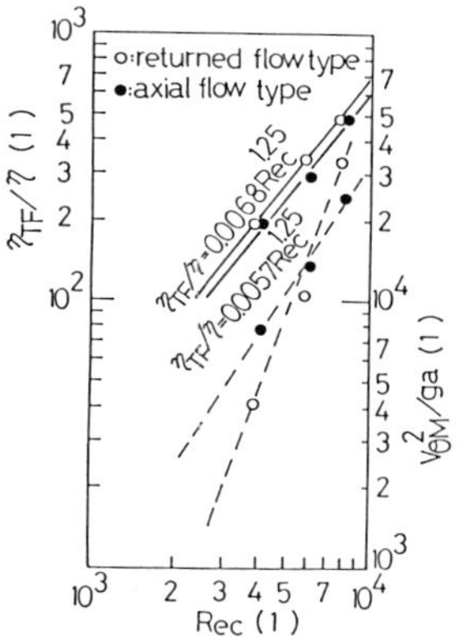

FIGURE 8–18. Relationship between η_{TF}/η, $V_{\theta M}{}^2/ga$, and R_{ec} for both types of the vortex chambers.

Figure 8–17 shows the relationship between ξ/ζ_C and R_{ec} for both types of vortex chambers. The value of ξ/ζ_C decreases with increasing Reynolds number; however ξ/ζ_C is about $\xi/\zeta_C \doteqdot 2.0$ for $R_{ec} = 1000 \sim 3000$. In this expression, Δp_c and Δp_{st} are defined as

$$\Delta p_c = \zeta_C\, \rho\frac{V_o^2}{2} \quad , \quad \Delta p_{st} = \xi\, \rho\frac{V_o^2}{2} \tag{8–41}$$

Figure 8–18 shows the relationship between the dimensionless eddy viscosity in the quasi-forced vortex region η_{TF}/η and R_{ec} and also between $V_{\theta M}^2/ga$ and R_{ec} for both types of vortex chambers. The value of η_{TF}/η can be estimated by Equation 8–33, and the empirical equation can be represented by

$$\eta_{TF}/\eta = 0.0068 \cdot R_{ec}{}^{1.25} \text{ (for the returned flow type),} \tag{8–42}$$

$$\eta_{TF}/\eta = 0.0057 \cdot R_{ec}{}^{1.25} \text{ (for the axial flow type)} \tag{8–43}$$

D. EQUIFLOW RATE LINES AND EQUIVORTICITY LINES

The equiflow rate line is defined as

$$\Psi(r,z) = \frac{\int_{R_1}^{r} 2\pi r \cdot dr \cdot V_z}{Q_o} \tag{8–44}$$

where V_z is an axial velocity and Q_o is the flow rate into the vortex chamber.

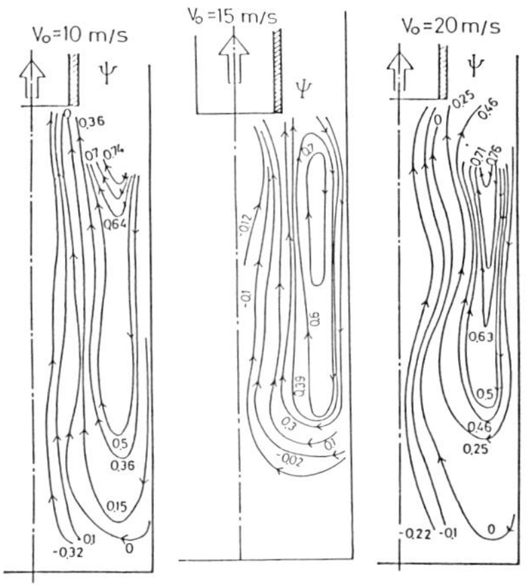

FIGURE 8–19. Equiflow rate lines in the returned flow type of the vortex chamber.

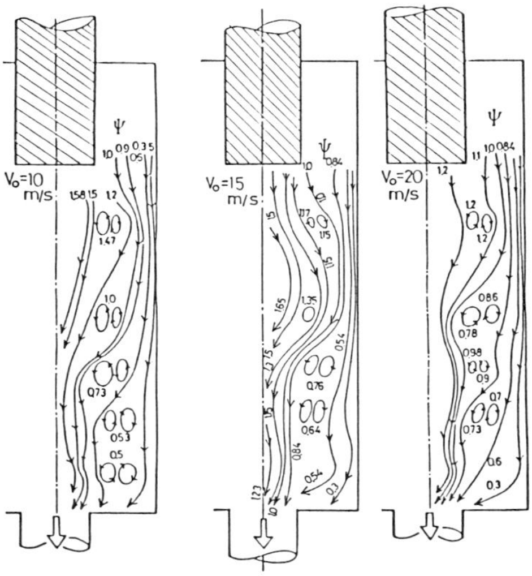

FIGURE 8–20. Equiflow rate lines in the axial flow type of the vortex chamber.

Figures 8–19 and 8–20 show the equiflow rate lines for V_o = 10, 15, and 20 m/s in the returned flow type of the vortex chambers. From these calculated results of the equiflow rate lines for the both types of vortex chambers, it is seen that there are very different characteristics of the flow patterns. For example, in the axial flow type it looks like the generation of the small double circulations. However, these results of the equiflow rate lines show the different flow patterns in comparison with those of the returned flow type of the vortex chamber (D_1 = 91.4 mm and H = 295 mm) by Smith.[3]

In order to recognize the characteristics of the pressure drop in the vortex chamber it is necessary to understand the distribution of the dimensionless vorticity lines Ω defined as

$$\Omega = \frac{\zeta}{I_a} \tag{8–45}$$

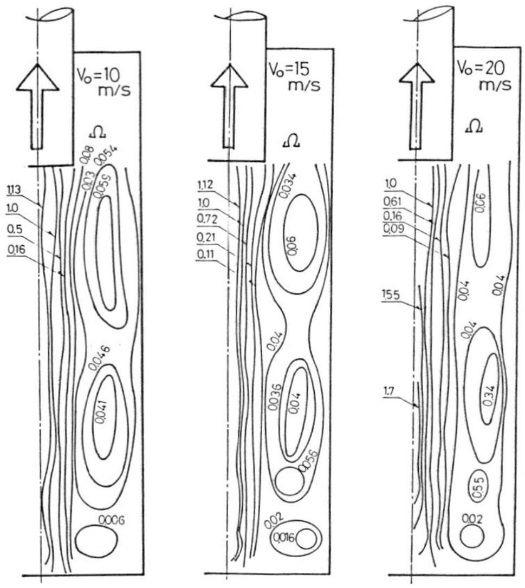

FIGURE 8–21. Equivorticity lines in the returned flow type of the vortex chamber.

where ζ is defined as

$$\zeta = \sqrt{\zeta_\theta^2 + \zeta_z^2} \tag{8–46}$$

$$\zeta_\theta = -\frac{\partial V_z}{\partial r} \quad , \quad \zeta_z = \frac{\partial V_\theta}{\partial r} + \frac{V_\theta}{r} \tag{8–47}$$

and also I_a is defined as

$$I_a = \frac{9V_o(D_1 - D_o)}{D_2^2} \tag{8–48}$$

Then Figures 8–21 and 8–22 show the dimensionless equivorticity lines in both types of vortex chambers.

From these figures it is seen that the strong vorticity is accumulated in the quasi-forced vortex region, while in the quasi-free vortex region the vorticity has, roughly speaking, disappeared except in the boundary layer on the concave wall surface. Figure 8–23 shows the radial distributions of the dimensionless vorticities which have the same characteristics for V_o = 10, 15, and 20 m/s in both types of vortex chambers. These calculated results of the dimensionless vorticities show the same distribution which is estimated by the calculated results of the Ogawa combined vortex model as shown in Figure 8–8. Accordingly, the main contribution to the pressure drop in the vortex chamber is based upon the high concentrated vorticity region in the quasi-forced vortex flow.

On the other hand the same phenomenon of the vorticity distribution, which was measured on the rotating flow of water in the cylindrical vortex chamber for the low Reynolds number $R_{ec} \doteqdot 30.0$, was published in the dissertation of Dr. R. G. Alan.[4]

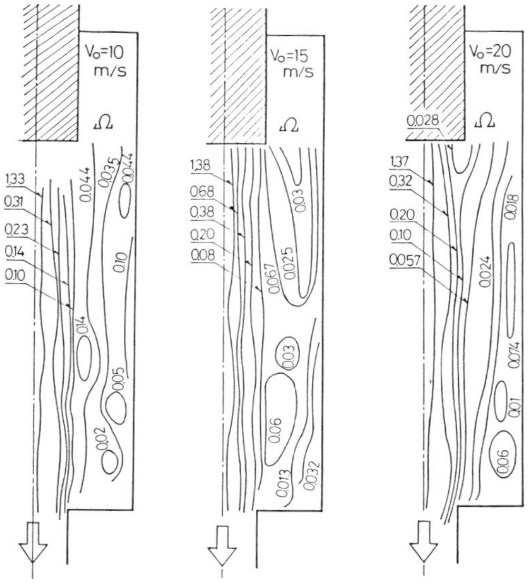

FIGURE 8–22. Equivorticity lines in the axial flow type of the vortex chamber.

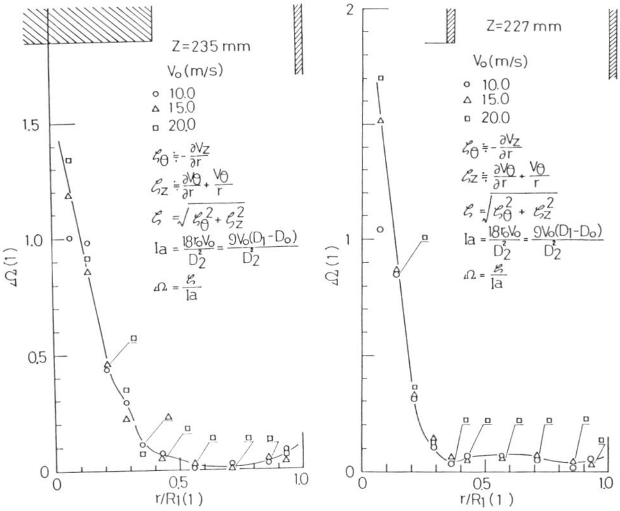

FIGURE 8–23. Radial distributions of the vorticity for both types of the vortex chamber.

E. DISTRIBUTION OF THE ANGULAR MOMENTUM FLUX

The angular momentum flux M_z can be defined as

$$M_z = 2\pi\rho \int_{R_1}^{r} V_\theta V_z r^2 dr \tag{8–49}$$

In this equation, a sign of the axial velocity V_z can be defined that the positive sign (+) of V_z is given when the flow direction is to be in the downstream direction, while the

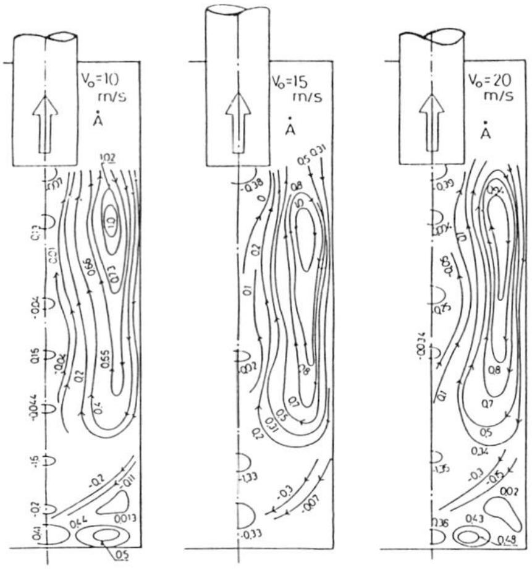

FIGURE 8–24. Distributions of the equiangular momentum flux in the returned flow type of the vortex chamber.

negative sign (−) of V_z is given when the flow direction is to be in the upstream direction. Here, in order to represent these quantities by the dimensionless forms, we introduce $\rho\, A_o V_o^2 r_o$ as a representative angular momentum flux. Therefore, the dimensionless angular momentum fluxes $\dot{A}$ and $\dot{A}_T$ are

$$\dot{A}(r) = \frac{2\pi}{A_o V_o^2 r_o} \int_{R_1}^{r} V_\theta V_z r^2 dr \tag{8–50}$$

$$\dot{A}_T = \frac{2\pi}{A_o V_o^2 r_o} \int_{R_1}^{0} V_\theta V_z r^2 dr \tag{8–51}$$

where a symbol r_o is defined as $r_o = R_1 - (D_o/2)$. These quantities can be numerically calculated by the distributions of the measured velocities V_θ and V_z at each section in the vortex chamber.[5]

Figure 8–24 shows the distributions of the equiangular momentum flux $\dot{A}(r)$ which was defined by Equation 8–50 for $V_o = 10$, 15, and 20 m/s in the returned flow type of the vortex chamber. The regions of the most strong values of $\dot{A}$ ($A = 0.94 \sim 1.02$) for $V_o = 10 \sim 20$ m/s were located in the bounds of $r/R_1 = 0.55 \sim 0.85$ and $Z/H_i = 0.07 \sim 0.345$ in the closed circulating flow. The pattern of the equiangular momentum flux is quite similar to that of the equiflow rate lines.

Figure 8–25 shows the distributions of the equiangular momentum flux $\dot{A}(r)$ for $V_o = 10$, 15, and 20 m/s in the axial flow type. These results are also quite like that of the equiflow rate lines of the axial flow type.

Figure 8–26 shows the distribution of the total angular momentum flux $\dot{A}_T$ which was defined by Equation 8–51. From this results it is clear that the total angular momentum flux flowed into the returned flow type is completely transported from the outer wall region into the vortex core region in the domain of $Z/H_i = 0 \sim 0.65$. Therefore, the total angular momentum flux in this region keeps at nearly $\dot{A}_T = 0$.

Figure 8–27 shows the distribution of $\dot{A}_T$ in the axial flow type. The value of $\dot{A}_T$ was decreased gradually from $\dot{A}_T = 1.5$ at $Z/H_i = 0$ to $\dot{A}_T = 1.0$ at $Z/H_i = 1.0$.

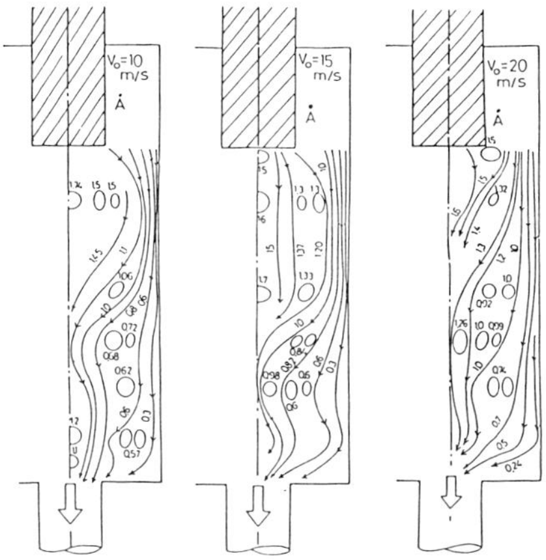

FIGURE 8–25. Distributions of the equiangular momentum flux in the axial flow type of the vortex chamber.

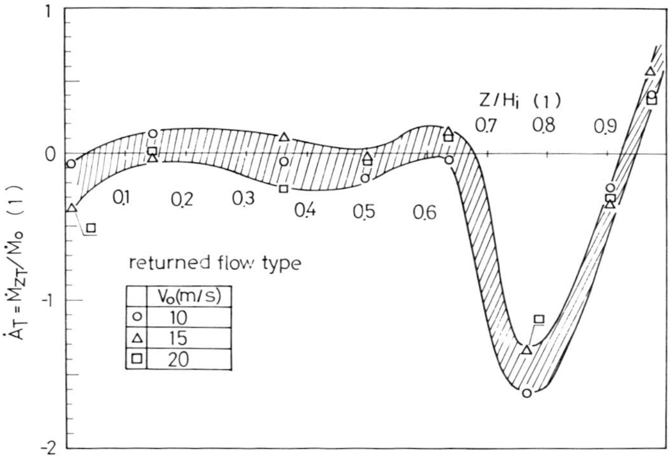

FIGURE 8–26. Distributions of the total angular momentum flux $\dot{A}_T$ in the returned flow type of the vortex chamber (D_1 = 140 mm).

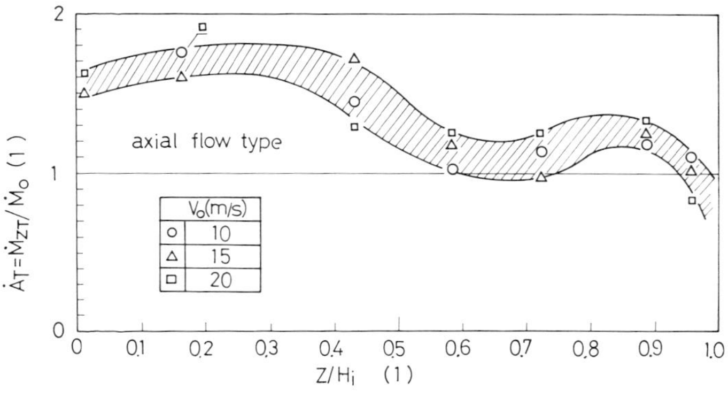

FIGURE 8–27. Distributions of the total angular momentum flux $\dot{A}_T$ in the axial flow type of the vortex chamber (D_1 = 140 mm).

This result can be explained simply by Equation 8–50, since the dimensionless angular momentum flux can be written as

$$\dot{A}(r) = \frac{2\pi}{A_o V_o^2 r_o} \int_{R_1}^{r} V_\theta V_z r^2 dr$$

In this equation, the distributions of the axial velocities in both types show quite different flow patterns, where in the returned flow type the axial velocity V_z is zero at about $r = b = 2R_1/3$, which is nearly independent of the Reynolds number. Therefore in the region of $b \leqslant r \leqslant R_1$, V_z directs to the downstream, while in the region of $0 \leqslant r \leqslant b$, V_z directs to the upstream. Therefore, when we calculate $\dot{A}(r)$ by integrating from $r = R_1$ to $r = 0$, it is physically possible to cancel the integration. Accordingly, the results $\dot{A}_T$ of Figure 8–26 of the returned flow type can be recognized. On the other hand, in the axial flow type, V_z directs to the downstream in the whole region from $r = 0$ to $r = R_1$ and V_z show nearly the same value along the Z-axis; however, the tangential velocity V_θ shows the sinuous increment to the downstream as

$$V_\theta(r,z) = V_{\theta o}(r) \cdot (1 + kz) \tag{8–52}$$

Substituting the above equation into Equation 8–51 of $\dot{A}_T$, we obtain

$$\dot{A}_T = \frac{2\pi}{A_o V_o^2 r_o} \int_{R_1}^{0} V_{\theta o}(r) \cdot (1 + kz) \cdot V_z \cdot r^2 \cdot dr =$$

$$= \frac{2(1 + kz)\pi}{A_o V_o^2 r_o} \int_{R_1}^{0} V_{\theta o}(r) \cdot V_z \cdot r^2 \cdot dr \tag{8–53}$$

This yields the result $\dot{A}_T$ of Figure 8–27 of the axial flow type.

F. DISTRIBUTION OF THE AXIAL MOMENTUM FLUX

The axial momentum flux $\dot{Z}(r)$ and the total axial momentum flux $\dot{Z}_T$ can be defined as

$$\dot{Z} = \frac{2\pi}{A_o V_o^2} \int_{R_1}^{r} V_z^2 \cdot r \cdot dr \tag{8–54}$$

$$\dot{Z}_T = \frac{2\pi}{A_o V_o^2} \int_{R_1}^{0} V_z^2 \cdot r \cdot dr \tag{8–55}$$

Figures 8–28 and 8–29 show the dimensionless distribution of the equiaxial momentum flux $\dot{Z}$ defined by Equation 8–54 for $V_o = 10$, 15, and 20 m/s in both types of the vortex chambers.

Figures 8–30 and 8–31 show the distributions of $\dot{Z}_T$ in both types of the vortex chambers. In the returned flow type, the value of $\dot{Z}_T$ decreases monotonically to the downstream direction, while in the axial flow type, the value of $\dot{Z}_T$ keeps nearly constant value. These differences are based upon the distributions of the axial velocities along the downstream direction (Z-direction) as shown in Figures 8–32 and 8–33.

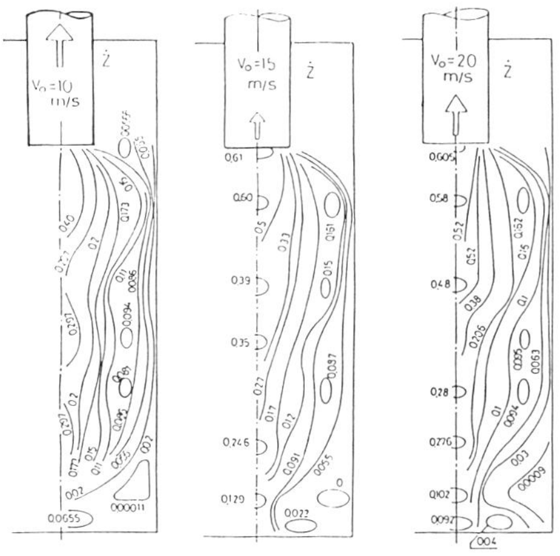

FIGURE 8–28. Equiaxial momentum flux in the returned flow type of the vortex chamber.

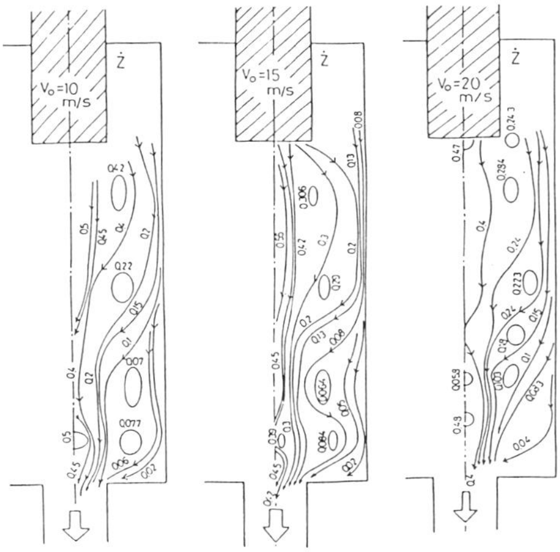

FIGURE 8–29. Equiaxial momentum flux in the axial flow type of the vortex chamber.

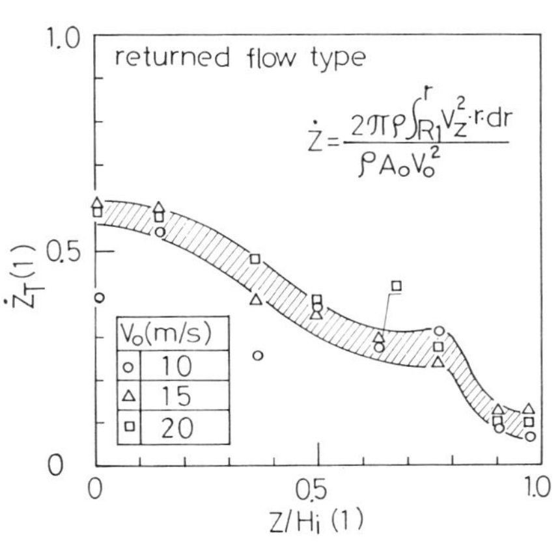

FIGURE 8–30. Distributions of the total axial momentum flux $\dot{Z}_T$ in the returned flow type of the vortex chamber (D_1 = 140 mm).

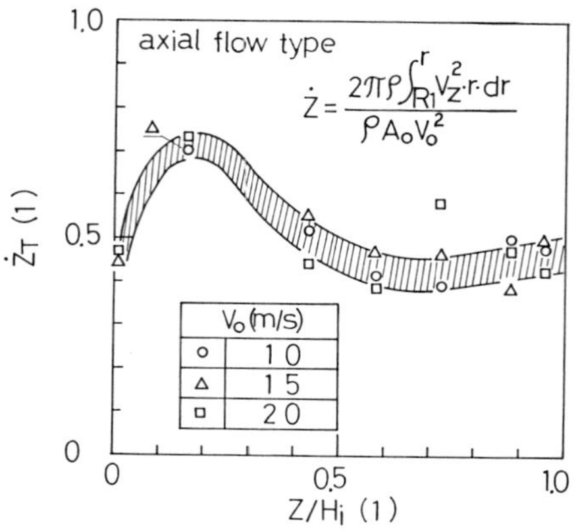

FIGURE 8–31. Distributions of the total axial momentum flux $\dot{Z}_T$ in the axial flow type of the vortex chamber (D_1 = 140 mm).

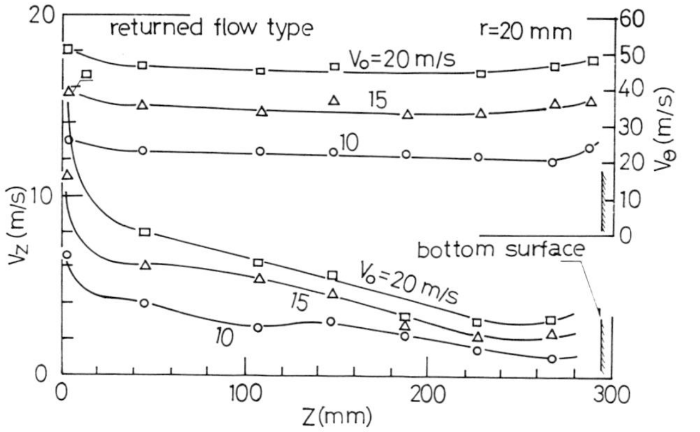

FIGURE 8–32. Distributions of V_θ and V_z at r = 20 mm along the Z-axis in the returned flow type of the vortex chamber (D_1 = 140 mm).

G. SWIRL PARAMETER

Here in order to make a representation of how to define a strength of the rotation of the turbulent rotational flow in the vortex chamber, we can introduce the following dimensionless quantity Γ as a swirl parameter

$$\Gamma = \frac{\dot{A}_T}{\dot{Z}_T} = \frac{\int_{R_1}^{0} V_\theta \cdot V_z \cdot r^2 \cdot dr}{r_o \int_{R_1}^{0} V_z^2 \cdot r \cdot dr} \tag{8–56}$$

Then Figures 8–34 and 8–35 show the calculated results by Equation 8–56 for both types of the vortex chambers. Comparing the characteristics of Γ for both types shows that there are very different characteristics of the three-dimensional turbulent rotational flow between them.

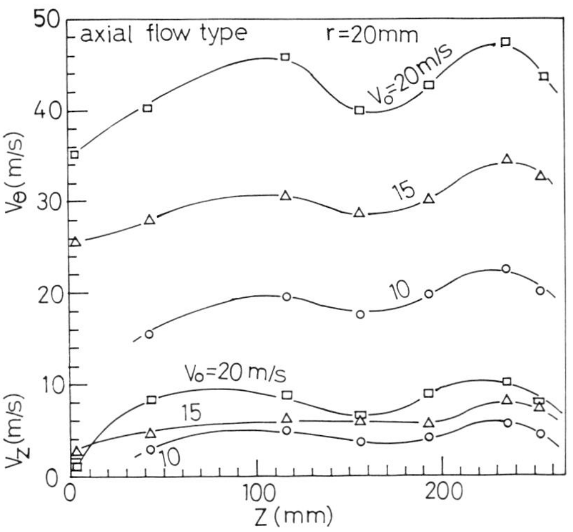

FIGURE 8–33. Distributions of V_θ and V_z at r = 20 mm along the Z-axis in the axial flow type of the vortex chamber (D_1 = 140 mm).

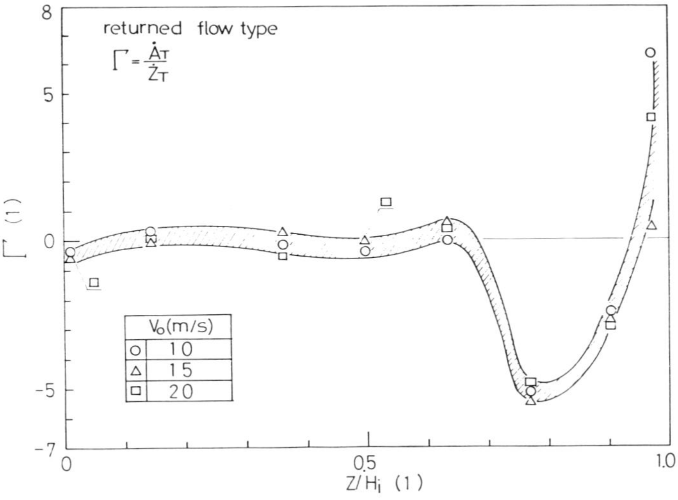

FIGURE 8–34. Distributions of the swirl parameters Γ in the returned flow type of the vortex chamber of D_1 = 140 mm.

MAIN SYMBOLS

$\dot{A}(1)$;	dimensionless angular momentum flux
$\dot{A}_T(1)$;	total angular momentum flux
F(s);	residue time in the vortex chamber
$H_i(m)$;	imaginary length
$I_a(1/s)$;	a representative vorticity
$\dot{M}_z(N \cdot m)$;	angular momentum flux
n(1);	velocity exponential on the quasi-free vortex
p(Pa);	static pressure

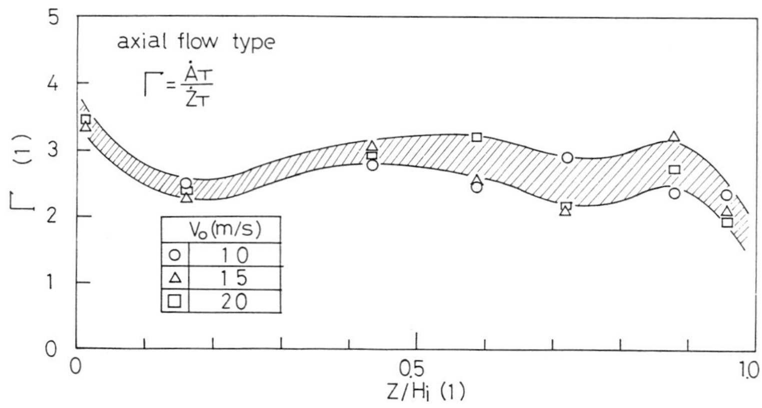

FIGURE 8–35. Distributions of the swirl parameters Γ in the axial flow type of the vortex chamber of $D_1 = 140$ mm.

$Q_o(m^3/s)$;	flow rate
$R_{ec}(1)$;	Reynolds number
$T_2(s)$;	spin-up time
$\mathbf{V}(m/s)$;	velocity vector
$V_o(m/s)$;	mean inlet velocity in the pipe
$V_{\theta M}(m/s)$;	maximum tangential velocity
$V_\theta(m/s)$;	tangential velocity
$V_z(m/s)$;	axial velocity
$W(N \cdot m/s = W)$;	energy dissipation
$\dot{Z}(1)$;	dimensionless axial momentum flux
$\dot{Z}_T(1)$;	total angular momentum flux
$\Gamma(1)$;	dimensionless quantity defined by $\dot{A}_T/\dot{Z}_T$
$\zeta(1/s)$;	vorticity component
$\zeta_C(1/s)$;	coefficient of the pressure drop
$\eta(Pa \cdot s)$;	viscosity
$\rho(kg/m^3)$;	density
$\Psi(1)$;	equiflow rate line
$\mathfrak{N}(1/s)$;	vorticity

REFERENCES

1. **Fujita, Y. and Ogawa, A.,** On the velocity distributions in the boundary layer on the bottom surface of the returned flow type of the vortex chamber, *J. Soc. Mech. Eng.*, preprint No. 790-10 (July 1979), 303, On the velocity distributions in the boundary layer in a returned flow type of the confined vortex chamber, *Nihon Univ., J. Coll. Eng.*, A-25 (March) 31, 1984.
2. **Bolchkob, E. P. and Smulybckii, I. I.,** Aerodinamika Bihreboi kamery s toltsebyi i bobobym bdubom, *Teor. Osn. Khim. Tekhnol.*, (2) 214, 1983.
3. **Smith, J. L.,** An experimental study of the vortex in the cyclone separator, *Trans. A. S. M. E.* (J. Basic Eng.), December-1962, 602.
4. **Alan, G. R.,** A forced vortex flow, Ph.D. thesis, University of Maryland, College Park, 1970.
5. **Ogawa, A.,** Experimental study of the angular and axial momentum fluxes in the cylindrical vortex chambers, World Congress III of Chemical Engineering, Tokyo, September 21 to 25, 1986, Vol. II, 208.

Chapter 9

VORTEX RING

I. INTRODUCTION

The motion of a vortex ring shows interesting phenomena for the physical process for which the fundamental investigation was already established by Helmholtz. When we consider the viscous action of fluid, the vortex ring is one example for the production mechanism of the discontinuous surfaces which are generated by the velocity gradients in the fluid layer. This is one of the most interesting problems in the study of the coherent structure of a viscous fluid.

Various kinds of behavior of vortex rings were investigated theoretically and experimentally in detail. For example, Helmholtz,[1] Lamb,[2] and Sommerfeld et al. (1950) had studied the mathematical analysis of the vortex ring. The subjects of this chapter include the morphological structure of the vortex ring, vibration phenomena, the joining of the vortex rings,[3,4] and also related effects of vortex rings.[5–9]

The main purposes of this study are to learn the fundamental characteristics of turbulent flow of the vortex ring. However, concerning the phenomena of the vibrations and breakdown, it is exceedingly difficult to get the exact solutions mathematically. Therefore, the characteristics which govern the motion of a progressive vortex ring in the domain of high Reynolds number ($R_e = U_fD/\nu = 1000 \sim 4000$) and also the the disruptions of the star-like vortex ring which takes place in the process of the progressive motion are described.

II. EXPERIMENTAL APPARATUS AND ITS METHOD

In order to alter the quantity of the momentum or the kinetic energy of a vortex ring over the extensive domain, two kinds of apparatus generating a vortex ring were used as shown in Figure 9–1. That is to say, one equipment is a plunger-type method which has changed the position of impact angle of a plummet and another one is the speaker-type method which has changed the output voltage as shown in Figure 9–2. The kinetic energy of a vortex ring is determined mostly by the advanced velocity of a vortex ring and its diameter. The reason for using a plunger method is that a vortex ring is created with the large quantity of acting fluid in a relatively long impact time under the sluggish impaction force, and then the reason for using a speaker method is that a vortex ring of small volume and of high advanced velocity is created by the impact force during a short acting time.

The experimental apparatus is composed of a cylinder of a transparent synthetic resin (inner diameter 150 mm, outer diameter 170 mm, and length 400 mm) and of a plunger in which the outer edge of the plunger plate is enveloped by the felt cloth and of a transparent square casing (749 mm × 749 mm, total length 2006 mm) in which a motion of a progressive vortex ring can be observed. Then in the case of the plunger method, a plumb which gives the impact force to the plunger edge has the weight of 340 g lead, and the swing angle θ is defined as shown in Figure 9–1. In order to take photographs of the motion of the progressive vortex ring for the perpendicular direction to the main axis, a camera was used.

In order to observe the phenomena of the star-like breakup for the main direction, two 8 mm movie cameras were set. Now, in the case of the production of the vortex ring by the speaker type as shown in Figure 9–2, a hi-fi speaker of 160 mm in diameter was adapted. In order to insure protection for the leakage of air through the cone surface, paint was

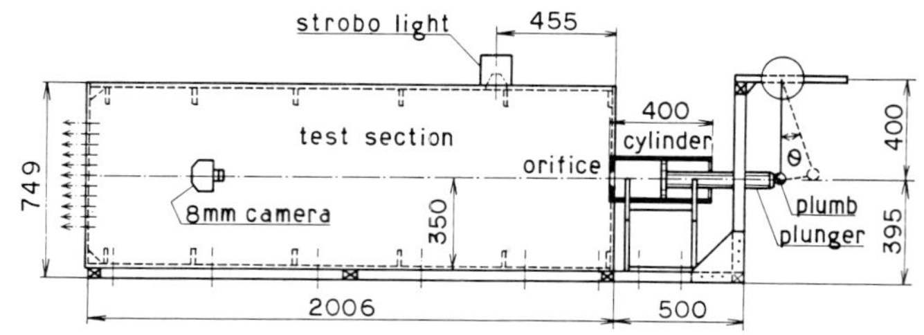

FIGURE 9–1. Experimental apparatus for producing the vortex ring by the plunger type.

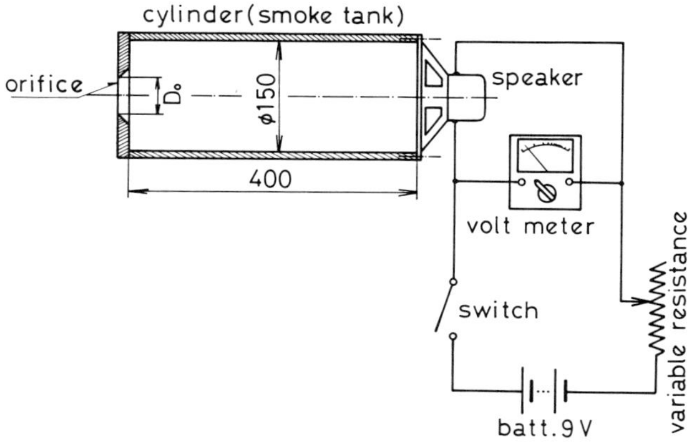

FIGURE 9–2. Apparatus producing the vortex ring by the speaker type.

sprayed on the cone surface. The input power to speaker was varied by the method of changing the resistance of the variable resistor. And by a compass needle of the voltmeter which was connected in parallel with this electrical circuit, the value of the input voltage was read. Then stroboscope equipment as a source of the intermittent light was continuously operated with a frequency 10 Hz.

In order to observe the motion of the vortex ring, the cylinder was filled with tobacco smoke. In creating the vortex motion under various conditions, we have employed five kinds of orifice diameter. For the plunger type, we have set up the relationship between the orifice diameter D_o and the swing angle θ of a plumb as written in Table 9–1, and also for the speaker type the relationship between the orifice diameter D_o and the voltage E of a speaker is also written in Table 9–1. Then the main diameter D, the tube diameter d, and the advanced velocity U of a vortex ring which characterized the behavior of a vortex ring were estimated by enlargement of the photographs.

The physical process of the formation and of the unstable breakdown of the star-like type was observed by the 8-mm (64 frames per s) camera which was aligned in the main direction of a vortex motion as shown in Figure 9–1. The specification of the speaker used in these experiments is written in Table 9–2.

III. RESULTS OF THE EXPERIMENTS

A. EXPLANATION OF THE SYMBOLS USED

Because of the difficulty of the measurements of jet velocity at the exit place of orifice and also because of the incorrectness of the relationship between the jet velocity at the

TABLE 9–1
Relationship between the Orifice Diameter D_o and the Swing Angle θ and between the Orifice Diameter D_o and the Voltage E

Plunger type	D_o (mm)	70		60			50		40		20
	θ (deg)	35	30	30	25	22	18	15	18	15	2

Speaker type	D_o (mm)		40			30		20	
	E (vol)	6.0	5.0	4.2	4.0	3.0	2.4	1.6	1.3

TABLE 9–2
Norm of Speaker (Koral Sound Co.)

Type 6A-7

Nominal size	160 mm
Sensitivity	90 dB
Rated input power	3 W
Voice coil impedance	8Ω
Bottle opening	148 mm
Depth	88.5 mm

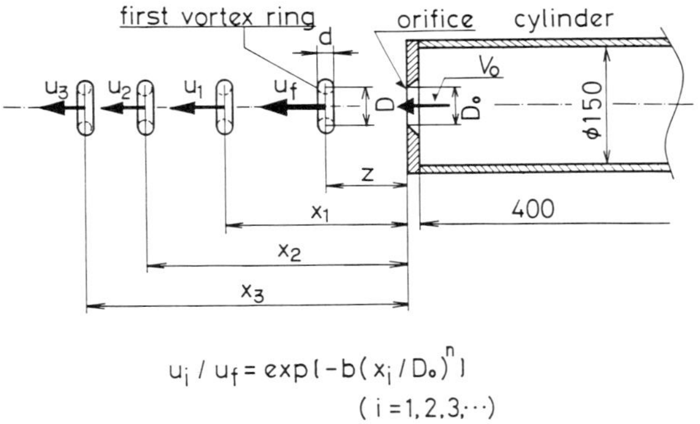

FIGURE 9–3. Flow visualization of vortex ring by stroboscope light.

orifice and the progressive velocity of the vortex ring, we have selected the primary photographic vortex ring as a representative vortex ring in both cases of plunger and speaker types. Then defining U_f as the moving advanced velocity of the primary vortex ring, the moving velocity U_i at the arbitrary distance X_i from the exit place of an orifice was nondimensionalized as shown in Figure 9–3. Now the decreasing velocity of an advanced vortex ring can be described approximately by the following empirical equation:

$$\frac{U_i}{U_f} = \exp\left[-b\left(\frac{X_i}{D_o}\right)^n\right] \quad , \quad (i = 1, 2, 3, \ldots) \tag{9–1}$$

where D_o is a diameter of an orifice and b and n are the index number of the progressive velocity of the vortex ring. Here we have used two types of Reynolds number; that is to say, one is Reynolds number R_{ef} for the advanced distance of a vortex ring, the other is Reynolds number R_e for the diameter of the vortex ring. Therefore defining D as a representative diameter of a primary photographic vortex ring, Z as an advanced distance from the orifice to the primary vortex ring as shown in Figure 9–3, two kinds of Reynolds number can be defined as

$$R_{ef} = \frac{U_f \cdot Z}{\nu} \tag{9–2}$$

$$R_e = \frac{U_f \cdot D}{\nu} \tag{9–3}$$

where ν is the kinematic viscosity of air.

B. PREDICTION OF THE PROGRESSIVE VELOCITY BY THE SIMPLIFIED EQUATION OF MOTION

When the Reynolds number R_e is a high value, the coefficient of the drag force of this vortex ring may be assumed to be a constant value of ζ_d. Therefore the equation of motion can be described as

$$m \frac{dU}{dt} = -\pi \cdot d \cdot D \cdot \zeta_d \frac{\rho U^2}{2} \tag{9–4}$$

where m is a mass of the vortex ring defined as $m = \rho\pi^2 d^2 D/4$ and πDd is the frontal area of the vortex ring. Here integrating this equation, we obtain the relation between U and x

$$\frac{U}{U_o} = \exp\left(-\frac{2 \cdot \zeta_d \cdot D_o}{\pi d} \frac{x}{D_o}\right) \tag{9–5}$$

where U_o is the mean progressive velocity at $x = 0$. On this experimental result, the value of ζ_d for $R_e = 600 \sim 4500$ is estimated as

$$\zeta_d = 0.05 \sim 0.062$$

On the other hand Baird, Wairegi, and Loo[10] had investigated the mechanical characteristics of the vortex ring in the water vessel. The vortex ring in the water was produced from the orifice of the diameter $D_o = 25.4$ mm ~ 63.5 mm. The orifice Reynolds number R_{eo} defined as $R_{eo} = D_o \cdot V_o/\nu$ was $3 \times 10^4 \sim 3 \times 10^5$, where V_o is the injection velocity of jet from the orifice. Then applying Bernoulli equations, they derived the relationship between the progressive velocity U_f of the nascent vortex ring and the injection velocity V_o as

$$U_f = \frac{1}{2} V_o$$

And also the empirical equation between U_f and V_o in the water vessel was

$$U_f(\text{m/s}) = 0.6 \times V_o^{0.8}$$

C. CHARACTERISTICS OF THE PROGRESSIVE VORTEX RING

1. Observations of the Vortex Rings

In Figures 9–4 through 9–9, some examples are shown of the effects of a change in the advanced velocity U, of the tube diameter d, and of the main diameter D of the vortex

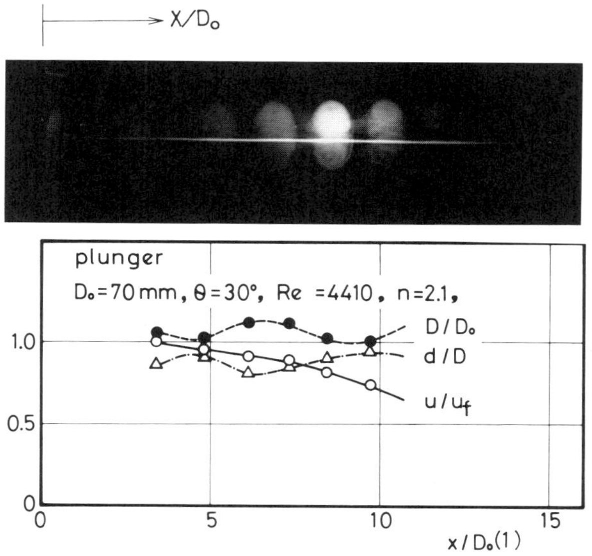

FIGURE 9–4. Vicissitudes of the vortex ring for D_o = 70 mm and R_e = 4410.

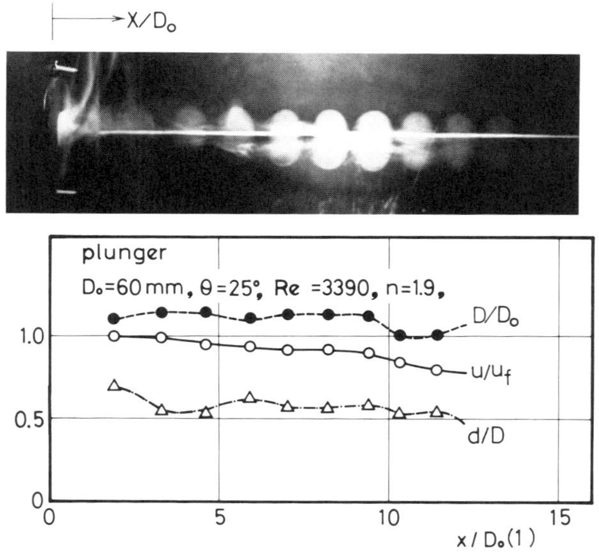

FIGURE 9–5. Vicissitudes of the vortex ring for D_o = 60 mm and R_e = 3390.

ring leaving from the orifice for the various orifice diameters. When we assume that the vortex ring is a homogeneous torus, the volume V_v and the surface area S of this vortex ring are given as

$$V_v = \pi^2 \cdot D \cdot d^2/4 \quad , \quad S = \pi^2 \cdot D \cdot d$$

From the results of the experiments, the curves of D/D_o and of d/D showed nearly axial symmetry along an axis x/D_o. On the other hand, Taylor[11] investigated the formation of a

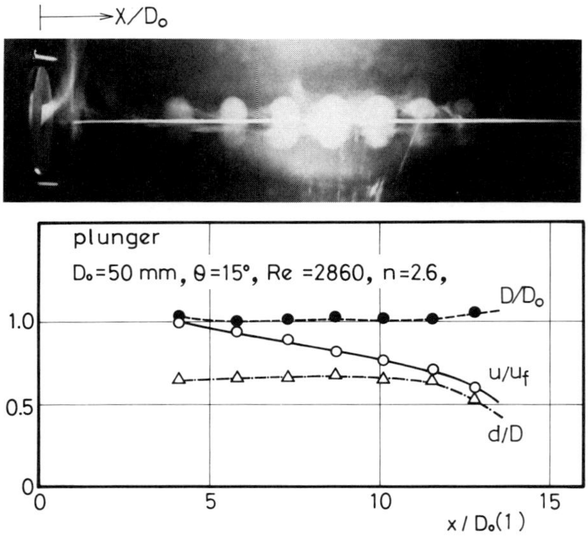

FIGURE 9–6. Vicissitudes of the vortex ring for D_o = 50 mm and R_e = 2860.

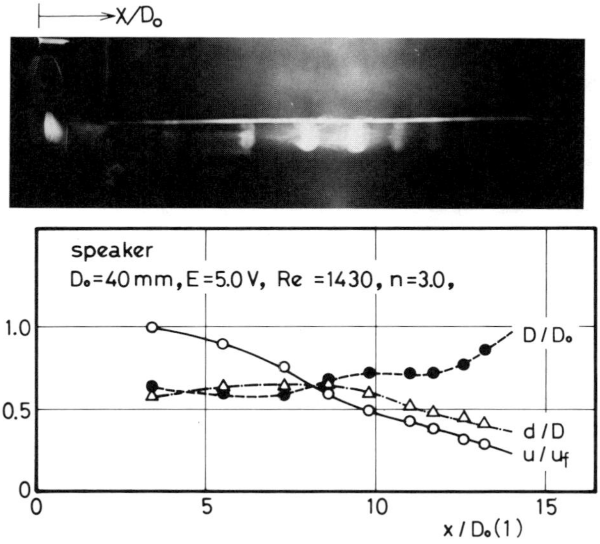

FIGURE 9–7. Vicissitudes of the vortex ring for D_o = 40 mm and R_e = 1430.

vortex ring by giving an impulse to a circular disc of the radius c. Then considering a flat circular disc of the radius c in perfect fluid of density ρ which suddenly moved forward with velocity U along the axis of the symmetry, he derived the theoretical relationship for the vortex formation in connection with Lamb's theory as shown in Figure 9–10.

2. Characteristics of the Index Number of the Advanced Velocity

Figure 9–11 shows the experimental results between the index numbers b and n which were used for Equation 9–1 of the advanced velocity for both types of equipment. From

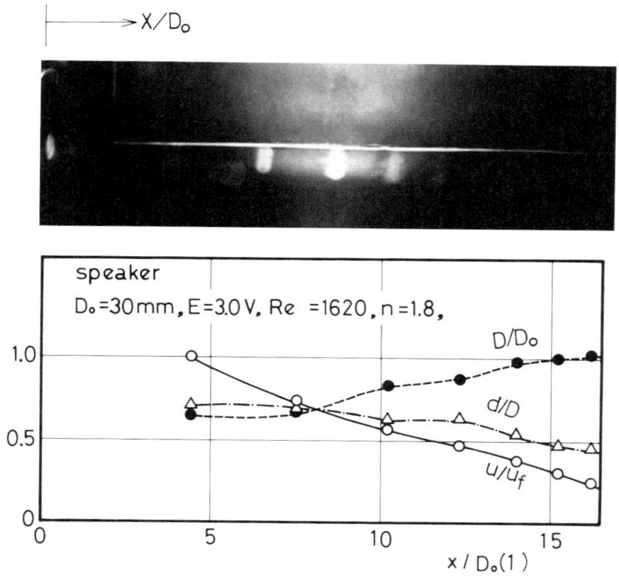

FIGURE 9–8. Vicissitudes of the vortex ring for D_o = 30 mm and R_e = 1620.

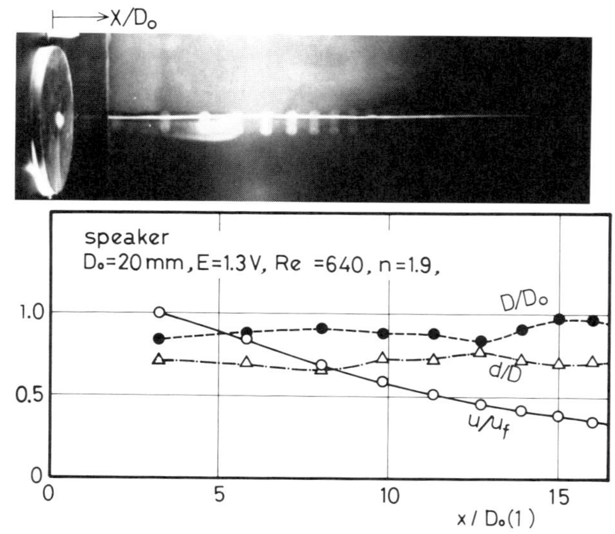

FIGURE 9–9. Vicissitudes of the vortex ring for D_o = 20 mm and R_e = 640.

this figure, the rate of magnitude of the advanced velocity of the vortex ring had little difference due to the features of the two kinds of equipment which generate the vortex ring.

When the rate of decreasing velocity for the two types of equipment is compared, contrasting the magnitudes of the indices b and n separately does not indicate all of the characteristics for the decreasing velocity. Then in order to make clear the effects of the index numbers b and n on the advanced velocity of the vortex ring, Figure 9–12 shows the curves of U/U_f for various values of b and n. From this figure, in spite of the two types of equipment, the rate of decrease of the advanced velocity of a vortex ring was largely

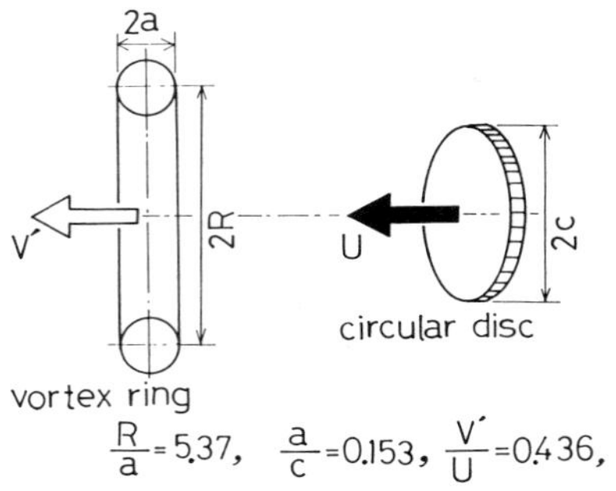

FIGURE 9–10. Generation of the vortex ring with a circular disk by G. I. Taylor.

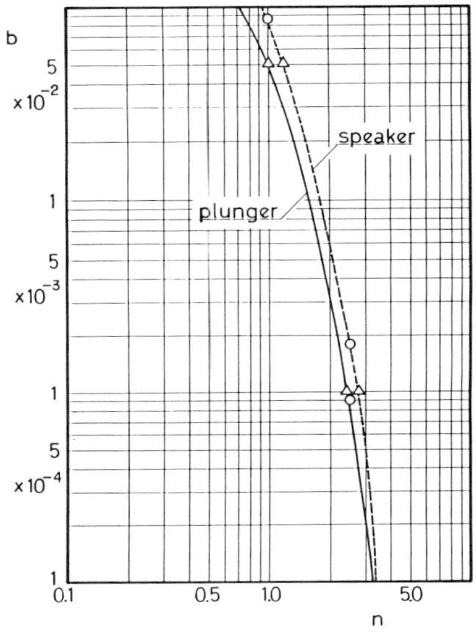

FIGURE 9–11. Relationship between b and n.

influenced by the index number n irrespective of the magnitude of index number b. Therefore from these experimental results we can describe the conclusions as follows:

1. The rate of magnitude of the decreasing advanced velocity is mainly influenced by the index number n.
2. Increasing the index number n, the rate of magnitude of the decreasing advanced velocity is also increased.
3. In the case of speaker type, the rate of decreasing is large. It is unfortunate that the physical process of the vortex ring formation near the exit place of the orifice could not be sized exactly by the photographic method and by the 8-mm moving picture (64 frames per s).

The values of index in these experiments were within the limit $n = 1.0 \sim 2.0$. However, sometimes we found large values of n. This problem was caused by the instantaneous photographs taken of the first nascent vortex ring.

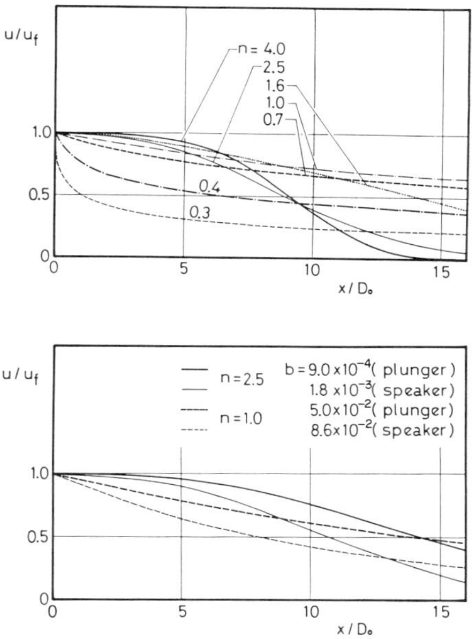

FIGURE 9–12. Dimensionless velocity of the vortex ring by Equation 9–1.

D. MOMENTUM AND KINETIC ENERGY OF A VORTEX RING

The well-known formula for the advanced velocity of the vortex ring can be expressed as follows:

$$U = \frac{\Gamma}{2\pi D}\left(\ln\frac{8D}{d_o} - \frac{1}{4}\right) \tag{9–6}$$

where D is the diameter of a vortex ring and d_o is the diameter of the vortex core. When we apply this formula to these experiments, there is a problem of how to estimate the diameter of the vortex core. As already described above, it was not possible to determine clearly the diameter of the vortex core in these visualization methods.

However, from the results of Prof. Sullivan, the value of d_o/D equals 0.27 in the case of $U = 79.3$ cm/s, $D = 5.56$ cm. In these experiments, the domain of advanced velocity U was 65 ~ 190 cm/s and that of the diameter D was 3.0 ~ 6.4 cm. And also the values of d_o/D for the five kinds of the orifices were nearly constant; therefore, the variation of the value $(\ln(8D/d_o) - 1/4)$ was minute. Therefore the advanced velocity U of the vortex ring was given as follows:

$$U = \frac{k_1\Gamma}{D} \tag{9–7}$$

The magnitude of the circulation Γ can be described deductively by Reynolds number $R_e(= U_fD/\nu)$. As an example, when the veloicty $U_f = 94$ cm/s and the diameter 5.5 cm was attained on the swing angle of plumb $\theta = 18°$ for the orifice diameter $D_o = 50$ mm, the magnitude of the circulation Γ for $d_o/D = 0.27$ was obtained from Equation 9–4 as follows:

$$\Gamma = \frac{2\pi \cdot D \cdot U}{\ln(8D/d_o) - 1/4} = \frac{2\pi \times 5.5 \times 94}{\ln(8/0.27) - 1/4} = 1035\ (\mathrm{cm^2/s})$$

Under the same conditions, given a value $d_o/D = 0.30$, the magnitude of circulation was

$$\Gamma = \frac{2\pi \times 5.5 \times 94}{\ln(8/0.30) - 0.25} = 1071 \text{ (cm}^2\text{/s)}$$

Therefore given a little different values of d_o/D, the value of Γ was not influenced so much by changing the value of d_o/D.

Next the kinetic energy E of a vortex ring can be written as follows:

$$E = \frac{\Gamma^2 \cdot D}{4} \rho\left(\ln \frac{8D}{d_o} - \frac{7}{4}\right)$$

where ρ is the density of air. As an example on the same numerical condition stated above, the kinetic energy E of a vortex ring becomes

$$E = \frac{\Gamma^2 \cdot D}{4} \rho\left(\ln \frac{8D}{d_o} - \frac{7}{4}\right) = 5.55 \times 10^{-4} (\text{N} \cdot \text{m})$$

E. COMPARISON OF THE MOMENTUM BETWEEN THE PLUNGER TYPE AND THE SPEAKER TYPE

In the case of a plunger type, the impulse for the generation of a vortex ring was supplied by the slow gradient of static pressure and by enough time in the cylinder. But in the case of the speaker type, the impulse was supplied in a short interval of time and by the abrupt gradient of static pressure, so the fluid which was compressed in the cylinder constituted the flow field of a high speed jet at the edge of the orifice. In order to compare the decreasing quantities of the vortex motion, it is necessary to compare the quantities of momentum preserving the vortex ring for both methods. Therefore we propose the following method.

In the case of the plunger type, the mean Reynolds number R_{ef} for the exit diameter of an orifice $D_o = 40$ mm and for the swing angle of plumb $\theta = 15°$ was 9.0×10^3. Similarly, in the case of the speaker, the mean Reynolds number R_{ef} for the voltage $E = 5.0$ V and for the orifice diameter $D_o = 40$ mm was 8.9×10^3.

Then (on the abscissa), the velocity U_f for the first vortex ring to appear and (on the ordinate) the velocity U_{20} at the distance $Z = 20$ cm from the orifice and the diameter D_{20} of the vortex ring at $Z = 20$ cm was plotted in Figure 9–13. From this figure, for the plunger type

$$U_{20}(\text{cm/s}) = 1.1\, U_f - 21$$

and for the speaker type

$$U_{20}(\text{cm/s}) = 0.3\, U_f + 53$$

were obtained by the experiments.

Figure 9–14 shows the values of the Reynolds number R_e (namely corresponding to the momentum) against the moving velocity of the first vortex ring for the plunger and speaker types. And then the relationships between the value of index n and the Reynolds number R_e are shown in Figure 9–15. Further the rate of decrease of an advanced vortex ring produced by the speaker was larger than that of the plunger type.

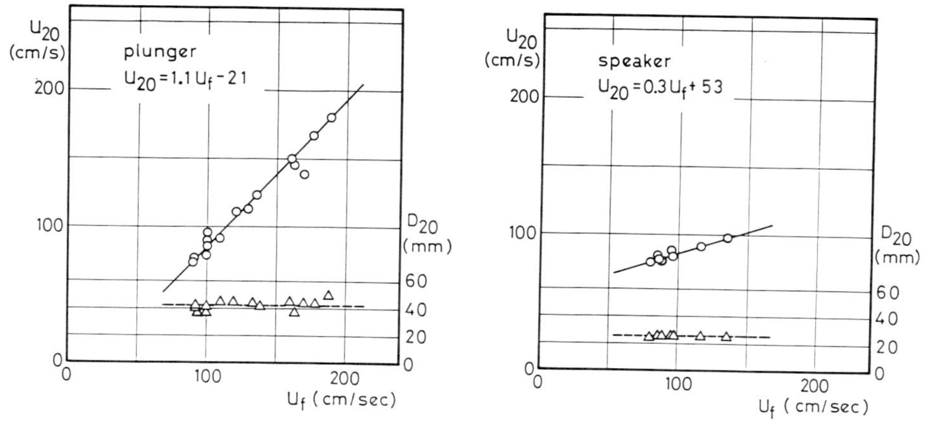

FIGURE 9–13. Relationship between U_{20} and U_f for both types of the vortex generators.

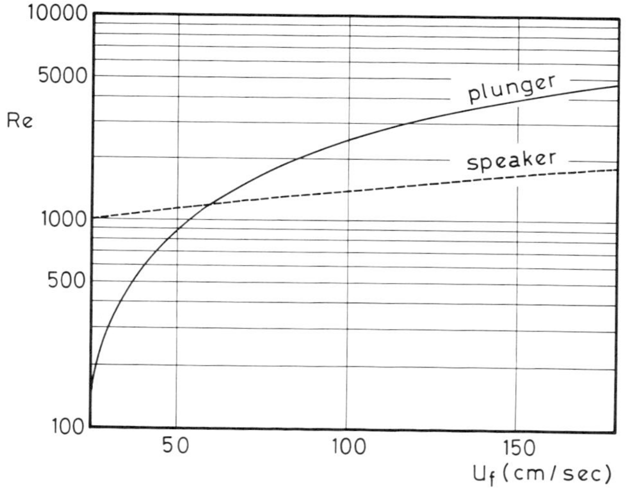

FIGURE 9–14. Relationship between R_e and U_f for both types of the vortex generators.

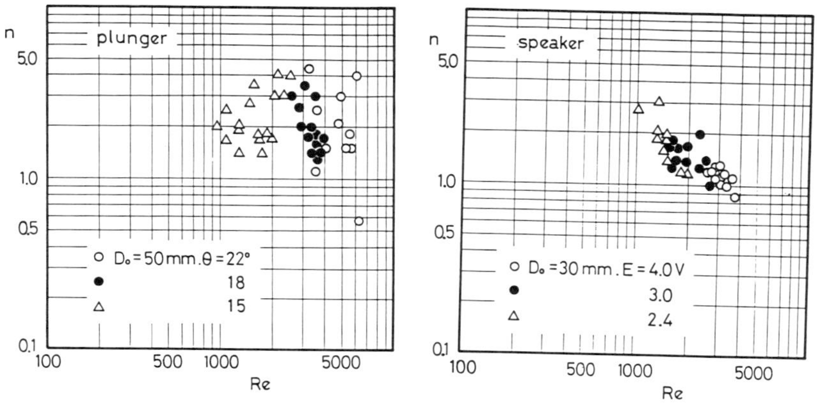

FIGURE 9–15. Relationship between n and R_e for both types of the vortex generators.

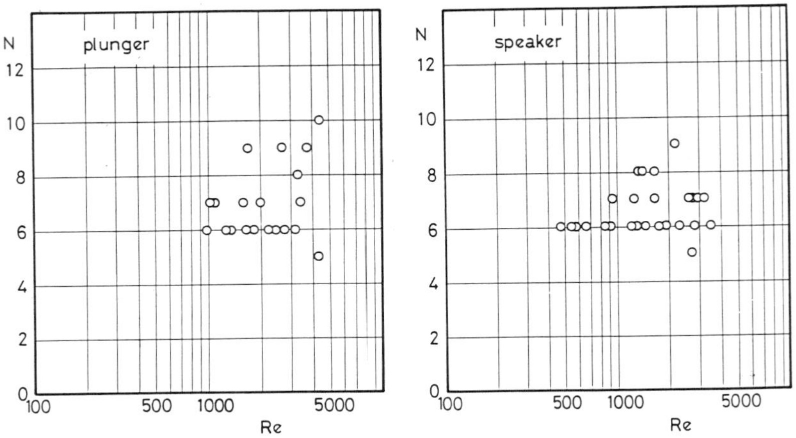

FIGURE 9–16. Relationship between N and R_e for both types of the vortex generators.

F. STAR-TYPE BREAKUP OF A VORTEX RING

The mechanisms of the production for the star type breakup of a vortex ring are as follows:

1. Concerning the exit orifice being truly circular or not
2. Whether the concentration of the smoke in the smoke chamber is homogeneous or not
3. The problems concerning the effect of fluid turbulence of the surrounding air

While observing the behaviors of a star-type breakup of a vortex ring, the regular distribution of the peaks around a vortex ring extended entirely. And also the kernel of the vortex ring was in a state of the rigid body. There was no place where the diameter of a vortex kernel was changed around the vortex ring.

The problem concerning the fluid turbulence was excluded by the observations of the many types of vortex rings. From the above-stated results, the vortex ring was not influenced by the shape of the exit orifice and by the local change of the smoke concentration. The star-type breakup of a vortex ring was governed by the total energy included in the vortex ring. The correlation between the Reynolds number R_e and the number N of the peaks for the star-type breakup is shown in Figure 9–16. However, from this figure, we could not find the clear relationship between N and R_e, because we have confirmed by the observation of the vortex ring that the number N of the star-type breakup was not determined as a whole by the given energy to the vortex ring at the exit of the orifice plate, but decreased with decay of the advanced velocity of a vortex ring. Figure 9–17 shows the behavior of the vicissitude of a vortex ring under the conditions $R_e = 2050$ and $n = 1.74$ by the speaker. From these photos, it was clear that the number N of peaks decreased with decay of the advanced velocity from equiseven angle shape to an equifive angle shape. Figure 9–18 shows the processes of the vicissitude of a vortex ring under the conditions $R_e = 1650$ and $n = 2.2$. These photos were taken within the time interval $0.1 \sim 0.15$ s after inputting the power to the speaker. From these photos, the star-like breakup of the vortex ring was produced by the cause and the process as follows:

1. A vortex ring was composed of the vortex kernel rotating like a rigid body and of the free vortex.

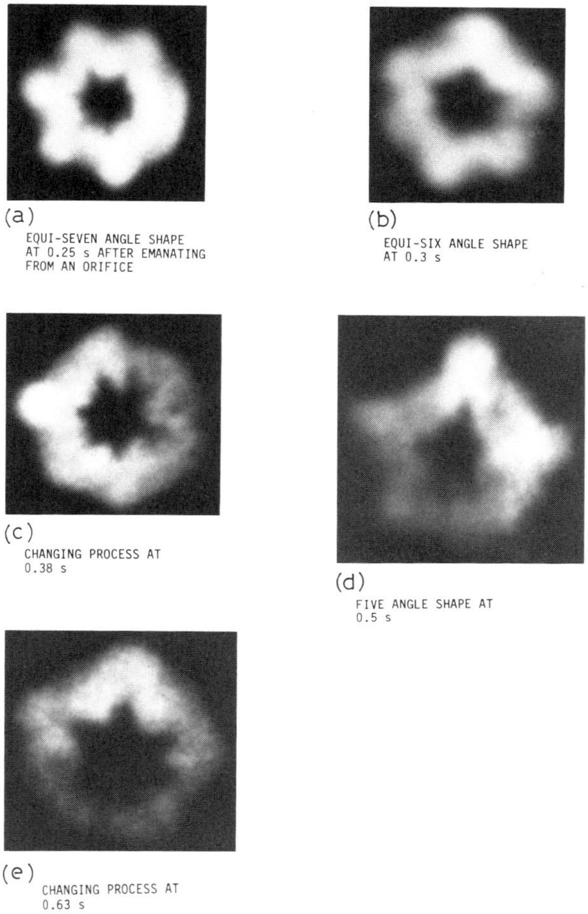

FIGURE 9–17. Behaviors of the vicissitude of a vortex ring under the conditions of $R_e = 2050$, $n = 1.74$ by the speaker.

2. A vortex ring emanating from an orifice was the stable circular form; however, the vortex kernel was a wave-like form due to the small disturbance of the vortex kernel itself. When the kinetic energy of a vortex kernel became high, and, on the other hand, when the kinetic energy of a vortex ring was a lower level, so the frequency decreased. Then when the frequency coincided with the critical frequency of the vortex kernel, so the vortex ring developed gradually to the star-like form.
3. The energy dissipation at each peak of the star-like vortex ring due to the viscous action for the vorticity became high; therefore, the momentum of the vortex ring containing large energy rapidly fell off.
4. The energy contained in the vortex ring was dissipated with a decrease of the advanced velocity, so that corresponding to the containing energy the number N of the peaks is also changed gradually from N to N-1 and N-2 and so on.
5. Therefore the vortex ring losing the kinetic energy became a stationary ring of smoke. Then this vortex ring disappeared into the atmosphere.[12–14]

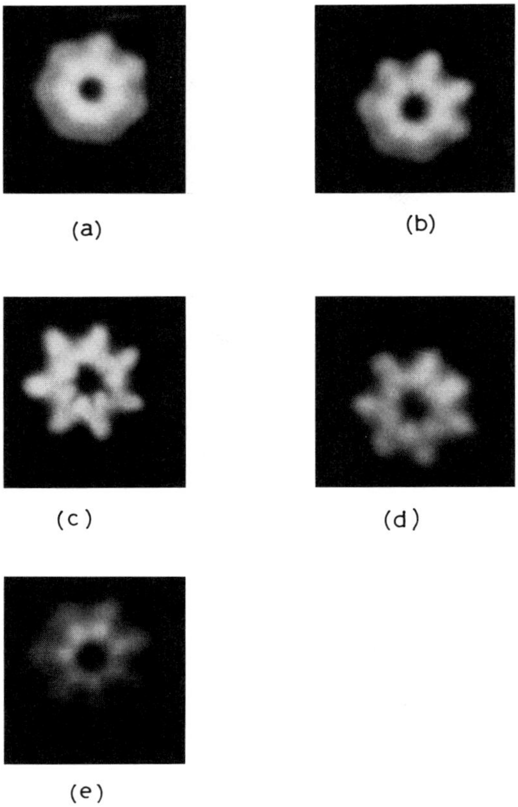

FIGURE 9–18. Processes of the vicissitude of a vortex ring under the conditions of R_e = 1650 and n = 2.2 by a speaker. These photos were taken within the time interval 0.1. ~ 0.15 s after inputting the power to the speaker.

MAIN SYMBOLS

b(1);	index of the advanced decreasing velocity
d(m);	tube diameter of the vortex kernel
D(m);	main diameter of the vortex ring
D_o(m);	diameter of the orifice
E(N · m);	kinetic energy of the vortex ring
n(1);	index of the advanced decreasing velocity
R_e(1);	Reynolds number for the diameter of a vortex ring
R_{ef}(1);	Reynolds number for the advanced distance Z of the nascent vortex ring
U(m/s);	advanced decreasing velocity of the vortex ring
U_f(m/s);	advanced velocity of a first nascent vortex ring
Z(m);	advanced distance of a first vortex ring
Γ(m^2/s);	circulation of the vortex ring
ρ(kg/m^3);	density of fluid

REFERENCES

1. **Helmholtz, H.,** Über Integrale der hydrodynamischen Gleichungen welche den Wirbelsbewegungen entsprechen, Crelle-Borchardt, 55, 25, 1858; translated into Japanese by Nagaoka, H., Kado ronbunshu, Maruzen, Taisho, 2nd ed.
2. **Lamb, H.,** *Hydrodynamics,* Cambridge, 6th ed., 1932, 236.
3. **Oshima, Y. and Asaka, S.,** Interaction of two vortex rings along parallel axis in air, *J. Phys. Soc. Jpn.,* 42(2), 708, 1977.
4. **Oshima, Y. and Asaka, S.,** Interaction of multi-vortex rings, *J. Phys. Soc. Jpn.,* 42(4), 1391, 1977.
5. **Didden, N. and Liess, C.,** Experimente zum Einfluß der Anfangsbedingungen auf die Instabilität von Ringwirbel, Z. A. M. M., 56, T206, 1976.
6. **Magarvey, R. H. and MacLatchy, C. S.,** The formation and structure of vortex ring, *Can. J. Phys.,* 42, 678, 1964.
7. **Magarvey, R. H. and MacLachy, C. S.,** The disintegration of vortex ring, *Can. J. Phys.,* 42, 684, 1964.
8. **Yamada, H.,** On the motions of the vortex ring (in Japanese), Collected papers of the preprints of *J. Soc. Mech. Eng.,* No. 740-9, 219, 1974–78.
9. **Boyarintsev, V. I., Levchenko, E. S., and Savin, A. S.,** Motion of two vortex rings, Fluid dynamics (translated from Russian), Mekh. Zhidkosti Gaza, 20(5), 818, 1985.
10. **Baird, M. H., Wairegi, T., and Loo, H. J.,** Velocity and momentum of vortex rings in relation to formation parameters, *Can. J. Chem. Eng.,* 55 (Feb.), 19, 1977.
11. **Taylor, G. I.,** Formation of a vortex ring by giving an impulse to a circular disc and then dissolving it away, *J. Appl. Phys.,* 24, 104, 1953.
12. **Igarashi, M. and Ogawa, A.,** Experiments of the vortex motion, Collected papers of preprint of *J. Soc. Mech. Eng.,* (1976–11), Tohoku-shibu, 64–66.
13. **Igarashi, M. and Ogawa, A.,** Experiments of the vortex motion, Collected papers of preprint of *J. Soc. Mech. Eng.,* No. 770–8, 37, 1977.
14. **Ogawa, A. and Igarashi, M.,** Behaviours of the vortex ring in the domain of the high Reynolds number, *Nihon Univ. J. Coll. Eng.,* Vol. A-20, 97, 1979.

Chapter 10

STABILITY OF ROTATING FLOW

I. INTRODUCTION

There are many papers concerning the stability problems of the Taylor vortex, of the Görtler vortex,[1] and of the vortex ring and also of rotating flow in the vortex chamber or in fluid machines.

In this chapter the physical explanations of the stability problems, which can be applied to chemical engineering and to powder engineering, are described in detail.

II. INSTABILITY AND TRANSITION OF A BOUNDARY LAYER ON A ROTATING CONE

It is very important to investigate the fundamental characteristics concerning the changing patterns from the laminar boundary layer, the transition layer, and the turbulent boundary layer of the complicated three-dimensional boundary layer on the axis-symmetrical body in the turbo-machine. Profs. Kobayashi, Kohama, Arai, and Ukaku[2] have studied the instability and the transition of the boundary layer on the rotating cones theoretically and experimentally.

Figure 10–1 shows the experimental apparatus of the rotating cone and the main symbols, where V_∞(m/s) means the free stream velocity and V_e(m/s) means the velocity at the brink of the boundary layer along the surface of the cone, ω(rad/s) means the angular velocity of the rotating cone, $\widetilde{x}_c$(m) means the starting point of the transition region and the distance from the summit of the cone angle 2θ(deg) to $\widetilde{x}_c$, $\widetilde{x}_t$(m) means the terminal point of the transition region and also the dimensionless quantities S, R_{ec}, and R_{et} are applied, and, in addition to this, a symbol ϵ(deg) means the angle of the spiral vortices in the transition region.

In these experiments the cone angle 2θ and the surface distance L along the cones are described in this figure. Then in order to make a flow visualization, they smeared the titanium-tetra chloride over the black painted cone surface. So, we can see the transition region and the turbulent boundary layer by the generation of the white smoke.

Figure 10–2 shows the examples of the sketch of the flow visualization of the boundary layer on the surface of the rotating cone of $2\theta = 15°$. The rotation of the cone is clockwise on seeing from the upstream to the downstream. In the upper sketch of the driving condition of $V_\infty = 0$ m/s, N = 500 rpm, the Taylor vortices are generated. In the middle sketch for $V_\infty = 1.8$ m/s and N = 2000 rpm in the transition region the spiral vortices are generated and after that the flow reaches the turbulent boundary layer. In the lower sketch for $V_\infty = 4.8$ m/s and N = 1000 rpm, there is the laminar boundary layer over the whole surface of the rotating cone.

Figure 10–3 shows the experimental results by Prof. Kobayashi et al. Figure 10–3(a) shows the relationship between V_e/V_∞ and $\widetilde{x}/L$ for $2\theta = 15°$, 30°, and 60°, respectively. Figure 10–3(b) shows the relationship between ϵ(deg) and $S = \omega R/V_e$. With increasing the value of S, the spiral vortices approach the peripheral vortices on the Taylor vortices. Figure 10–3(c) shows the relationship between $R_{ec} = \widetilde{x}_c V_e/\nu$, $R_{et} = \widetilde{x}_t V_e/\nu$, and S, where R_{ec} is the Reynolds number at the starting point of the transition layer and also R_{et} is the Reynolds number at the terminal point of the transition point.

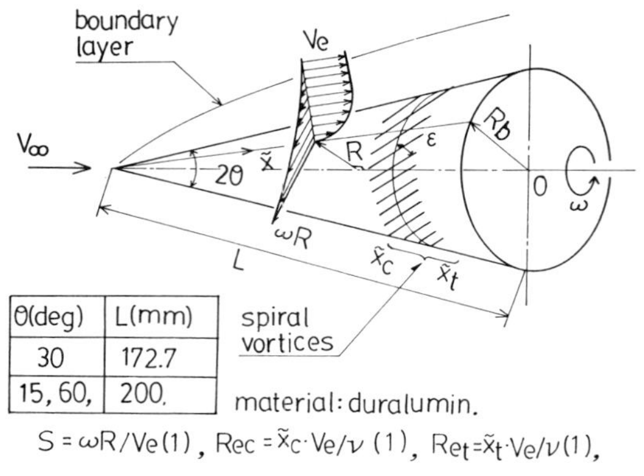

FIGURE 10–1. Rotating cone and the main symbols.

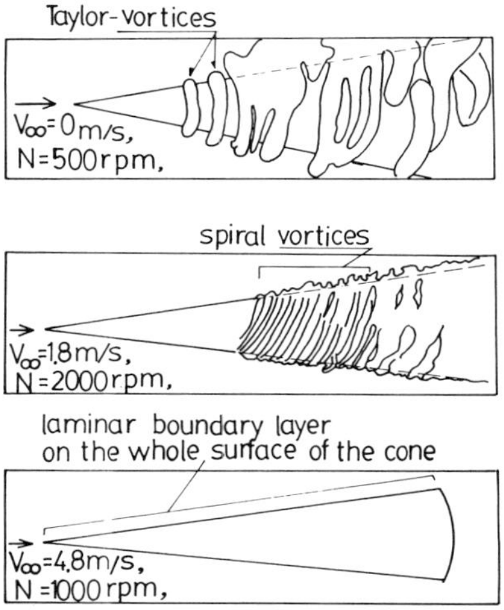

FIGURE 10–2. Sketch of the flow visualization of the boundary layer on the surface of the rotating cone of $2\theta = 15°$ (the direction of the rotation of the cone is clockwise on seeing from the upstream to the downstream).

III. TAYLOR-GÖRTLER VORTICES

A. EUTENEUER EXPERIMENT

Concerning the Taylor vortex problems, G. I. Taylor had done splendid theoretical studies and also experiments in the coaxial cylinders.[3] As shown in Figure 10–4, R_1 and R_2 are the radii of the two rotating cylinders and Ω_1 and Ω_2 are the angular velocities of both cylinders, where a symbol k is $k = \Omega_1/\Omega_2$. When the cylinders are rotating in the same direction with slightly different velocities, Taylor derived the stability criterion that the rotating motion of fluid changes from being a stable condition to being an unstable condition (appearance of Taylor vortices) passing through the value of Ω_2 given by the following equation

$$\frac{\pi^4\nu^2(R_1 + R_2)}{2\Omega_2^2(R_1 - R_2)^3 R_2^2 [1 - (kR_1^2/R_2^2)](1 - k)} \left(\frac{1 - k}{1 + k}\right) = 0.0571 \qquad (10\text{–}1)$$

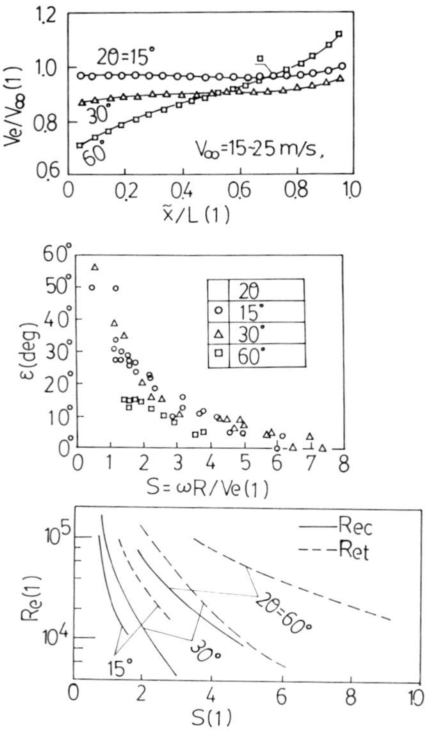

FIGURE 10–3. Experimental results of the flow characteristics on the rotating cone.

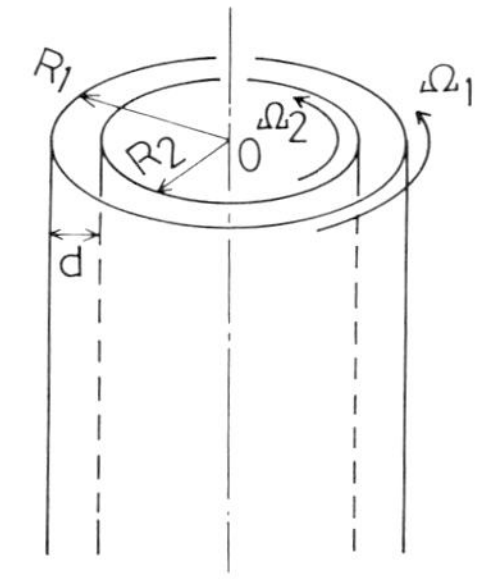

FIGURE 10–4. Schematic illustration of the rotating cylinders.

Therefore when the outer cylinder of the radius R_1 is at rest, so k is equivalent to zero. On the other hand when the direction of the outer rotating cylinder is the opposite direction to the inner rotating cylinder, so the value of k becomes negative. Here in the case of $k = 0$, the above equation becomes

$$\frac{V_{\theta 2} \cdot (R_1 - R_2)}{\nu} \sqrt{\frac{R_1 - R_2}{R_1 + R_2}} = 29.18 \tag{10–2}$$

where $V_{\theta 2}(= \Omega_2 R_2)$ means the rotating velocity of the inner rotating cylinder. When the gap between them is very narrow, the above equation transforms to

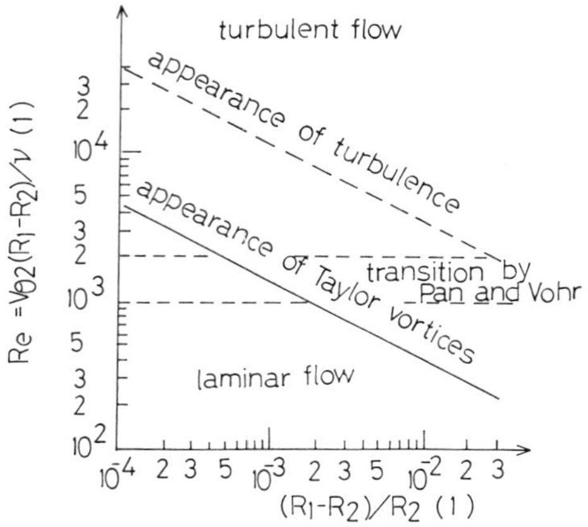

FIGURE 10–5. Appearance of the Taylor vortices in the narrower gap between the coaxial cylinders.

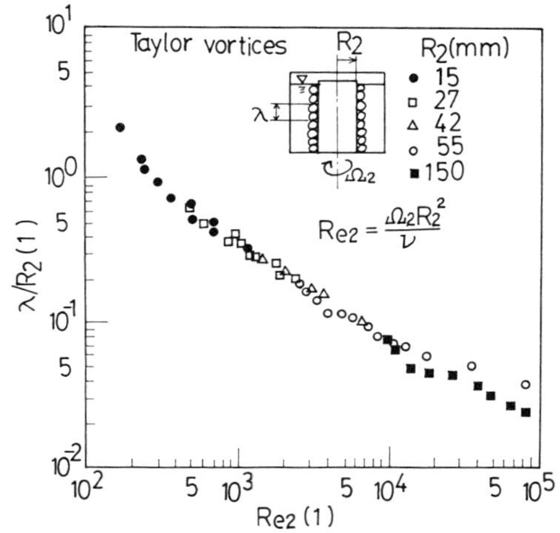

FIGURE 10–6. Relationship between the wavelength and the Reynolds number for the Taylor vortices in the wide gap.

$$\frac{V_{\theta 2}(R_1 - R_2)}{\nu}\sqrt{\frac{R_1 - R_2}{R_2}} = 41.3 \tag{10–3}$$

If the value of the left-hand side is higher than 41.3, the appearance of the Taylor vortices will occur. This theoretical relation was confirmed by the experimental results for the narrow gap between the cylinders.

Figure 10–5 shows the suggestion of the appearance of the Taylor vortices in the narrower gap between the coaxial cylinders by Frêne and Godet.[4] Here Figures 10–6 and 10–7 show the interesting Euteneuer experiments.[5] Figure 10–6 shows the experimental results of the Taylor vortices which appear on the rotating inner cylinder of the radii R_2 = 15, 27, 42, 55, and 150 mm, respectively. The wavelength λ is defined as the length of a pair of the Taylor vortices, and the Reynolds number R_{e2} is defined as

$$R_{e2} = \frac{\Omega_2 R_2^2}{\nu} = \frac{V_{\theta 2} \cdot R_2}{\nu} \tag{10–4}$$

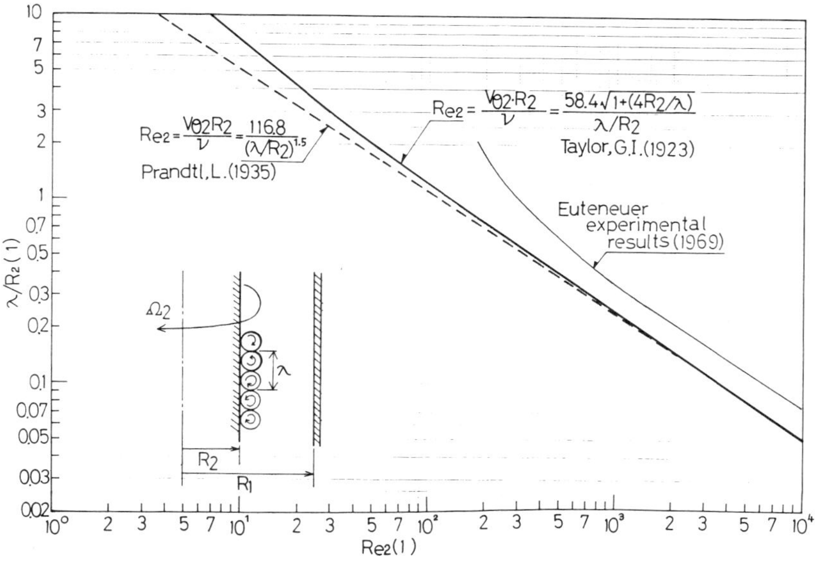

FIGURE 10–7. Prediction of the wavelength of the Taylor vortices by Taylor and by Prandtl.

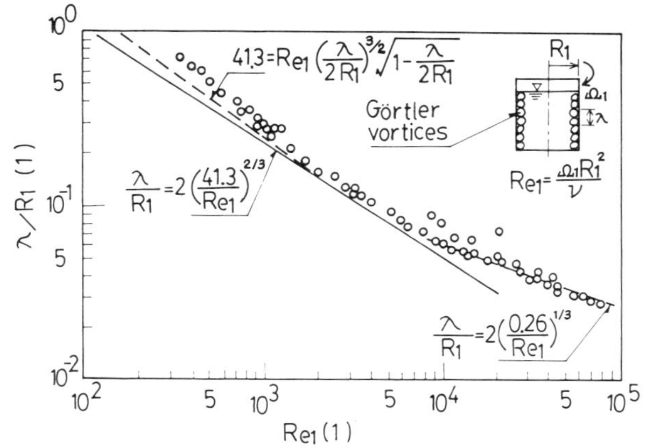

FIGURE 10–8. Relationship between the wavelength of the Görtler vortices and the Reynolds number.

The experimental results of λ/R_2 are longer than the predicted values by Equation 10–3 of Taylor, since the gap $(R_1 - R_2)$ between them in Euteneuer experiments was wider than that of the Taylor theoretical studies. So the flow conditions have very different characteristics in comparison with the assumed flow conditions by G. I. Taylor. Figure 10–8 shows the experimental results of Görtler vortices, which appear on the concave wall surface on the decelerating rotational velocity of the outer cylinder of the radius R_1.[6] The wavelength λ/R_1 depends on the Reynolds number R_{e1}. In this figure two empirical equations and an equation which was transformed from the Taylor theoretical equation are shown in comparison with the experimental values.

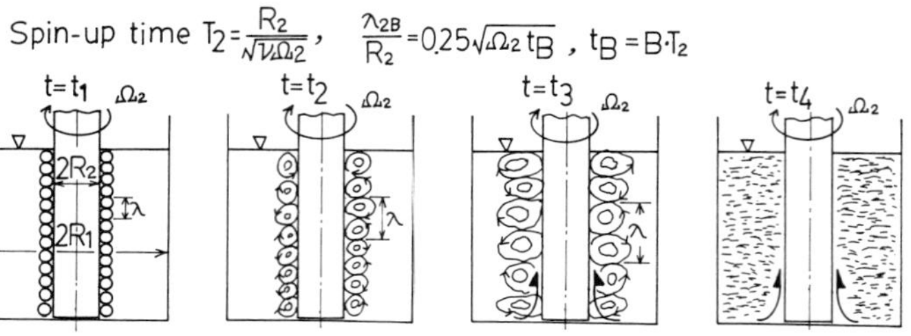

FIGURE 10–9. Schematic illustration of the development, transition, and collapse of the Taylor vortices in the wide gap of the coaxial cylinders.

B. DEVELOPMENT, TRANSITION, AND COLLAPSE OF TAYLOR VORTICES WITH LAPSE OF TIME

When the space between the outer rest cylinder and the inner rotating cylinder becomes too wide, for example $(D_1 - D_2)/D_2 \geqslant 1$, so the wavelength λ of the Taylor vortices becomes longer with lapse of time as shown in Figure 10–9.[7] When the inner cylinder rotates with a constant angular velocity Ω_2, with lapse of time $t = t_1$, the regular Taylor vortices suddenly appear on the surface of the rotating cylinder. With further lapse of time $t = t_2 > t_1$, the regular shaped Taylor vortices become larger. After time T_2 which is defined as

$$T_2 = \frac{R_2}{\sqrt{\nu \Omega_2}} \tag{10–5}$$

the wavelength of the Taylor vortices increases more rapidly. An inward flow toward the inner rotating cylinder in the boundary layer on the fixed bottom surface is generated. After that the unstable and larger Taylor vortices break down and the rotating flow in the coaxial cylinders becomes turbulent flow. This physical phenomenon is very important for considering the rotating flow of liquid in the wider span in the coaxial cylinders. However, when the Reynolds number R_{e2} defined as

$$R_{e2} = \frac{R_2^2 \Omega_2}{\nu} \tag{10–6}$$

is kept to the lower value as $R_{e2} \leqslant 500$, even if the lapse of time elapses the spin-up time T_2, the wavelength λ of the Taylor vortices is inclined to be constant. And then when the Reynolds number R_{e2} is within the limits of $500 \leqslant R_{e2} \leqslant 1300$, so the wavelength λ shows unstable behavior. Further when the Reynolds number R_{e2} is higher than $R_{e2} = 1300$, the wavelength λ further increases with lapse of time.[8,9]

The relationship between the time t_B of the breakdown of the Taylor vortices and the spin-up time T_2 can be represented as

$$t_B = B \cdot T_2 \tag{10–7}$$

where a constant B depends on the ratio (D_2/D_1).

C. GÖRTLER VORTEX

Three-dimensional disturbances generate in the boundary layer along the concave wall surface. This unstable effect is due to the centrifugal force on the concave wall surface.[10,11]

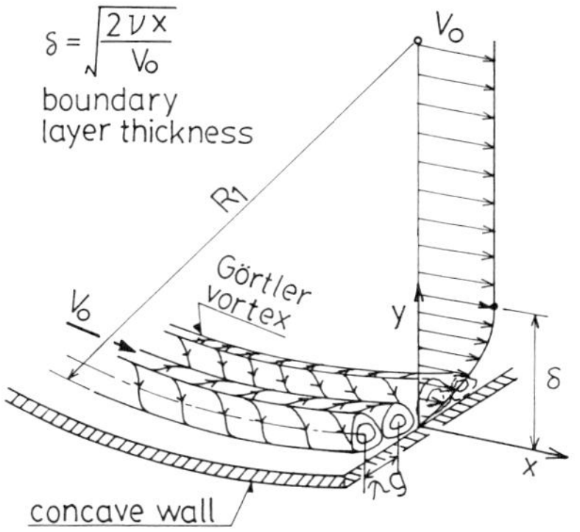

FIGURE 10–10. Schematic illustration of Görtler vortices in the boundary layer on the concave wall surface.

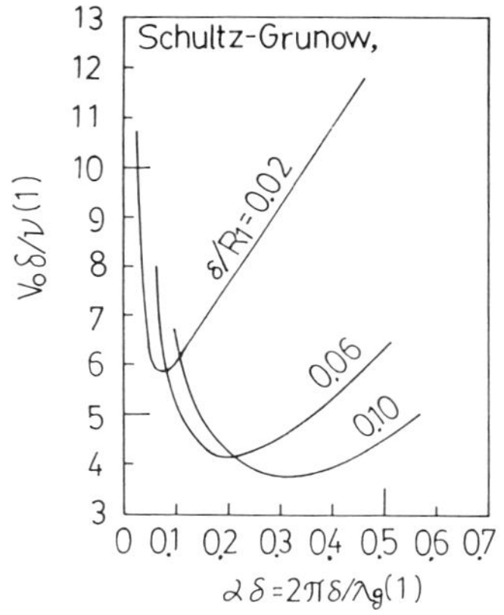

FIGURE 10–11. Limit of the stability for the boundary layer on the concave wall surface after Schultz-Grunow (in case of $\beta = 0$, where β is real and denotes the amplification factor).

The existence of this effect was first shown by H. Görtler.[12] Taking the x-direction to the main flow and the y-direction at right angles to the main flow as shown in Figure 10–10, the disturbance velocity to the x-direction v_x' can be represented as

$$v_x' = v_x(y)[\cos(\alpha z)] \exp(\beta t) \qquad (10\text{–}8)$$

where β is the real value and shows the amplification factor, in which $\lambda_g = 2\pi/\alpha$ gives the wavelength of the disturbance (Görtler vortices). Figure 10–11 shows the numerical values of the neutral curves of the relationship between $V_o\delta/\nu$ and $\alpha\delta = 2\pi\delta/\lambda_g$ after Schultz-Grunow,[13] where R_1 means the radius of the concave wall surface, and δ is the boundary layer thickness. Here the relationship between δ and the momentum thickness δ_2 can be represented as $\delta_2 = 0.047\delta$.

IV. STABILITY CRITERION IN A CIRCULAR CYLINDER BY RAYLEIGH

Consider the steady flow of viscous fluid between two coaxial circular cylinders of infinite extension in the z-direction in the limiting case of arbitrarily large Reynolds number as shown in Figure 10–12. We can use a simple method proposed by Rayleigh.[14]

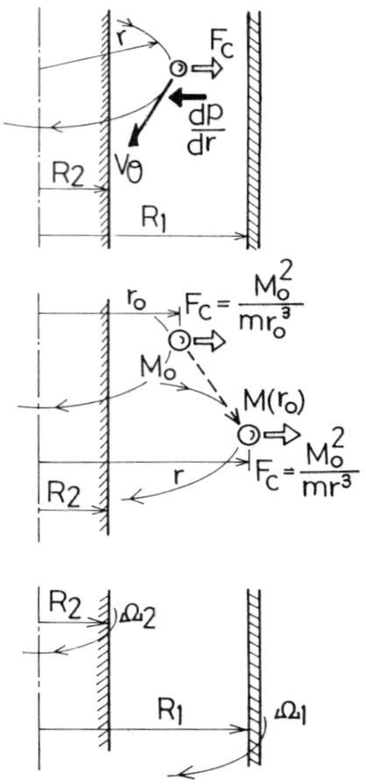

FIGURE 10–12. Schematic illustration of Rayleigh criterion.

Here we consider any arbitrarily small region of fluid and assume that it is displaced from the trajectory of its motion. In this process forces arise which act on the displaced portion of fluid. For the stability of the basic motion it is necessary that these forces should strive to return the displaced element to its original position. We denote that the angular momentum of the fluid element moving at the radius r = constant by $M(r) = mr^2\omega$. The centrifugal force $F_c = mV_\theta^2/r = mr\omega^2 = M^2/mr^3$ acting on this element balances the corresponding pressure gradient dp/dr.

Let us assume now that an element at $r = r_o$ is subjected to a small displacement from its trajectory on to a radius $r > r_o$. The angular momentum is conserved and will be equal to $M_o = M(r_o)$, while the centrifugal force becomes $F_c = M_o^2/mr^3$. In order that the element should tend to return into its initial position, it is necessary that this centrifugal force should be less than that existing at the distance r from the axis, i.e., less than $Fc(r) = M^2/mr^3$. Therefore a necessary condition for stability will be

$$\frac{M^2}{mr^3} - \frac{M_o^2}{mr^3} > 0 \quad \text{or} \quad M^2 - M_o^2 > 0 \tag{10–9}$$

Since

$$M^2 - M_o^2 \doteqdot 4m^2\omega^2 r_o^3\, dr \tag{10–10}$$

therefore,

$$\frac{M^2 - M_o^2}{dr} = 4m^2\omega^2(r - dr)^3 \doteqdot 4m^2\omega^2 r^3 = 2M\frac{dM}{dr} \tag{10–11}$$

The stable condition becomes

$$M\frac{dM}{dr} > 0 \quad \text{or} \quad \frac{1}{2}\frac{d}{dr}M^2 > 0 \tag{10–12}$$

that is to say,

$$m(\Omega_1 R_1^2 - \Omega_2 R_2^2) > 0 \tag{10–13}$$

for the stable condition.

When the cylinders rotate in opposite directions, Ω changes its sign in the space between the cylinders, so that the above condition is not satisfied in the whole volume of fluid, i.e., the motion is unstable. When both cylinders rotate in the same direction and $\Omega_1 > 0$ and $\Omega_2 > 0$, then $\omega > 0$; the above condition takes on the form

$$\Omega_1 R_1^2 > \Omega_2 R_2^2 \text{ (stable)} \tag{10–14}$$

When the outer cylinder is at rest ($\Omega_1 = 0$) and only the inner cylinder rotates, the motion is unstable. Conversely, when the inner cylinder is at rest ($\Omega_2 = 0$), the motion is stable. For example, the tangential velocity can be represented

$$V_\theta r^n = \Gamma_n = \text{constant} \tag{10–15}$$

so the stable condition can be described as

$$\frac{dM^2}{dr} = m^2\,\Gamma_n^2\,2(1 - n)\,r^{1-2n} > 0 \tag{10–16}$$

The more detailed, above condition can be represented as

$$n < 1 \text{ stable} \tag{10–17}$$

$$n = 1 \text{ critical} \tag{10–18}$$

$$n > 1 \text{ unstable (Taylor vortex)} \tag{10–19}$$

V. PRANDTL THEORY OF STABILITY OF TURBULENT ROTATIONAL FLOW

Consider the two-dimensional fluid motion in the space of the coaxial rotating cylinder.[15] Assume that the tangential velocity of fluid motion is dependent on the radius r. Now when the tangential velocity is increased with the radius r, the lump of fluid cannot penetrate from the inside region toward the outside region, since the centrifugal force of the fluid particle in the inside region is smaller than that of fluid particle in the outside region. As shown in Figure 10–13, when the tangential velocity can be written by the forced vortex as

$$V_\theta = \omega r \tag{10–20}$$

where ω is the angular velocity, the centrifugal force F_c at the radius r per unit volume is

$$F_c = \rho\frac{V_\theta^2}{r} = \rho\omega^2 r \tag{10–21}$$

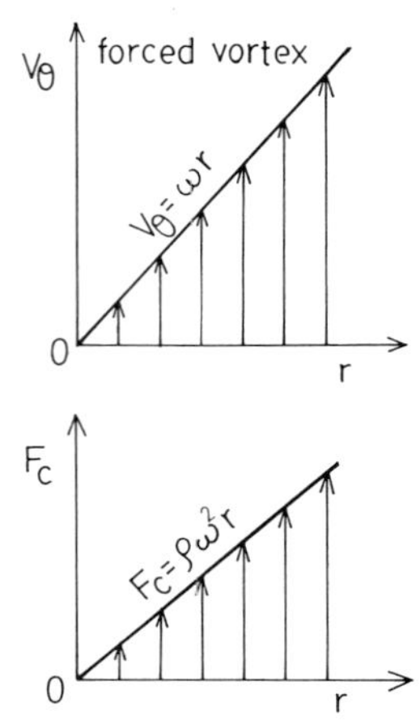

FIGURE 10–13. Distributions of the tangential velocity and of the centrifugal force of the forced vortex.

Therefore the centrifugal force F_c is proportional to the radius r; so this force in the inside region is smaller than that in the outside region. The lump of fluid in the outside region cannot penetrate into the inside region, since this lump of fluid is turned back to the original place by the stronger centrifugal force. This physical situation of the rotational flow is called the stable condition. On the other hand, assuming that the tangential velocity of the lump of fluid is decreased with increasing the radius r, and when the rotating fluid on a circular locus of the arbitrary radius r is moved toward the outside region by the action of the centrifugal force, so this outside protruded lump of fluid moves further out, since the restoration force to this lump of fluid being returned toward the inside original place becomes weaker with the increase of radius r. For example, the tangential velocity is

$$V_\theta = \frac{\Gamma}{r} \tag{10–22}$$

where Γ is a constant value. So the centrifugal force F_c per unit volume can be expressed as shown in Figure 10–14 as

$$F_c = \rho\frac{V_\theta^2}{r} = \frac{\rho\Gamma^2}{r^3} \tag{10–23}$$

In this case the centrifugal force F_c is inversely proportional to the third power of the radius r. Therefore the centrifugal force shows an extreme decrease with increasing the radius r. This physical situation is an unstable condition.

From these fundamental ideas we consider the lump of fluid rotating with the tangential velocity on the circular locus as shown in Figure 10–15. The centrifugal force per unit volume is $F_c = \rho V_\theta^2/r$.

Assuming that this lump of fluid on the vorticity layer of Ω is moved from the circular locus of the radius r toward the circular locus of the radius r + l by the same disturbance under the conservation of the angular momentum, where l is the small distance, the small increment v_θ of the tangential velocity at the radius r + l is

$$v_\theta = l\left(\frac{dV_\theta}{dr} + \frac{V_\theta}{r}\right) \tag{10–24}$$

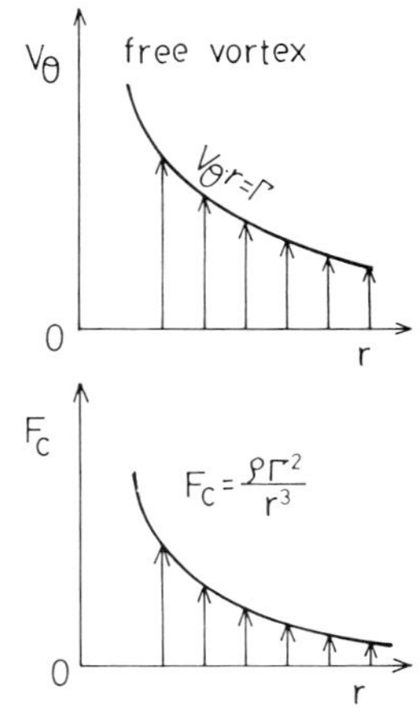

FIGURE 10–14. Distributions of the tangential velocity and of the centrifugal force of the free vortex.

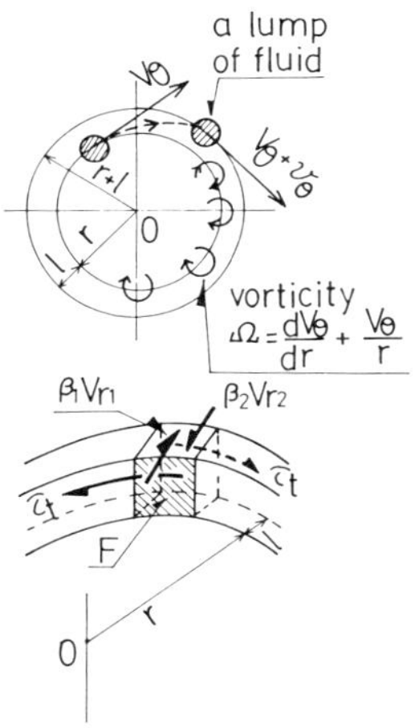

FIGURE 10–15. Schematic illustration of the fundamental idea of the Prandtl theory.

where $(dV_\theta/dr + V_\theta/r)$ is the vorticity Ω.

Then the difference of the centrifugal forces between the centrifugal force at the radius r + l and that at the radius r is

$$\rho \frac{[(V_\theta + v_\theta)^2 - V_\theta^2]}{r} \tag{10–25}$$

The first approximation of this quantity is

$$\frac{2\rho V_\theta v_\theta}{r} \tag{10–26}$$

Therefore the work done $W_f(N/m^2)$ of this lump of fluid per unit volume moving the distance $^l/_2$ to the outward direction against this centrifugal force can be obtained by Equation 10–24 and Equation 10–26 as

$$W_f = \frac{2\rho V_\theta v_\theta}{r} \cdot \frac{1}{2} = \frac{\rho V_\theta v_\theta l}{r} = \frac{\rho l V_\theta}{r} \cdot l\left(\frac{dV_\theta}{dr} + \frac{V_\theta}{r}\right) = \tag{10–27}$$
$$= \frac{\rho l^2 V_\theta}{r} \cdot \left(\frac{dV_\theta}{dr} + \frac{V_\theta}{r}\right) \quad , \quad (N/m^2)$$

The relative transition velocity (deformation velocity) between the two cross-sectional surfaces which are a small radial distance apart in the rotational flow is

$$\frac{dV_\theta}{dr} - \frac{V_\theta}{r} \tag{10–28}$$

So the work done W_r to the surface F by the turbulent Reynolds stress ($\tau_t = -\rho\overline{v_r v_\theta}$) based upon the turbulent rotational flow on the cylindrical side surface is

$$W_r' = Fl\tau_t\left(\frac{dV_\theta}{dr} - \frac{V_\theta}{r}\right) \tag{10–29}$$

It is assumed that the Reynolds stress τ_t may be expressed as

$$\tau_t = -\rho\overline{v_r v_\theta} \fallingdotseq \rho l\left(\frac{dV_\theta}{dr} + \frac{V_\theta}{r}\right)(\beta_1 V_{r1} + \beta_2 V_{r2}) \tag{10–30}$$

where $\beta_1 V_{r1}$ is the outward radial velocity and $\beta_2 V_{r2}$ is the inward radial velocity. Then, divide W_r' by flow rate $F(\beta_1 V_{r1} + \beta_2 V_{r2})$ as

$$W_r = \rho l^2\left(\frac{dV_\theta}{dr} + \frac{V_\theta}{r}\right)\left(\frac{dV_\theta}{dr} - \frac{V_\theta}{r}\right)$$

and assume that the transition condition from laminar flow to the turbulent rotational flow is based upon the rotating flow (the unstable condition). The work done W_f by the centrifugal force to the unit of volume of fluid is less than that W_r by the Reynolds stress. Consequently, the following condition can be obtained

$$\rho l^2\left(\frac{dV_\theta}{dr} + \frac{V_\theta}{r}\right)\left(\frac{dV_\theta}{dr} - \frac{V_\theta}{r}\right) > \frac{\rho l^2 V_\theta}{r}\left(\frac{dV_\theta}{dr} + \frac{V_\theta}{r}\right) \tag{10–31}$$

The above equation can be transformed to

$$\frac{dV_\theta}{dr} - \frac{V_\theta}{r} > \frac{V_\theta}{r} \tag{10–32}$$

Therefore when a term $(\partial V_\theta/\partial r + V_\theta/r)$ is positive Equation 10–32 becomes

$$\frac{dV_\theta}{dr} > \frac{2V_\theta}{r} \tag{10–33}$$

While a term $(\partial V_\theta/\partial r + V_\theta/r < 0)$ is negative, Equation 10–32 becomes

$$\frac{dV_\theta}{dr} < 2\,\frac{V_\theta}{r} \tag{10–34}$$

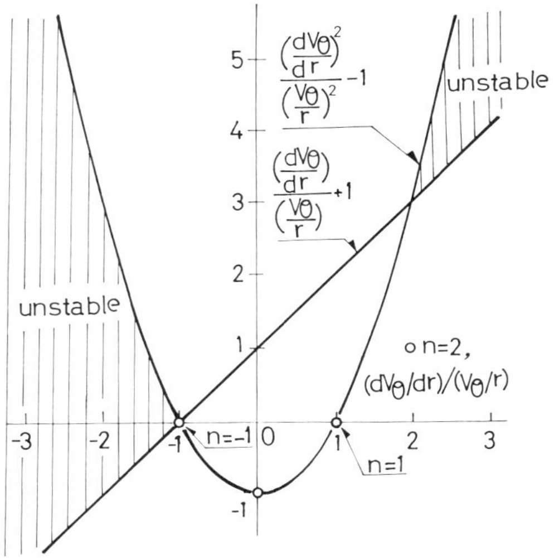

FIGURE 10–16. The domains of the stable and unstable conditions by the Prandtl theory.

If the tangential velocity is the positive value and also $\partial V_\theta/\partial r + V_\theta/r < 0$, the following condition can be obtained

$$\frac{dV_\theta}{dr} < -\frac{V_\theta}{r} \tag{10–35}$$

Here in order to express more concretely the above considerations of Equation 10–31, dividing both sides by a term $(V_\theta/r)^2$ and after that taking $[(dV_\theta/dr)/V_\theta/r)]$ as an abscissa, the possible regions of the unstable turbulent production are shown by the shaded lines in Figure 10–16.

If the distribution of the tangential velocity is

$$V_\theta = ar^n \tag{10–36}$$

the following condition can be obtained

$$\begin{aligned} n &< -1, && \text{unstable (Taylor vortex)} \\ n &= -1, && \text{critical point (free vortex)} \\ -1 < n &< 2, && \text{stable (laminar flow)} \\ n &= 1, && \text{stable (forced vortex)} \\ n &> 2, && \text{stable (turbulent flow).} \end{aligned}$$

VI. VIBRATION OF VORTEX FLOW

A. FUNDAMENTAL EQUATION

Here the simplified equation of the vortex vibration will be explained by using the Ogawa combined vortex model. Let us assume that a small lump of fluid rotating steadily at the arbitrary radius r is in the mechanical equilibrium between the centrifugal force to the outward direction and the pressure gradient to the inward direction. The following equation can be stated

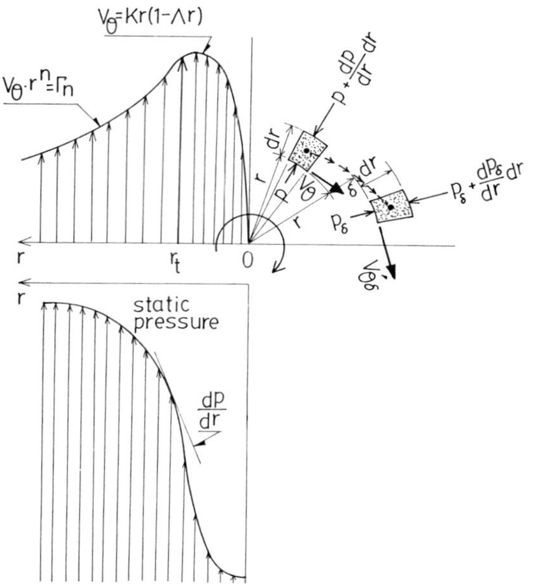

FIGURE 10–17. Schematic illustration for the stability criterion by the Ogawa combined vortex model.

$$\frac{dp}{dr} = \rho\frac{V_\theta^2}{r} \tag{10–37}$$

Now assuming that this lump of fluid which is rotating in the mechanical equilibrium on the circle of the radius r is moved a small radial displacement δ toward the outward direction by some disturbance as shown in Figure 10–17, then the equation of motion to the radial direction for this lump of fluid which was displaced only the small radial distance δ is

$$\frac{d^2\delta}{dt^2} = \frac{V_{\theta\delta}'^2}{r+\delta} - \frac{1}{\rho}\cdot\frac{dp_\delta}{dr} \tag{10–38}$$

where $V'_{\theta\delta}$ is the tangential velocity at the radius r + δ corresponding to the conservation of the angular momentum between the radius r and the radius r + δ as follows

$$V_\theta \cdot r = V'_{\theta\delta}\cdot(r+\delta) \tag{10–39}$$

The pressure p_δ is the static pressure at the radius r + δ and the pressure gradient is

$$\frac{dp_\delta}{dr} = \rho\frac{V_{\theta\delta}^2}{r+\delta} \tag{10–40}$$

where $V_{\theta\delta}$ is the steady tangential velocity at the radius r + δ. Therefore $V_{\theta\delta}$ does not always coincide with $V'_{\theta\delta}$ and further the relationship between V_θ and r can be related to

$$V_\theta \cdot r^n = \Gamma_n = \text{constant} \tag{10–41}$$

When the flow reaches its potential flow, the exponent n becomes one (n = 1). Consequently substituting Equations 10–30 and 10–40 into Equation 10–38, we obtain the fundamental equation

$$\frac{d^2\delta}{dt^2} = \left(\frac{r}{r + \delta}\right)^2 \cdot \frac{V_\theta^2}{r + \delta} - \frac{V_{\theta\delta}^2}{r + \delta} \tag{10–42}$$

B. QUASI-FORCED VORTEX

In the quasi-forced vortex region the tangential velocity is

$$V_\theta = Kr(1 - \Lambda r) \tag{10–43}$$

Then the tangential velocity $V_{\theta\delta}$ at the radius $r + \delta$ can be expressed nearly as

$$V_{\theta\delta} \doteqdot V_\theta + K(1 - 2\Lambda r)\delta \tag{10–44}$$

Substituting the above equation into Equation 10–43, we obtain

$$\frac{d^2\delta}{dt^2} = -\frac{2V_\theta^2}{r}\left[1 + \frac{Kr(1 - 2\Lambda r)}{V_\theta}\right]\delta = -\Omega^2\delta \tag{10–45}$$

This equation is the equation of simple harmonic motion which gives the frequency f. The frequency is

$$f = \frac{K}{\sqrt{2}\pi}\sqrt{(1 - \Lambda r)(2 - 3\Lambda r)} \tag{10–46}$$

Denote the angular velocity of the vortex core at $r = 0$ by Ω_o. From Equation 10–46 Ω_o is equal to

$$\Omega_o = K \tag{10–47}$$

Consequently the frequency f_c at the vortex core of $r = 0$ becomes

$$f_c = 2f_o \tag{10–48}$$

where f_o is the frequency for Ω_o. This equation is in agreement with the theory of Yeh.[16] According to these results, the rotating flow in the quasi-forced vortex is in a stable condition.

C. QUASI-FREE VORTEX

In the quasi-free vortex region the distribution of the tangential velocity is

$$V_\theta \cdot r^n = \Gamma_n = \text{constant} \tag{10–49}$$

Substituting Equation 10–49 into Equation 10–39, we obtain the equation of $V'_{\theta\delta}$

$$V'_{\theta\delta} = \left(\frac{r}{r + \delta}\right)\frac{\Gamma_n}{r^n} \tag{10–50}$$

And also the relationship between the pressure gradient and the tangential velocity can be expressed as

$$\frac{dp_\delta}{dr} = \rho\left(\frac{1}{r + \delta}\right) \cdot \left(\frac{r}{r + \delta}\right)^{2n} V_\theta^{2n} \tag{10–51}$$

So substituting Equations 10–50 and 10–51 into Equation 10–42, we obtain

$$\frac{dp^2_\delta}{dt^2} = -2(1 - n)\frac{V_\theta^2}{r^2} \cdot \delta \tag{10–52}$$

Therefore the stable condition in the quasi-free vortex region must satisfy the following condition

$$1 - n > 0 \tag{10–53}$$

This condition is in agreement with the theories of Rayleigh and Prandtl. Then the equation of the frequency of the stable condition is

$$f = \frac{V_\theta}{\pi r}\sqrt{\frac{1 - n}{2}} \tag{10–54}$$

In the unstable condition of $1 - n < 0$, the equation is

$$\frac{d^2\delta}{dt^2} = 2(n - 1)\frac{\Gamma_n^2}{r^{2(1+n)}}\delta = \xi^2 \cdot \delta \tag{10–55}$$

The solution of this equation is

$$\delta = C_1 \sinh\xi t + C_2 \cosh\xi t \tag{10–56}$$

where C_1 and C_2 are constants. Consequently when the lapse of time t approaches infinity, the displacement δ is also increased. Therefore this condition is called an unstable condition. This condition is in agreement with the theory of Prandtl.

D. MEASUREMENT OF THE VIBRATION IN THE VORTEX CHAMBER

In order to compare the theoretical analysis with the experimental results of the vortex vibration in the vortex chamber, a returned flow type of a vortex chamber made of transparent plastic was used as shown in Figure 10–18. The diameter of the vortex chamber was D_1 = 150 mm, that of the inner pipe was D_2 = 50 mm, and the imaginary cylindrical length was H_i = 365 mm. The measurements of the tangential and axial velocities and also of the static pressure for V_o = 5.0, 10.0, 15.0, and 20.0 m/s at z = 30 mm beneath the edge of the inner pipe were done by a cylindrical pitot tube. While the measurements of the frequency of the vibration on the turbulent rotational flow were done at the radii r = 10, 20, 30, 40, 50, 60, and 70 mm by the semiconductor pressure transducer at the bottom of the vortex chamber.

Figure 10–19 shows the distributions of the tangential and axial velocities and of the static pressures. The exponent n of the quasi-free vortex was n = 0.58, 0.68, 0.76, and 0.80 for V_o = 5.0, 10.0, 15.0, and 20.0 m/s, respectively. Then the exponent increases

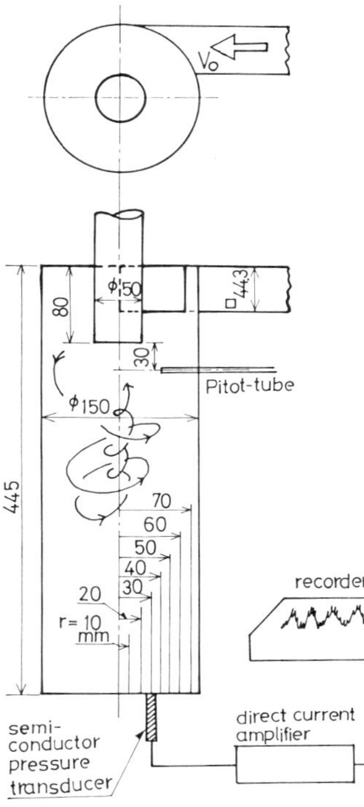

FIGURE 10–18. The main sizes of the vortex chamber.

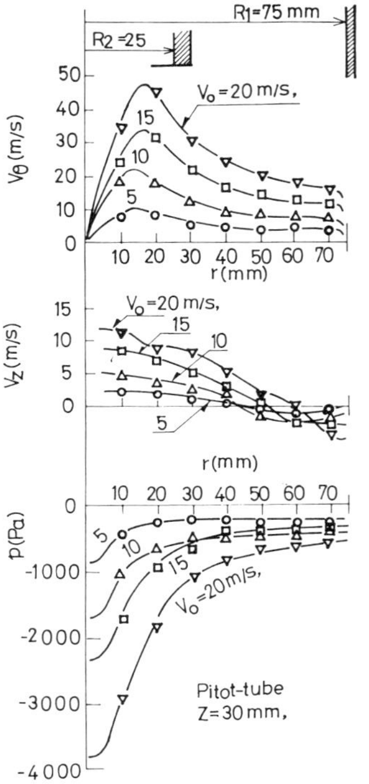

FIGURE 10–19. Distributions of the tangential and axial velocities and of the static pressure.

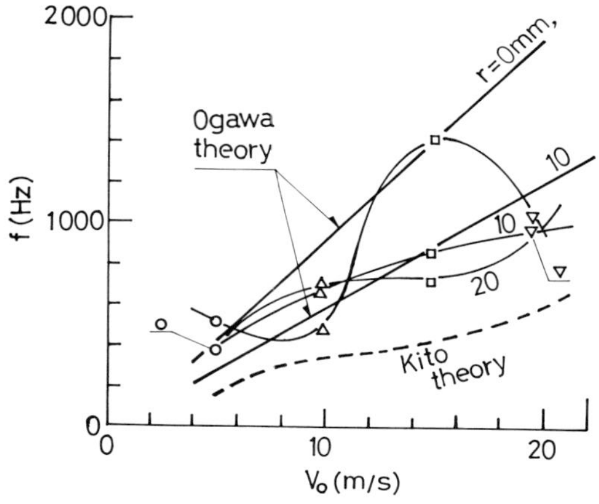

FIGURE 10–20. Relationship between the frequency and the mean inlet velocity of the vortex chamber.

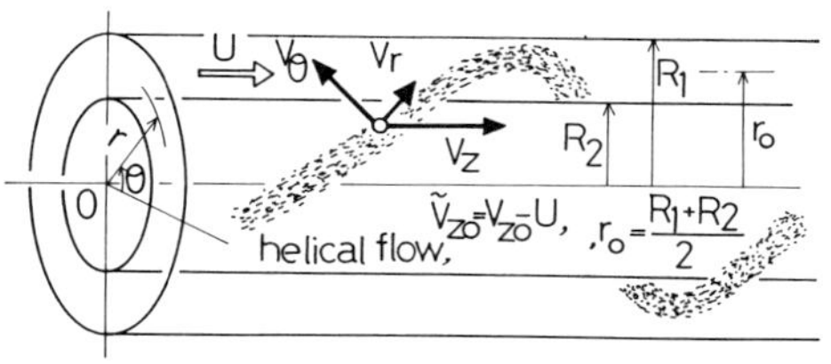

FIGURE 10–21. Helical flow and the main symbols of the coaxial cylinders.

little by little with increasing Reynolds number. Since the inertia force dominates the viscous force with increasing the Reynolds number, so the tangential velocity profile becomes a free vortex (n = 1.0). Also the radii a corresponding to the maximum tangential velocities are located nearly at $a \doteqdot (2/3)R_2$, and also this value is nearly independent of the Reynolds number. In addition to this, in the quasi-forced vortex region the depression of the static pressure is remarkable. Therefore 80% of the total energy dissipation of the turbulent rotational flow is generated in the quasi-forced vortex region, and also 15% is generated in the quasi-free vortex region and by the wall friction on the concave wall surface in the vortex chamber.

Figure 10–20 shows the relationship between the calculated results of the frequency of the vibration and the experimental results of that for r = 0, 10, and 20 mm, respectively. The numerical results in the quasi-forced vortex region coincide with the experimental results.[17] In this figure the dotted line is the theoretical value, which was derived by Prof. F. Kito for the application the exhaust pipe of the Kaplan turbine.[18]

VII. STABILITY OF ROTATING FLOW IN THE COAXIAL CYLINDERS BY H. LUDWIEG

A. EULER'S DIFFERENTIAL EQUATIONS

Euler's differential equation for cylindrical coordinates can be written as shown in Figure 10–21[19]

$$\frac{\partial w_r}{\partial t} + w_r\frac{\partial w_r}{\partial r} + w_\theta\frac{\partial w_r}{r\partial\theta} - \frac{w_\theta^2}{r} + w_z\frac{\partial w_r}{\partial z} = -\frac{1}{\rho}\frac{\partial p'}{\partial r} \tag{10–57}$$

$$\frac{\partial w_\theta}{\partial t} + w_r\frac{\partial w_\theta}{\partial r} + w_\theta\frac{\partial w_\theta}{r\partial\theta} + \frac{w_r w_\theta}{r} + w_z\frac{\partial w_\theta}{\partial z} = -\frac{1}{\rho}\frac{\partial p'}{r\partial\theta} \tag{10–58}$$

$$\frac{\partial w_z}{\partial t} + w_r\frac{\partial w_z}{\partial r} + w_\theta\frac{\partial w_z}{r\partial\theta} + w_z\frac{\partial w_z}{\partial z} = -\frac{1}{\rho}\frac{\partial p'}{\partial z} \tag{10–59}$$

and the continuity equation is

$$\frac{\partial w_r}{\partial r} + \frac{w_r}{r} + \frac{\partial w_\theta}{r\partial\theta} + \frac{\partial w_z}{\partial z} = 0 \tag{10–60}$$

Assuming that the velocity components are composed of the basic flow with the small disturbed flow of v_r, v_θ, v_z in the narrow gap between the coaxial cylinders, the velocity components and the static pressure are

$$w_r = v_r(r, \theta, z, t) \tag{10–61}$$

$$w_\theta = V_\theta(r) + v_\theta(r, \theta, z, t) \tag{10–62}$$

$$w_z = V_z(r) + v_z(r, \theta, z, t) \tag{10–63}$$

$$p' = P(r) + p(r, \theta, z, t) \tag{10–64}$$

Substituting Equations 10–61 to 10–64 into Equations 10–57 to 10–60, and neglecting the higher order terms of v_r, v_θ, and v_z, the following equations are obtained

$$\frac{\partial v_r}{\partial t} + V_\theta\frac{\partial v_r}{r\partial\theta} - \frac{2V_\theta}{r}v_\theta + V_z\frac{\partial v_r}{\partial z} = -\frac{1}{\rho}\frac{\partial p}{\partial r} \tag{10–65}$$

$$\frac{\partial v_\theta}{\partial t} + \frac{\partial V_\theta}{\partial r}v_r + V_\theta\frac{\partial v_\theta}{r\partial\theta} + \frac{V_\theta}{r}v_r + V_z\frac{\partial v_\theta}{\partial z} = -\frac{1}{\rho}\frac{\partial p}{r\partial\theta} \tag{10–66}$$

$$\frac{\partial v_z}{\partial t} + \frac{\partial V_z}{\partial r}v_r + V_\theta\frac{\partial v_z}{r\partial\theta} + V_z\frac{\partial v_z}{\partial z} = -\frac{1}{\rho}\frac{\partial p}{\partial z} \tag{10–67}$$

$$\frac{\partial v_r}{\partial r} + \frac{v_r}{r} + \frac{\partial v_\theta}{r\partial\theta} + \frac{\partial v_z}{\partial z} = 0 \tag{10–68}$$

where

$$\rho\frac{V_\theta^2}{r} = -\frac{\partial P}{\partial r} \tag{10–69}$$

For the small disturbances, we assume the following equations

$$v_r = \hat{v}_r(r) \exp[i(\alpha z + \gamma\theta - \beta t)] \tag{10–70}$$

$$v_\theta = \hat{v}_\theta(r) \exp[i(\alpha z + \gamma\theta - \beta t)] \tag{10–71}$$

$$v_z = \hat{v}_z(r) \exp[i(\alpha z + \gamma\theta - \beta t)] \tag{10–72}$$

$$p = \hat{p}(r) \exp[i(\alpha z + \gamma\theta - \beta t)] \tag{10–73}$$

where α and γ are real numbers, and $\beta(\beta = \beta_r + i\beta_i)$ is a complex number. Substituting Equations 10–70 to 10–73 into Equations 10–65 to 10–68 we obtain

$$\left(-\beta i + V_\theta \frac{\gamma}{r} i + V_z \alpha i\right)\hat{v}_r - \frac{2V_\theta}{r}\hat{v}_\theta = -\frac{1}{\rho}\frac{\hat{p}}{r} \tag{10–74}$$

$$\left(-\beta i + V_\theta \frac{\gamma}{r} i + V_z \alpha i\right)\hat{v}_\theta + \left(\frac{dV_\theta}{dr} + \frac{V_\theta}{r}\right)\hat{v}_r = -\frac{1}{\rho}\frac{\gamma}{r} i\hat{p} \tag{10–75}$$

$$\left(-\beta i + V_\theta \frac{\gamma}{r} i + V_z \alpha i\right)\hat{v}_z + \frac{dV_z}{dr}\hat{v}_r = -\frac{1}{\rho}\alpha i\hat{p} \tag{10–76}$$

and the continuity equation

$$\frac{d\hat{v}_r}{dr} + \frac{\hat{v}_r}{r} + \frac{\gamma}{r} i\hat{v}_\theta + \alpha i\hat{v}_z = 0 \tag{10–77}$$

Here we consider the stationary ring form vortex in the basic flow without z-component in the narrow gap; so, we can make the following assumption

$$V_z = 0 \quad , \quad dV_z/dr = 0 \quad , \quad \gamma = 0 \quad , \quad \beta_r = 0 \tag{10–78}$$

In addition to this the gap in the coaxial cylinders is very narrow ($\Delta r/r_o << 1$). So we may neglect $\hat{v}_r/r$ in the continuity equation and also we may put the following approximate quantities as

$$V_\theta \fallingdotseq V_{\theta o} \quad , \quad r \fallingdotseq r_o \quad , \quad dV_\theta/dr \fallingdotseq C_\theta \quad , \quad dV_z/dr \fallingdotseq C_z \tag{10–79}$$

Therefore Equations 10–74 to 10–77 can be transformed to

$$\beta_i\hat{v}_r - \frac{2V_{\theta o}}{r_o}\hat{v}_\theta = -\frac{1}{\rho}\frac{d\hat{p}}{dr} \tag{10–80}$$

$$\beta_i\hat{v}_\theta + \left(C_\theta + \frac{V_{\theta o}}{r_o}\right)\hat{v}_r = 0 \tag{10–81}$$

$$\beta_i\hat{v}_z = -\frac{1}{\rho}\alpha i\hat{p} \tag{10–82}$$

$$\frac{d\hat{v}_r}{dr} + \alpha i\hat{v}_z = 0 \tag{10–83}$$

Eliminating the components $\hat{v}_\theta$, $\hat{v}_r$, and $\hat{p}$, we obtain the equation for $\hat{v}_r$

$$\frac{d^2\hat{v}_r}{dr^2} - \alpha^2\left[1 + \frac{2}{\beta_i^2}\cdot\frac{V_{\theta o}}{r_o}\left(C_\theta + \frac{V_{\theta o}}{r_o}\right)\right]\hat{v}_r = 0 \qquad (10\text{–}84)$$

Consequently when the value in the parentheses of the second term on the left-hand side becomes negative a value, the solution of Equation 10–84 can be obtained as the sine and cosine curves (sinusoidal waves) as follows

$$\begin{aligned}\hat{v}_r = {} & A \cos\alpha\sqrt{1 + \frac{2}{\beta_i^2}\cdot\frac{V_{\theta o}}{r_o}\left(C_\theta + \frac{V_{\theta o}}{r_o}\right)}\, r + \\ & + B \sin\alpha\sqrt{1 + \frac{2}{\beta_i^2}\cdot\frac{V_{\theta o}}{r_o}\left(C_\theta + \frac{V_{\theta o}}{r_o}\right)}\, r\end{aligned} \qquad (10\text{–}85)$$

In order to keep the stable condition in the narrow gap, the value in the parentheses of the second term on the left-hand side becomes positive

$$\frac{V_{\theta o}}{r_o}\left(C_\theta + \frac{V_{\theta o}}{r_o}\right) = \frac{V_{\theta o}}{r_o}\left(\frac{dV_{\theta o}}{dr_o} + \frac{V_{\theta o}}{r_o}\right) = \frac{V_{\theta o}}{r_o^2}\frac{d}{dr_o}(V_{\theta o}r_o) > 0 \qquad (10\text{–}86)$$

B. STABILITY OF THE HELICAL FLOW

Next we consider the stability of the helical flow in the narrow gap in the coaxial cylinders. Now let us imagine the flow in the moving coordinate system with the velocity U to the z-axis and also let us assume that the flow is an axis-symmetry and the symbols of the velocity components are as follows:

$$V_{\theta o} \quad , \quad \tilde{V}_{zo} = V_{zo} - U \quad , \quad C_\theta \quad , \quad C_z \qquad (10\text{–}87)$$

Here we may select the velocity U which is independent of radius r. When the stream-surface which is constituted by the same stream lines moves to the perpendicular direction for the cylinder surface, the following condition is obtained

$$\left(\frac{V_{zo} - U}{V_{\theta o}}\right) 2\pi r_o = \left(\frac{V_{zo} - U + C_z\cdot\Delta r}{V_{\theta o} + C_\theta\cdot\Delta r}\right) 2\pi(r_o + \Delta r) \qquad (10\text{–}88)$$

Neglecting the higher term of $(\Delta r)^2$, Equation 10–88 becomes

$$\frac{U}{V_{\theta o}} = \frac{V_{zo}}{V_{\theta o}} + \frac{(C_z r_o/V_{\theta o})}{1 - (C_\theta r_o/V_{\theta o})} \qquad (10\text{–}89)$$

$$\tan\delta = \frac{V_{zo} - U}{V_{\theta o}} = \frac{\tilde{V}_{zo}}{V_{\theta o}} = -\frac{(C_z r_o/V_{\theta o})}{1 - (C_\theta r_o/V_{\theta o})} \qquad (10\text{–}90)$$

where δ is the slope angle of the helical stream line on the radius r_o in the relative system with the velocity U. Then we may apply the stability criterion of Rayleigh in the narrow gap in the coaxial cylinders.

We introduce the following quantities

$$r_{oeff} = \frac{r_o}{\cos^2\delta} \qquad (10\text{–}91)$$

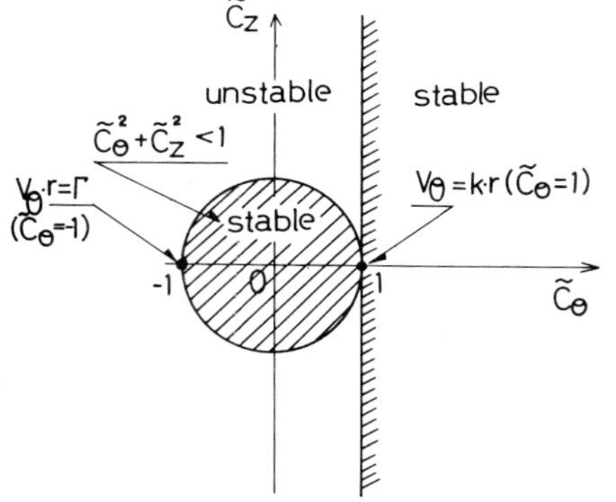

FIGURE 10–22. The domains of the stable and unstable conditions by the theory of Ludwieg.

$$V_{\theta oeff} = \sqrt{V_{\theta o}^2 + V_{zo}^2} = \frac{V_{\theta o}}{\cos\delta} \tag{10–92}$$

$$C_{\theta eff} = C_\theta \cos\delta + C_z \sin\delta \tag{10–93}$$

Substituting these quantities into Equation 10–90, the following equation is derived

$$\frac{C_{\theta eff} \cdot r_{oeff}}{V_{\theta oeff}} = \frac{r_o(C_\theta + C_z \cdot \tan\delta)}{V_{\theta o}} > -1 \tag{10–94}$$

Further transforming Equation 10–94, the stable criterion is obtained

$$\frac{C_\theta r_o}{V_{\theta o}} - \frac{(C_z r_o/V_{\theta o})^2}{1 - (C_\theta r_o/V_{\theta o})} > -1 \tag{10–95}$$

Here in order to make clear the stability criterion of Equation 10–95, we introduce the dimensionless variables

$$\tilde{C}_\theta = \frac{C_\theta r_o}{V_{\theta o}} \quad , \quad \tilde{C}_z = \frac{C_z r_o}{V_{\theta o}} \tag{10–96}$$

Then Equation 10–95 can be transformed

$$\tilde{C}_\theta - \frac{\tilde{C}_z^2}{1 - \tau_\theta} > -1 \tag{10–97}$$

The stable and unstable regions are shown in Figure 10–22 of the $\tilde{C}_\theta$, $\tilde{C}_z$ planes. In this figure a point ($\tilde{C}_\theta = -1$, $\tilde{C}_z = 0$) is corresponding to the free vortex (potential flow) and a point ($\tilde{C}_\theta = 1$, $\tilde{C}_z = 0$) is corresponding to the forced vortex.

Moreover it may be possible to apply the criterion of the stability of Equation 10–95 into the stability problem on the wider gap in the coaxial cylinders as follows:

$$\frac{dV_\theta}{dr}\frac{r}{V_\theta} - \frac{(dV_z/dr)^2(r/V_\theta)^2}{1 - (dV_\theta/dr)(r/V_\theta)} > -1 \tag{10–98}$$

R. S. Scorer[20] has investigated the stability criterion in curved flow based upon the fundamental conception of Richardson.

MAIN SYMBOLS

a(m);	radius of the maximum tangential velocity
f(Hz);	frequency of the vortex
l(m);	small length
n(1);	exponent of the quasi-free vortex
R_1(m);	radius of the outer cylinder
R_2(m);	radius of the inner cylinder
R_{e2}(1);	Reynolds number
R_{et}(1);	Reynolds number
R_{ec}(1);	Reynolds number
S(1);	velocity ratio
T_2(s);	spin-up time
U(m/s);	mean velocity to the z-axis
V_e(m/s);	velocity at the brink of the boundary layer
$V_{\theta 2}$(m/s);	tangential velocity of the inner cylinder
V_θ(m/s);	tangential velocity
V_r(m/s);	radial velocity
V_z(m/s);	axial velocity
V_o(m/s);	mean inlet velocity in the inlet pipe
V_∞(m/s);	free stream velocity
$\tilde{X}_c$(m);	starting point of the transition region
Γ(m^2);	circulation
δ(m);	thickness of the boundary layer and the small radial distance
λ(m);	wavelength
τ_t(Pa);	Reynolds stress
ν(m^2/s);	kinematic viscosity
Ω_1(rad/s);	angular velocity of the outer cylinder
Ω_2(rad/s);	angular velocity of the inner cylinder
ω(rad/s);	angular velocity of the rotating cone

REFERENCES

1. **Cognet, G.,** 3eme Rencontre sur les Ecoulements à Vortex de Taylor, 5-7, Avril (1983), Institut Naitonal Polytechnique de Lorraine, Universite de Nancy, France.
2. **Kobayashi, R., Kohama, Y., Arai, R., and Ukaku, M.,** The instability and transition of a boundary layer on a rotating cone (4th Report)(in Japanese), *Trans. J. Soc. Mech. Eng.,* Paper No. 86-0082A, 52(484), 3912, 1986.
3. **Taylor, G. I.,** Stability of a viscous liquid contained between two rotating cylinders, *Phil. Trans. R. Soc.,* A-223, 1923.
4. **Frêne, J. and Godet, M.,** La portance et le couple ræsistant d'un palier lisse fonctionnant en régime non laminaire, Mécanique, Matériaux, Electricité, *GAMI,* No. 274 (Oct.), 18, 1972.
5. **Euteneuer, G. A.,** Einige Ergebnisse experimenteller Untersuchungen an instätionären Görtler-Tayor Wirbeln, *Z. A. M. M.,* 50, T-177, 1970.
6. **Euteneuer, G. A., Heynatz, J. T., and Siedenkersting, H.,** Der Bodeneinfluß bei Anlauf-bei Bremsströmungen zäher Flüssigkeiten in rotierenden Zylindergefäßen, *Z. A. M. M.,* 48, T190, 1968.
7. **Ogawa, A., Fujita, Y., and Nagabayashi, H.,** Development and collapse of Taylor vortex on the rotating inner cylinder in the coaxial rest outer cylinder, *Chem. Eng. Comm.,* 37, 209, 1985.
8. **Ogawa, A.,** Time dependent phenomena of development processes of Taylor-vortices in the co-axial cylinders, *Nihon Univ. J. Coll. Eng.,* 26-A, (March), 117, 1985.
9. **Ogawa, A., Fujita, Y., and Nagabayashi, H.,** Experimental study of Taylor-Görtler vortex (characteristic time up to the break down of Taylor vortices), Flow Visualization and Digital Image Processing, VKIFD (Belgium) June 9–13, 1986.
10. **Schlichting, H.,** *Boundary Layer Theory,* 7th ed., McGraw-Hill, New York, 1979.
11. **Tani, I.,** Production of longitudinal vortices in the boundary layer along a concave wall, *J. Geophys. Res.,* 67(8), 3075, 1962.
12. **Görtler, H.,** Dreidimensionales Stabilitätstheorie zur laminarer Grenzschichten, *Z. A. M. M.,* 35, 362, 1955.
13. **Schultz-Grunow, F.,** Exakte Zugänge zu hydrohynamischen Problemen, *Z. Flugwiss.,* 23(5), 175, 1975.
14. **Lord Rayleigh,** Scientific Papers, Vol. VI (1911–1919), On the dynamics of revolving fluids, 447–453.
15. **Tollmien, W., Schlichting, H., Görtler, H., and Riegels, F. W.,** *Gesammelte Abhandlungen, Ludwig Prandtl, Zweiter Teil,* Springer-Verlag, 1961, 778.
16. **Traugott, S. C.,** Influence of solid-body rotation on screen-produced turbulence, *NACA*-TN-4135, August 1958.
17. **Ogawa, A., Hagiwara, T., and Fujita, Y.,** Collected papers of J. S. M. E., No. 761-2 (June), 45, 1976
18. **Kito, F. and Uchimaru, S.,** *J. Inst. Electrical Eng.,* No. 503, 713, 1930.
19. **Ludwieg, H.,** Stabilität der Strömung in einem zylindrischen Ringraum, *Z. Flugwiss.,* 8(5), 135, 1960.
20. **Scorer, R. S.,** Local instability in curved flow, *J. Inst. Math. Appl.,* 3, 250, 1967.

Chapter 11

VORTEX MODELS

I. INTRODUCTION

In this chapter the author introduces the three kinds of vortex models which are very interesting for the application to engineering. However there are many assumptions in these vortex models and one should take care to apply them in conditions that are restricted by the assumptions in these models.

II. BURGERS VORTEX MODEL

Let us assume that the fluid flow is the steady state, has axis-symmetry, and also the axial gradient (d/dz) of the physical quantities is small. So the Reynolds equation can be described as follows:

$$V_r \frac{\partial V_r}{\partial r} - \frac{V_\theta^2}{r} = -\frac{1}{\rho}\frac{\partial p}{\partial r} - \frac{\partial \overline{v_r^2}}{\partial r} - \frac{\overline{v_r^2}}{r} + \frac{\overline{v_\theta^2}}{r} \tag{11–1}$$

$$V_r \frac{\partial V_\theta}{\partial r} + \frac{V_r \cdot V_\theta}{r} = -\frac{\partial(\overline{v_r v_\theta})}{\partial r} - \frac{2(\overline{v_r v_\theta})}{r} \tag{11–2}$$

$$V_r \frac{\partial V_z}{\partial r} = -\frac{\partial(\overline{v_r v_\theta})}{\partial r} - \frac{(\overline{v_r v_\theta})}{r} \tag{11–3}$$

Here a term of the Reynolds stress $(-\overline{v_r v_\theta})$ can be changed into the following forms

$$-\overline{v_r v_\theta} = \nu_t\left(\frac{dV_\theta}{dr} - \frac{V_\theta}{r}\right) \tag{11–4}$$

where ν_t means the kinematic eddy viscosity. Therefore substituting Equation 11–4 into Equation 11–2, Equation 11–2 can be transformed to

$$V_r \frac{\partial V_\theta}{\partial r} + \frac{V_r V_\theta}{r} = \nu_t\left(\frac{d^2 V_\theta}{dr^2} + \frac{1}{r}\frac{dV_\theta}{dr} - \frac{V_\theta}{r^2}\right) \tag{11–5}$$

From the practical point of view, the kinematic eddy viscosity is not constant in the radial direction, but it is a function of the radial, axial, and tangential directions depending upon the intensity of turbulence. For example, according to the results by Ustimenko et al., the value of ν_t/ν in the quasi-forced vortex region reaches $\nu_t/\nu = 1000 \sim 10{,}000$, while that in the quasi-free vortex region reaches $\nu_t/\nu = 100 \sim 1000$, respectively. However in order to simplify the analysis, the value of ν_t/ν may be assumed to be constant. Then transforming Equation 11–5, we obtain

$$\frac{V_r}{r}\frac{d}{dr}(r\,V_\theta) = \nu_t \frac{d}{dr}\left[\frac{1}{r}\cdot\frac{d}{dr}(r\,V_\theta)\right] \tag{11–6}$$

This equation can be integrated

$$\frac{d}{dr}(r \cdot V_\theta) = C_1 \cdot r \cdot \exp\left(\int \frac{V_r}{\nu_t} dr\right) \qquad (11\text{–}7)$$

where C_1 is a constant of integration. Further integrating Equation 11–7, we obtain

$$V_\theta = \frac{C_1}{r} \int r \cdot \exp\left(\frac{V_r}{\nu_t} dr\right) \cdot dr + \frac{C_2}{r} \qquad (11\text{–}8)$$

where C_2 is a constant of integration. In this equation in order to take a finite value of V_θ at $r = 0$, a constant of integration C_2 must vanish. Further in order to transform Equation 11–8 into the dimensionless form, let us introduce $V_\theta = V_{\theta b}$ at the radius $r = b$, so we can obtain the following equation

$$V'_\theta = \frac{V_\theta}{V_{\theta b}} = \frac{\int_0^{r'} r' \cdot \exp\left(\int_0^{r'} \frac{V_r b}{\nu_t} dr'\right) \cdot dr'}{r' \int_0^1 r' \cdot \exp\left(\int_0^{r'} \frac{V_r b}{\nu_t} dr'\right) \cdot dr'} \qquad (11\text{–}9)$$

where a dimensionless radius r' is $r' = r/b$. The distribution of radial velocity V_r is a function of the radius r; so, we can obtain the distribution of the tangential velocity. Then from the continuity equation of the axis-symmetry we obtain

$$\frac{1}{r} \cdot \frac{\partial(r\, V_r)}{\partial r} + \frac{\partial V_z}{\partial z} = 0 \qquad (11\text{–}10)$$

Integrating the above equation,

$$r \cdot V_r = -\int_0^T \left(\frac{\partial V_z}{\partial z}\right) \cdot r \cdot dr \qquad (11\text{–}11)$$

When we assume that the radial velocity only depends on the radius r, the axial velocity V_z must be a linear function of z. Therefore the axial velocity is

$$V_z = f_o + f(r) \cdot z \qquad (11\text{–}12)$$

Substituting Equation 11–12 into Equation 11–11, Equation 11–11 becomes

$$r \cdot V_r = -\int_0^r f(r) \cdot r \cdot dr \qquad (11\text{–}13)$$

Again assuming that a function f(r) is to be a constant value Λ, Equation 11–13 can be written as

$$r \cdot V_r = -\frac{\Lambda}{2} \cdot r^2 \qquad (11\text{–}14)$$

And also introducing a boundary condition of $V_r = V_{rb}$ at $r = b$, Equation 11–14 can be transformed to

$$V'_r = \frac{V_r}{V_{rb}} = \frac{r}{b} = r' \tag{11–15}$$

Substituting Equation 11–15 into Equation 11–9 and after that integrating once more, we obtain the following distributions of the dimensionless tangential velocities V'_θ which depends on the Reynolds number R_{eb} defined as

$$R_{eb} = -\frac{b \cdot V_{rb}}{\nu_t} \tag{11–16}$$

$$V'_\theta = \frac{1}{r'} \cdot \frac{1 - \exp(-R_{eb} \cdot r'^2/2)}{1 - \exp(-R_{eb}/2)} \tag{11–17}$$

To the positive value of R_{eb}, we must add a negative sign to Equation 11–16, since the radial velocity V_{rb} is the inward direction (negative value). In this equation, when the Reynolds number R_{eb} becomes large, that is to say, when the inertia force acting on the fluid particle is stronger than the turbulent viscous stress, the tangential velocity distribution gradually approaches the free vortex or the potential flow. For example when R_{eb} reaches the higher value, the value of $\exp(-R_{eb} \cdot r'^2/2)$ approaches zero. So Equation 11–17 can be transformed to

$$V'_\theta = \frac{1}{r'} \tag{11–18}$$

On the other hand when the Reynolds number R_{eb} reaches the lower value, the term $\exp(-R_{eb} \cdot r'^2/2)$ can be transformed to

$$\exp\left(-\frac{R_{eb} \cdot r'^2}{2}\right) \fallingdotseq 1 - \frac{R_{eb}}{2} r'^2 \tag{11–19}$$

Therefore Equation 11–17 can be transformed to

$$V'_\theta = \frac{R_{eb}}{2[1 - \exp(-R_{eb}/2)]} \cdot r' \tag{11–20}$$

This equation shows the distribution of the forced vortex which is proportional to the radius r. This equation 11–17 was first derived by J. M. Burgers. Therefore, this equation can be called the Burgers Vortex Model after his contribution.[1]

In addition to this, we can define the two representative radii r'_a and r'_o, where r'_a is the radius corresponding to the maximum tangential velocity $V'_{\theta M}$, and also r'_o is the radius corresponding to 99% of V_θ of the free vortex ($V'_\theta = 1/r'$) as shown in Figure 11–1. These radii can be represented as follows:

$$r'_a = \frac{1.59}{\sqrt{R_{eb}}} \qquad r'_o = \frac{3.04}{\sqrt{R_{eb}}} \tag{11–21}$$

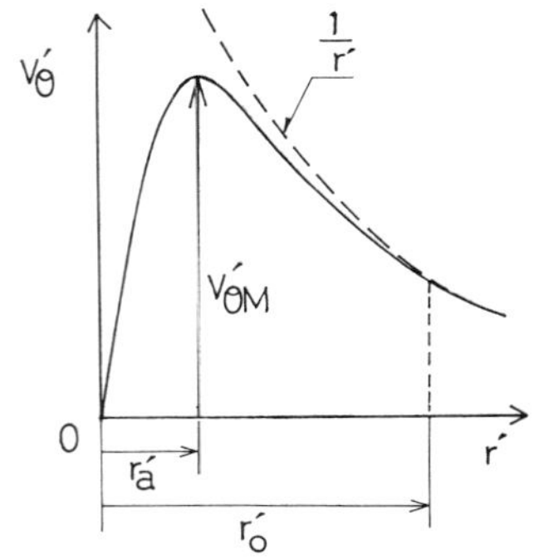

FIGURE 11–1. Schematic illustration of the main symbols of the vortex model.

Therefore, the ratio of r_o' to r_a' is about

$$r_o'/r_a' = 1.94 \tag{11–22}$$

While the distribution of the dimensionless vorticity ζ' can be obtained as

$$\zeta' = \frac{dV_\theta'}{dr'} + \frac{V_\theta'}{r'} = \frac{\text{Reb} \cdot \exp\left(-\frac{R_{eb}}{2} r'^2\right)}{\left[1 - \exp\left(-\frac{R_{eb}}{2}\right)\right]} \tag{11–23}$$

The dimensionless vorticity ζ' at the center ($r' = 0$) is equivalent to the Reynolds number R_{eb}, when the Reynolds number is larger than $R_{eb} \fallingdotseq 10$. These dimensionless distributions of the tangential velocity and of the vorticity for $R_{eb} = 1.0$, 5.0, 10.0, 20.0, 30.0, and 50.0 are shown in Figure 11–2.

III. SULLIVAN VORTEX MODEL

Burgers (one cell vortex) derived an exact solution of the Navier-Stokes equation for three-dimensional vortex flow. In addition to this, R. D. Sullivan[2] derived the closed solution with the two cells during the course of a study of a family of solutions in which V_r and V_θ are functions of the radius r only, and V_z is a function of r multiplied by z. Here the vortex of two cells means that the three-dimensional spiral flow does not simply spiral toward the central axis and out along it as shown in Figure 11–3, but has a region of reverse flow near the central axis as shown in the Sullivan model of Figure 11–3.

Then the solutions of V_r, V_z, and V_θ can be expressed as follows:

$$V_r = -a \cdot r + 6\left(\frac{\nu}{r}\right)\left[1 - \exp\left(-\frac{a \cdot r^2}{2\nu}\right)\right] \tag{11–24}$$

$$V_z = 2 \cdot a \cdot z\left[1 - 3 \cdot \exp\left(-\frac{a \cdot r^2}{2\nu}\right)\right] \tag{11–25}$$

$$V_\theta = \frac{\Gamma}{2\pi r}\left[\frac{H(a \cdot r^2/2\nu)}{H(\infty)}\right] \tag{11–26}$$

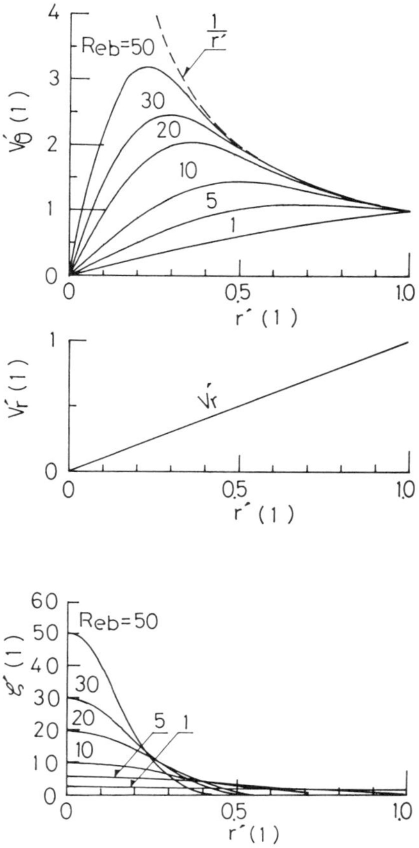

FIGURE 11–2. Distributions of the tangential and radial velocities and of the vorticities by Burgers.

where H is defined as

$$H(x) = \int_0^x \exp\left\{-t + 3\int_0^t \left[\frac{1 - \exp(-\tau)}{\tau}\right] d\tau\right\} dt \tag{11–27}$$

Further the distribution of the static pressure can be written as

$$p(r, z) = p_o - \left(\frac{\rho}{2}\right)\left\{4a^2 \cdot z^2 + a^2 \cdot r^2 + 36\left(\frac{\nu^2}{r^2}\right)\left[1 - \exp\left(-\frac{a \cdot r^2}{2\nu}\right)\right]^2\right\} + \rho\int_0^r \frac{V_\theta^2}{r} \cdot dr \tag{11–28}$$

Here it is noted that the axial gradient of the static pressure can be expressed as

$$\frac{\partial p}{\partial z} = -4\rho a^2 \cdot z \tag{11–29}$$

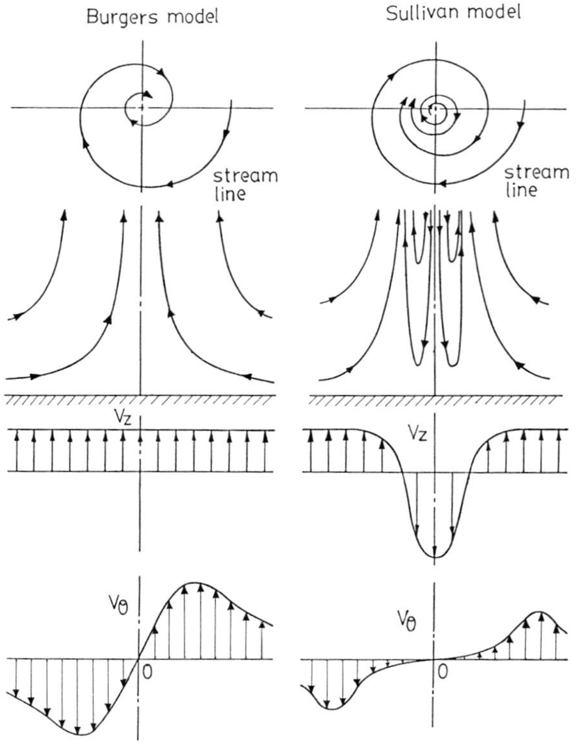

FIGURE 11–3. Comparison of Burgers model with Sullivan model.

Now the numerical results of the stream line, the axial velocity, the radial velocity, and the tangential velocity for the three kinds of Reynolds numbers R_{e1} by Donaldson and Sullivan[3] are shown in Figure 11–4, where R_{e1} is defined as

$$R_{e1} = \frac{V_{r1}R_1}{\nu} \tag{11–30}$$

Here a symbol V_{r1} is the radial velocity of gas flowing through the porous tube of the radius R_1. The positive value of V_{r1} is defined as a condition of flowing toward the outer direction; on the other hand, the negative value of V_{r1} is defined as a condition of flow towards the inner direction.

IV. CHEPURA VORTEX MODEL

Chepura derived the equation of the tangential velocity on the radial blade mixer in the cylindrical container.[4] As a result of the experiment, the distribution of the vorticity

$$\zeta = \frac{1}{r}\frac{d(r \cdot V_\theta)}{dr} = \frac{dV_\theta}{dr} + \frac{V_\theta}{r} \tag{11–31}$$

in the radial region $0 \leq r \leq r_t$ could be represented by a parabolic form as shown in Figure 11–5 as

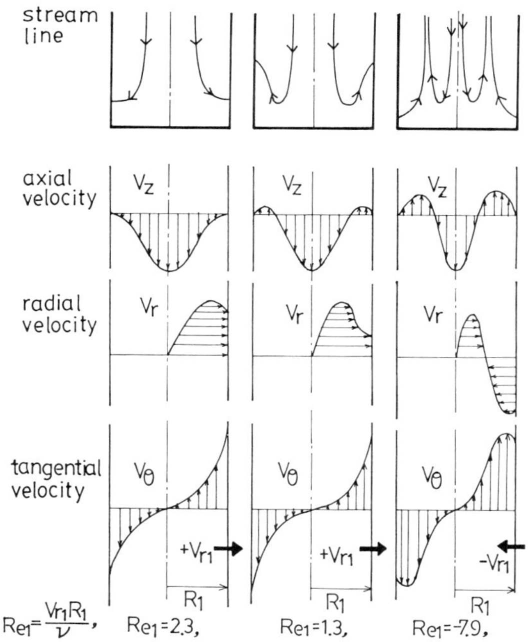

FIGURE 11–4. Schematic illustration of velocity distributions for various Reynolds numbers by Sullivan.

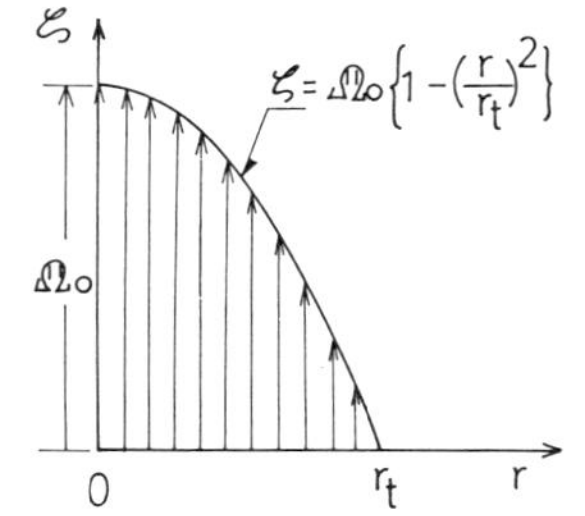

FIGURE 11–5. Vorticity distribution of Chepura model.

$$\zeta = \Omega_o\left[1 - \left(\frac{r}{r_t}\right)^2\right] \tag{11–32}$$

Substituting Equation 11–32 into Equation 11–31, the general solution of Equation 11–31 is obtained:

$$V_\theta = \frac{k \cdot r}{2}\left[2 - \left(\frac{r}{r_t}\right)^2\right] \tag{11–33}$$

While in this mixer, the distribution of the tangential velocity in the radial region $r_t \leqslant r \leqslant R_c$ can be represented by the free vortex

$$V_\theta = \frac{\Gamma}{2\pi r} \tag{11–34}$$

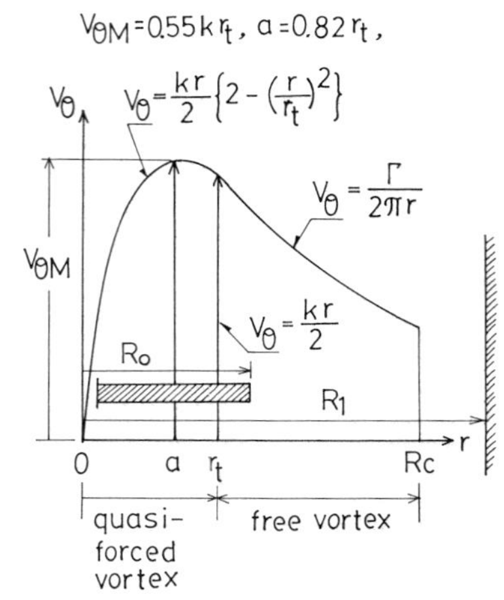

FIGURE 11–6. Schematic illustration of Chepura vortex model.

Then the tangential velocity is equal at the contact radius $r = r_t$. So the following equation is obtained

$$\frac{k \cdot r_t}{2} = \frac{\Gamma}{2\pi r_t} \tag{11–35}$$

The value of Γ is

$$\Gamma = \pi k \cdot r_t^2 \tag{11–36}$$

Therefore Equation 11–34 is

$$V_\theta = \frac{\pi k \cdot r_t^2}{2\pi r} = \frac{k \cdot r_t^2}{2 \cdot r} \tag{11–37}$$

The radius a corresponding to the maximum tangential velocity $V_{\theta M}$ is derived by differentiating Equation 11–33

$$a = \sqrt{2/3}\, r_t = 0.82 \cdot r_t \tag{11–38}$$

$$V_{\theta M} = \frac{2}{3}\sqrt{\frac{2}{3}}\, k \cdot r_t = 0.55 \cdot k \cdot r_t \tag{11–39}$$

The schematic illustration of this vortex model is shown in Figure 11–6.

Here as a result of the experiment, a relationship between a and r_t was obtained by Chepura[5] as follows:

$$a/r_t = 0.745, \quad r_t = 0.37\, D_o \tag{11–40}$$

where $D_o(= 2R_o)$ was the diameter of the rotating blade. In addition to this, he assumed that a constant k was $k \risingdotseq 1.05\, \omega_o$, where ω_o is the angular velocity of the vortex core in

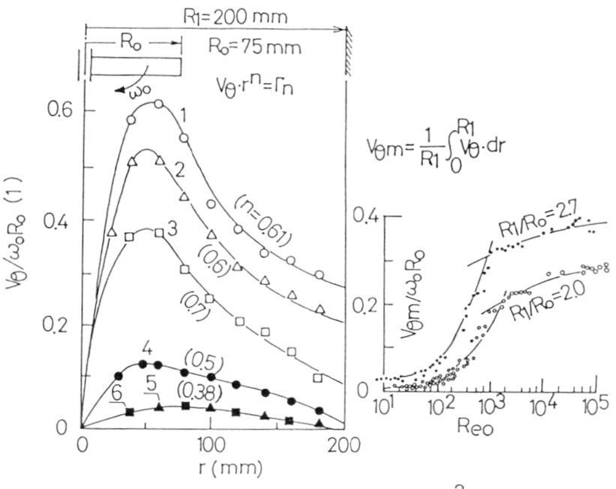

	η (Pa·S)	Reo (1)
1	0.001	45000
2	0.035	1260
3	0.0875	550
4	0.406	129
5	1.715	29
6	8.2	5.9

(ex. water η≒0.001 Pa·s at 20 °C.)

$R_{eo} = \frac{N(2R_o)^2}{\nu}$

N(rev/s), ω_o(rad/s), speed of paddle

FIGURE 11–7. Experimental result of the tangential velocity distributions in the mixer.

the quasi-forced vortex region. So the maximum tangential velocity $V_{\theta M}$ can be estimated by Equation 11–39 as

$$V_{\theta M} = 0.55 \cdot k \cdot r_t = 0.213 \cdot D_o \cdot \omega_o \tag{11–41}$$

In connection with Equation 11–40, a relationship between a and r_t can be estimated by the Ogawa combined vortex model as

$$\frac{a}{r_t} = \frac{2 + n}{2(1 + n)} \tag{11–42}$$

where n is the exponential of the quasi-free vortex. Here assuming that the quasi-free vortex ($n < 1$) becomes the free vortex ($n = 1$), Equation 11–42 becomes

$$a/r_t = 0.75 \tag{11–43}$$

Therefore Equation 11–42 is in good agreement with Equation 11–40.

Figure 11–7 shows the experimental results of the distributions of the tangential velocities for the effects of the various kinds of viscosities in the container of the diameter $D_1 = 2R_1 = 400$ mm with the diameter $D_o = 2R_o = 150$ mm of the rotating blade. It is very interesting to note that the property of the rotating fluid around the quasi-forced vortex flow is not always the free vortex, but it becomes the quasi-free vortex with the exponential n which depends on the Reynolds number R_{eo} defined as

$$R_{eo} = \frac{N \cdot (2R_o)^2}{\nu} \tag{11–44}$$

Further in this figure the experimental results of the mean tangential velocity $V_{\theta m}$ defined as

$$V_{\theta m} = \frac{1}{R_1} \int_0^{R_1} V_\theta \cdot dr \qquad (11\text{–}45)$$

for $R_1/R_o = 2.7$ and $R_1/R_o = 2.0$ are shown in dimensionless form by Blaginskij et al.[6]

MAIN SYMBOLS

a(m);	radius corresponding to the maximum tangential velocity
H(x)(1);	specially defined function
p(Pa);	static pressure
r(m);	radius
r′(1);	dimensionless radius
R_{e1}(1);	Reynolds number
R_{eb}(1);	Reynolds number
R_1(m);	radius of a cylinder
V_r(m/s);	radial velocity
V_θ(m/s);	tangential velocity
$V_{\theta M}$(m/s);	maximum tangential velocity
V_z(m/s);	axial velocity
v_r(m/s);	turbulent radial velocity
v_θ(m/s);	turbulent tangential velocity
v_z(m/s);	turbulent axial velocity
V'_r(1);	dimensionless radial velocity
V'_θ(1);	dimensionless tangential velocity
V'_z(1);	dimensionless axial velocity
z(m);	axial distance
Γ(m^2/s);	circulation
ζ'(1);	dimensionless vorticity component
ν_t(m^2/s);	kinematic eddy viscosity
ρ(kg/m^3);	density
Ω_o(rad/s);	vorticity component at the center ($r = 0$)

REFERENCES

1. **Burgers, J. M.,** *Advances in Applied Mechanics,* Vol. 1, Academic Press, New York, 1948, 171.
2. **Sullivan, R. D.,** A two-celled vortex solution of the Navier-Stokes equations, *J. Aero/Space Sci.,* Read Forum 26(11), 767, 1959.
3. **Donaldson, C. D. and Sullivan, R. D.,** Behaviour of solution of the Navier-Stokes equations for a complete class of three-dimensional viscous vortices, Proc. 1960 Heat Transfer and Fluid Mechanics, 1960, 16.
4. **Chepura, I. B., Solobjeb, A. B., Tumanob, I. U. B., Planobskij, A. N.,** O Tangentsialnoi Sostabliaiushehej Poliu Skolostej B Gladkostennom Sosude C Radiajno-Lopastnoj Meshalokoj, *Teor. Osn. Khim. Teknol.,* 3, 404, 1969.
5. **Chepura, I. B.,** O Nekotoroj Zakonomernosti Izmeneniik S Radialjno-Lopastnymi Meshalkami, *Teor. Osn. Khim. Teknol.,* 5(5), 764, 1971.
6. **Blaginskij, L. N., Begacheb, B. I., Glukhob, B. N., and Bolchkoba, L. N.,** O Bliianii Biazkosti Na Okruzhniik Skoroctj Zhidkosti B Apparatakh S Meshalkami, *Teor. Osn. Khim. Teknol.,* 5(3), 446, 1971.

Chapter 12

TRAILING VORTEX

I. INTRODUCTION

The vortices generated at the edges of the wings of an airplane are called the trailing vortex or the wing tip vortex. This vortex is generated by the intrusion of the stream around the wing tip from the high static pressure on the lower surface of the wing in to the low static pressure on the upper surface due to the static pressure difference between both surfaces of the wing during flight. The schematic illustration is shown in Figure 12–1. The trailing vortex contributes not only a decline of the lift coefficient of the wing but also an uninvited accident of the airplane which is swallowed up by this vortex. One example of an airplane accident in Japan is introduced.

An investigation committee of airplane accidents of the Ministry of Transport of Japan made a statement concerning the causality of the accident of the Cessna airplane which crashed and was destroyed by fire at the Nagoya airport (Japan) at 5:00 p.m. on July 28, 1983. The committee came to a conclusion that the Cessna airplane making the preparation for landing at a height of about 10 m was caused to fall by rolling into the turbulent trailing vortex of a diameter of about 20 m and the swirling velocity $V_\theta = 12$ m/s which was produced at the tip of the wing of a Boeing 727 landing on the landing strip about one minute before.[1] A schematic illustration of this accident is shown in Figure 12–2.

But according to the interesting study of Hitzel,[2] the demands of high speed performance on the modern transport and fighter airplane require slender wings, whose leading edge vortex flow results in increased lift at a high angle of attack. And also Maxworthy described a theory of the aircraft trailing vortices by using the knowledge of the motion of the turbulent vortex ring.[3]

Then, before introducing the theoretical analysis of the trailing vortex, the measured distributions of the tangential velocities of the trailing vortex by Kaufmann[4] is shown in Figure 12–3. He used a wing section of GP-387 with an aspect ratio $\Lambda = 5.0$ and a half wing length $S = 250$ mm. The free stream velocity in the wind tunnel was $V_\infty = 24.6$, 30.2, and 30.4 m/s and the measuring position behind the wing was $Z = 1270$ mm.

From this figure it is evident that there is a vortex core in the center region of the vortex, and around the vortex core there is a quasi-free vortex. Here in the case of $V_\infty = 30.4$ m/s, the radius r_0 corresponding to the maximum tangential velocity was nearly $r_\theta \fallingdotseq 1.3$ cm and the circulation Γ was nearly $\Gamma \fallingdotseq 9400$ cm^2/s. Further the analytical results of the viscous trailing vortex by Newman[5] are shown in Figure 12–4. He derived the theoretical equations of the tangential, radial, and axial velocities of a laminar viscous trailing vortex. The tangential V'_θ and radial V'_R velocities of the dimensionless form can be expressed as

$$V'_\theta = \frac{1}{R}(1 - e^{-R^2}) \tag{12–1}$$

$$V'_R = 2.33\,R \cdot e^{-R^2} \tag{12–2}$$

where the symbols V'_θ and V'_R are defined as $V'_\theta = V_\theta/V_{\theta M}$, $V'_R = V_r/V_R$, and $R = r/r_\theta$.

The radius r_θ corresponding to the maximum tangential velocity $V_{\theta M}$ is defined as

$$r_\theta = 2z(\nu/V_\infty \cdot z)^{1/2}$$
$$V_{\theta M} = (\Gamma/4\pi z)(V_\infty \cdot z/\nu)^{1/2}$$

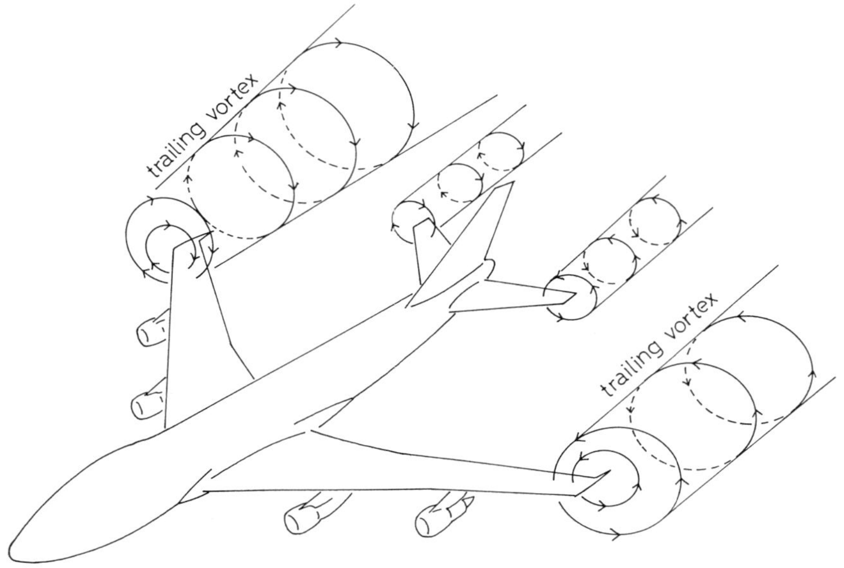

FIGURE 12–1. Schematic illustration of the trailing vortices generating at the edges of the wings.

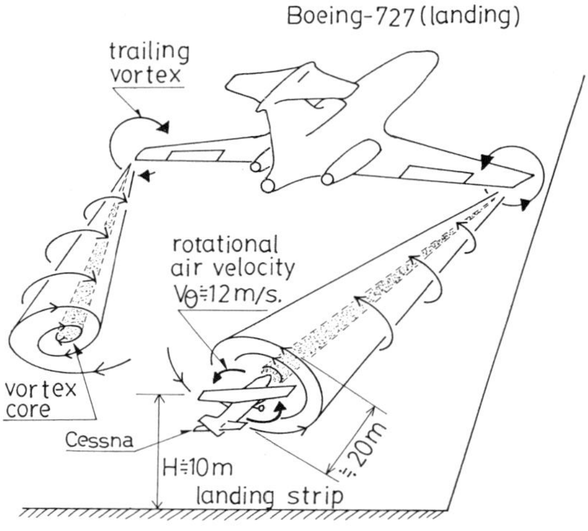

FIGURE 12–2. Cessna accident on the landing strip on July 1983 at Nagoya airport in Japan.

When the value of R^2 is small, V'_θ of Equation 12–1 can be transformed to

$$V'_\theta = \frac{R}{1 + R^2} \tag{12–3}$$

II. THEORETICAL ANALYSIS OF THE TRAILING VORTEX MODEL

A. THEORETICAL ANALYSIS

An experimental investigation of the viscous decay of a steady, three-dimensional, helical vortex is considered by Dosanjh et al.[6] The trailing vortex was generated by a rectangular,

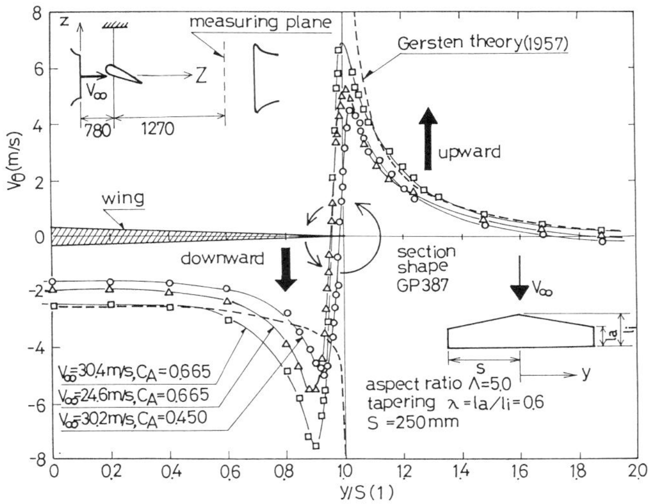

FIGURE 12–3. Distributions of the measured tangential velocities for a wing section of GP-387 by Kaufmann.

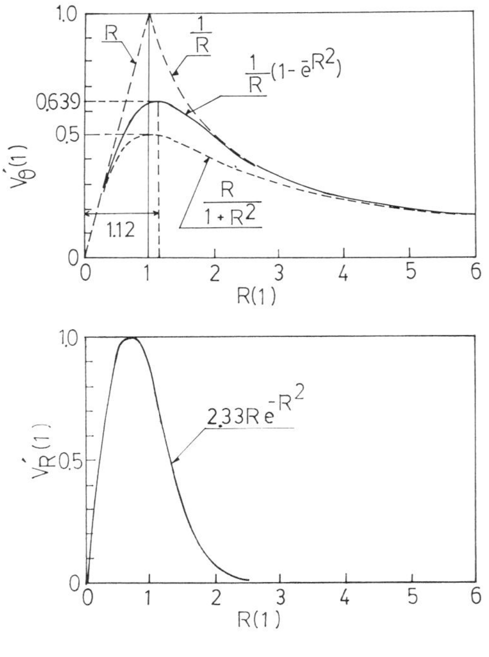

FIGURE 12–4. Distributions of the dimensionless tangential and radial velocities by Newman.

symmetrical half-aerofoil (NACA-0009) of 266.7 mm span, 76.2 mm chord, cantilevered from the wall of the circular test section of a low speed wind tunnel. All tests were performed at a Reynolds number $R_e = CV_\infty/\nu$ based on the free stream velocity and chord length of the aerofoil.

In order to make the mathematical analysis, the flow in the single trailing vortex may be assumed to be axis-symmetric. This assumption is actually satisfied at the distances far downstream of the aerofoil. The symbols used are defined in Figure 12–5.

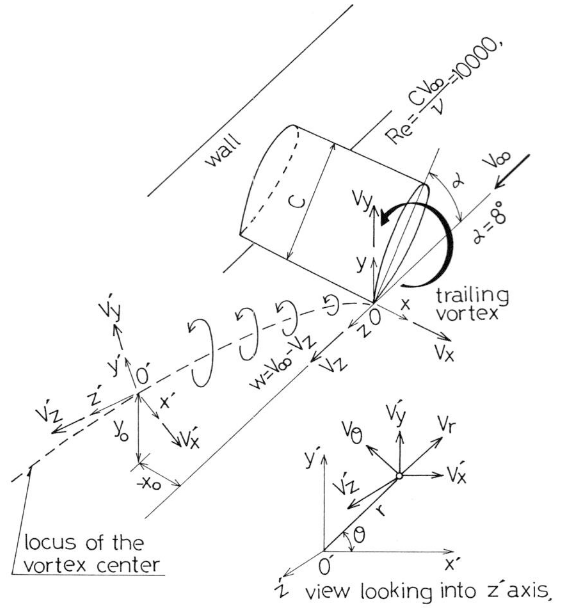

FIGURE 12–5. Schematic illustration of the main symbols.

Newman[5] linearized and solved these equations by introducing the following assumptions:

1. The axial velocity deficiency $W = V_\infty - v_z$ and the peripheral velocity V_θ are small compared with the free stream velocity V_∞.
2. The radial velocity V_r is very small compared with V_∞.
3. The Reynolds number of the main flow defined as $V_\infty \cdot z/\nu$ is large.

Then the Navier-Stokes equation can be reduced into the following forms

$$\frac{V_\theta^2}{r} = \frac{1}{\rho}\frac{\partial p}{\partial r} \tag{12–4}$$

$$V_\infty \cdot \frac{\partial V_\theta}{\partial z} = \nu\left(\frac{\partial^2 V_\theta}{\partial r^2} + \frac{1}{r}\cdot\frac{\partial V_\theta}{\partial r} - \frac{V_\theta}{r^2}\right) \tag{12–5}$$

$$V_\infty \frac{\partial W}{\partial z} = \nu\left(\frac{\partial^2 W}{\partial r^2} + \frac{1}{r}\frac{\partial W}{\partial r}\right) \tag{12–6}$$

$$\frac{1}{r}\cdot\frac{\partial}{\partial r}(rV_r) - \frac{\partial W}{\partial z} = 0 \tag{12–7}$$

The boundary and initial conditions to be satisfied by these linearized equations are $z > 0$, $V_\theta \rightarrow 0$, and $W \rightarrow 0$ as $r \rightarrow \infty$, $z \rightarrow \infty$, $V_\theta \rightarrow 0$, and $W \rightarrow 0$ for all r, $z = 0$, $V_\theta = \Gamma_o/2\pi r$; $W = 0$, where Γ_o is the circulation in the initial vortex (assumed potential) generated at $z = 0$. Under these boundary conditions the solutions are as follows:

$$V_\theta = \frac{\Gamma_o}{2\pi r}\left[1 - \exp\left(-\frac{V_\infty r^2}{4\nu z}\right)\right] \tag{12–8}$$

$$V_r = -\frac{Ar}{2z^2}\cdot\exp\left(-\frac{V_\infty\cdot r^2}{4\nu z}\right) \tag{12–9}$$

$$W = V_\infty - v_z = \frac{A}{z}\exp\left(-\frac{V_\infty\cdot r^2}{4\nu z}\right) \tag{12–10}$$

Here the solution of V_θ which satisfies these boundary conditions is identical with that given by Lamb[7] for the development with lapse of time of a two-dimensional viscous vortex when the time t is replaced by z/V_∞, and also it has been given previously by Squire.[8,9]

The constant A of Equation 12–10 can be determined by the considerations of momentum. The rotational, axial, and radial flows are independent of one another and thus Equations 12–9 and 12–10 represent the distributions of W and V_r for all values of the vortex strength Γ which is circulating around the vortex at large r. Thus if D is the drag force of the body which is producing the vortex, it is apparent that the limit of D as $\Gamma \rightarrow 0$, D_o say, is such that

$$D_o = \int_0^\infty \rho(V_\infty - W)\cdot W\cdot 2\pi r\cdot dr = \int_0^\infty \rho V_\infty\cdot W 2\pi r\cdot dr$$

since W is small in comparison with V_∞. Therefore we can obtain the following relation by integrating the above equation

$$A = D_o/4\pi\rho\nu \tag{12–11}$$

where D_o is the profile drag of the aerofoil. Then the circulation $\Gamma = 2\pi r V_\theta$ of the vortex can be expressed by Equation 12–8 as

$$\Gamma = \Gamma_o\left[1 - \exp\left(-\frac{V_\infty\cdot r^2}{4\nu Z}\right)\right] \tag{12–12}$$

Further in order to determine the radial positions r_θ and r_r corresponding to the maximum tangential and radial velocities, taking the derivatives $\partial V_\theta/\partial r$ and $\partial V_r/\partial r$ and equating them to be zero, the following equations are obtained

$$r_\theta = \sqrt{\frac{1.26\times 4\nu z}{V_\infty}} \fallingdotseq 1.12\sqrt{\frac{4\nu z}{V_\infty}} \tag{12–13}$$

$$r_r = \sqrt{\frac{2\nu z}{V_\infty}} \tag{12–14}$$

The relationship between r_θ and r_r is

$$r_\theta/r_r = 1.58 \tag{12–15}$$

This theoretical value is in good agreement with the experimental values for $1.0 < z/C < 6.0$. Therefore the maximum tangential and radial velocities can be expressed as

$$V_{\theta M} = 0.638 \cdot \frac{\Gamma_o}{2} \cdot \sqrt{\frac{V_\infty}{4\nu z}} \tag{12–16}$$

$$V_{rM} = -\frac{0.303A}{z^2} \cdot \sqrt{\frac{2\nu z}{V_\infty}} \tag{12–17}$$

For comparing the experimental data with the preceding analysis, the dimensionless representations of V_θ and V_r can be expressed as

$$\frac{V_\theta}{V_{\theta M}} = \frac{1.4}{(r/r_0)} \cdot \left\{1 - \exp\left[-1.26 \cdot \left(\frac{r}{r_\theta}\right)^2\right]\right\} \tag{12–18}$$

$$\frac{V_r}{V_{rM}} = 1.65\left(\frac{r}{r_r}\right) \cdot \exp\left[-0.5 \cdot \left(\frac{r}{r_r}\right)^2\right] \tag{12–19}$$

Now to put Equation 12–10 into the dimensionless form, we define a wake width b. When b/2, which is called the half-width, is the distance from its center to where $W = W_M/2$, and if Equation 12–10 is normalized so that $W/W_M = 1/2$ at $r/(b/2) = 1$, Equation 12–10 becomes

$$\frac{W}{W_M} = \exp\left[-0.693\left(\frac{r}{b/2}\right)^2\right] \tag{12–20}$$

where

$$\frac{b}{2} = 2\sqrt{\frac{0.693\nu z}{V_\infty}} \tag{12–21}$$

and

$$W_M = \frac{A}{z} = \frac{D_o}{4\pi\rho\nu z} \tag{12–22}$$

B. VELOCITY DISTRIBUTIONS

1. Basic Form of Symmetrical Wing Section of NACA-0009

Figure 12–6 shows the basic thickness form of NACA-0009 and also the square of the dimensionless velocity V/V_∞ from the leading edge, where V is the velocity along the surface of the wing section.[10] In the experiments of Dosanjh et al. they used the cantilevered NACA-0009 wing of 266.7 mm span and of 76.2 mm chord from the wall of the circular test section as shown in Figure 12–6. All experiments were performed as closely as possible to the uniform free stream velocity $V_\infty \fallingdotseq 2.74$ m/s for giving the Reynolds number $R_e = V_\infty \cdot C/\nu$ of 10,000. Here all measurements taken in the trailing vortex were performed at the angle of attack $\alpha = 8°$.

2. Axial Velocity Deficiency

Figure 12–7 shows the distribution of the dimensionless axial velocity deficiency in the single trailing vortex at the various given position of z/C. In this figure a solid line shows the theoretically predicted value of Equation 12–20 which gives good agreement with the

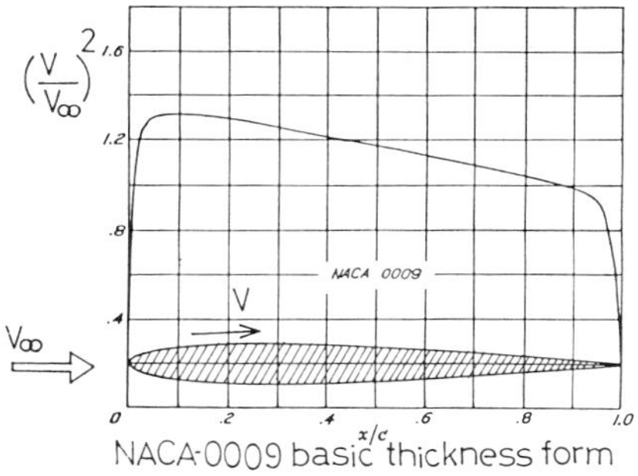

FIGURE 12–6. Wing section of NACA-0009.

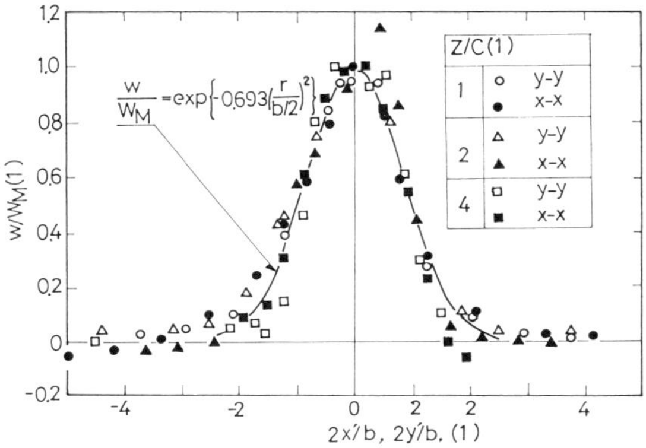

FIGURE 12–7. Distributions of the dimensionless axial velocity deficiency.

measured values of W/W_M for $z/C \leqslant 4.0$. Further they describe that the axial velocity is essentially constant throughout the wake within the accuracy of the experimental measurement for $z/C \geqslant 6.0$.

3. Tangential Velocity

Figure 12–8 shows the distributions of the dimensionless tangential velocity at the axial positions of $z/C = 1.0$, 4.0, and 6.0, respectively. In this figure the solid lines show the theoretically predicted values of Equation 12–18, which is in agreement with the measured values of $V_\theta/V_{\theta M}$ in the region of the quasi-forced vortex flow. But outside of this vortex core, the theoretically predicted value gives a little deviation from the measured value.

4. Radial Velocity

Figure 12–9 shows the distributions of the dimensionless radial velocity in the trailing vortex at $z/C = 1.0$, 4.0, and 6.0. In this figure the inward flow is corresponding to the positive value of V_r/V_{rM}, and also the solid lines show the theoretically predicted value of Equation 12–19. The measured values show the unsymmetrical distribution, since the vortex center is not always completely aligned with the measuring axis, and also the radial velocity component is small in comparison with V_θ and W.

5. Circulation

The circulation Γ in a circular region of the radius r is defined as

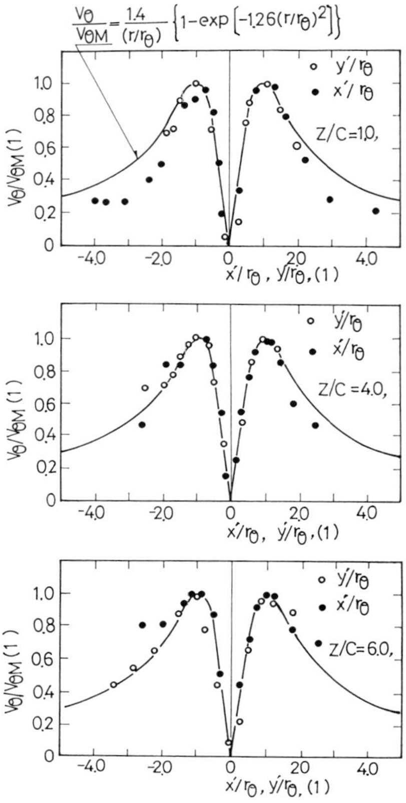

FIGURE 12–8. Distributions of the dimensionless tangential velocity.

$$\Gamma = 2\pi r\, V_\theta \tag{12–23}$$

or the dimensionless circulation can be defined as

$$\frac{\Gamma}{2\pi C V_\infty} = \frac{V_\theta}{V_\infty} \cdot \frac{r}{C} \tag{12–24}$$

Figure 12–10 shows the distributions of the dimensionless circulation at z/C = 1.0, 4.0, and 6.0, respectively.

Substituting Equation 12–13 into Equation 12–16, we obtain

$$\frac{\Gamma_o}{2\pi C V} = 1.397\, \frac{V_{\theta M}}{V_\infty} \cdot \frac{r_\theta}{C} \tag{12–25}$$

For example, from the experimental values of $V_{\theta M}$, V_∞, and r_θ at z/C = 1.0 to 8.67, an average value of $\Gamma_o/2\pi C V_\infty = 282 \times 10^{-4}$ was estimated, and therefore for $V_\infty = 2.73$ m/s and C = 76.2 mm, we obtain the average initial circulation $\Gamma_o = 0.0372\ m^2/s$.

Owen[11] studied the decay of a trailing vortex containing the turbulent vortex core. The region separating the vortex core from the outer irrotational flow (free vortex), in which the interaction between the turbulence and the lapse of time like $(time)^{0.5}$, is a characteristic of

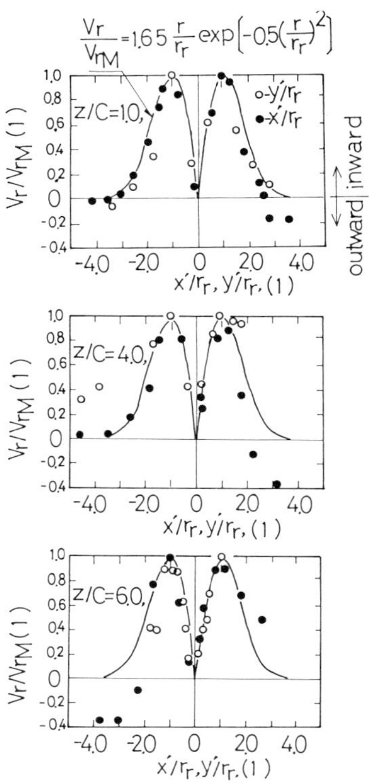

FIGURE 12–9. Distributions of the dimensionless radial velocity.

viscous diffusion. An important practical conclusion is that the maximum tangential velocity $V_{\theta M}$ in the trailing vortex generated by the wing of an airplane in flight can be represented by

$$V_{\theta M} = 0.065(\Gamma/\nu)^{0.75} (\nu/t)^{0.5} \tag{12–26}$$

where $\Gamma(m^2/s)$ is the circulation about the large circuits enclosing the vortex, $\nu(m^2/s)$ is the kinematic viscosity, and t(s) is the lapse of time of the vortex. Figure 12–11 shows the relationship between $V_{\theta M}/\sqrt{\nu/t}$ and Γ/ν.

For example, assuming that the circulation Γ of the Vulcan airplane is $\Gamma = 200\ m^2/s$ and that the lapse of time is $t = 60$ s and also the kinematic viscosity of air is $\nu = 1.5 \times 10^{-5}\ m^2/s$, the maximum tangential velocity at $t = 60$ s can be estimated by Equation 12–26 as

$$V_{\theta M} = 7.17\ m/s$$

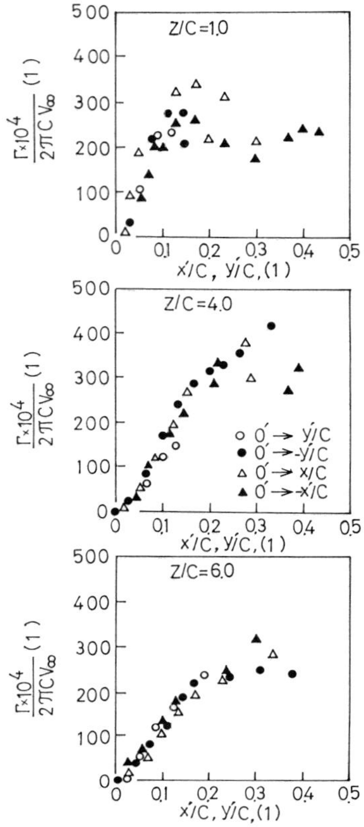

FIGURE 12–10. Distributions of the dimensionless circulation.

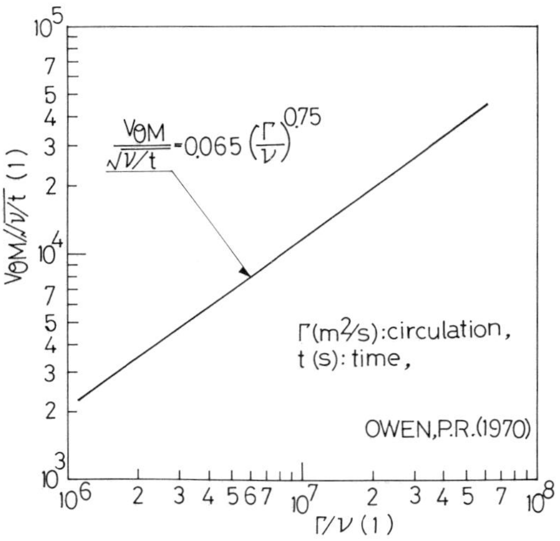

FIGURE 12–11. Relationship between $V_{\theta M}/\sqrt{\nu/t}$ and Γ/ν for the trailing vortex.

MAIN SYMBOLS

$A(m^2/s)$;	constant
$C(m)$;	chord length of wing
$D_o(N)$;	drag force
$p(Pa)$;	static pressure
$R(1)$;	dimensionless radius (r/r_θ)
$r_\theta(m)$;	radius of the maximum tangential velocity
$r_r(m)$;	radius of the maximum radial velocity
$t(s)$;	time
$V_\infty(m/s)$;	free stream velocity
$V_r(m/s)$;	radial velocity
$V_\theta(m/s)$;	tangential velocity
$V_z(m/s)$;	axial velocity
$V_{\theta M}(m/s)$;	maximum tangential velocity
$V_{rM}(m/s)$;	maximum radial velocity
$W(m/s)$;	axial velocity deficiency ($= V_\infty - V_z$)
$W_M(m/s)$;	maximum axial velocity deficiency
$\Gamma(m^2/s)$;	circulation
$\Gamma_o(m^2/s)$	circulation in the initial vortex
$\Lambda(1)$;	aspect ratio
$\nu(m^2/s)$;	kinematic viscosity of fluid

REFERENCES

1. Sankei Newspaper of Japan on September 6, 1984.
2. **Hitzel, S. M.,** *Slender Wings with Leading Edge Vortex-Separation, Recent contributions to fluid mechanics,* Haase, W., Ed., Springer-Verlag, 1982, 138.
3. **Maxworthy, T.,** The motion of aircraft trailing vortices, *J. Appl. Mech.,* (Trans. Am. Soc. Mech. Eng.), 42, 279, 1975.
4. **Kaufmann, W.,** Über die Ausbreitung kreiszlindrischer Wirbel in zähen Flüssigkeiten, *Ing. Arch.,* 31(1), 1, 1962.
5. **Newman, B. G.,** Flow in a viscous trailing vortex, *Aeronaut. Q.,* May-1959, 149.
6. **Dosanjh, D. S., Gasparek, E. P., and Eskinazi, S.,** Decay of a viscous trailing vortex, *Aeronaut. Q.,* May-1962, 167.
7. **Lamb, H.,** *Hydrodynamics,* 6th ed., Cambridge University Press, 1932.
8. **Squire, H. B.,** The growth of a vortex in turbulent flow, *A. R. C.,* 16, 666, 1954.
9. **Squire, H. B.,** The growth of a vortex in turbulent flow, *Aeronaut. Q.,* August-1965, 302.
10. **Abbott, I. H. and Doenhoff, A. E.,** *Theory of Wing Section,* Dover Pub. Inc., New York,
11. **Owen, P. R.,** The decay of a turbulent trailing vortex, *Aeronaut. Q.,* Feb.-1970, 69.

Chapter 13

BATHTUB VORTEX

I. INTRODUCTION

There are many interesting phenomena concerning bathtub vortex motion.[1,2] We observe the flow behavior of the water surface which is gradually lowering the depth Z of water surface from the bottom of the bathtub due to the flowing down through a drain hole. The water surface is in a stationary state early on. However the depth $Z = Z_1$ of water surface becomes shallower with lapse of time; so, the water surface starts to rotate and the strength of rotation of this water surface becomes stronger and then this rotation mode changes to a vortex motion. This vortex motion which is generated in water just above the drain hole is generally called a bathtub vortex.

The generation of this bathtub vortex originates by the results of Coriolis' force due to the rotation of the earth and also due to the secondary flow in the boundary layer on the bottom surface of the bathtub or of the vessel.[3] And further when a depth Z of the water surface at the end of the draining process becomes shallower, $Z = Z_2$, suddenly the direction of this vortex rotation changes to a contrary-wise direction. This interesting observed phenomenon of vortex reversal was explained physically by Sibulkin.[4] A schematic illustration of this phenomenon is shown in Figure 13–1.

Here I should like to introduce a manuscript of *Bath-Tub Vortex* by A. H. Shapiro.[5] Regarding the direction of the bathtub vortex or the kitchen sink, some claim that the direction of swirling is always the same in the northern hemisphere, and that in the southern hemisphere it is always opposite that for the northern hemisphere. Others claim that there is no unique direction of swirl in either hemisphere.

Shapiro said that under well-controlled conditions of experimentation, the observer looking downward at a drain in the northern hemisphere will always see a counterclockwise vortex, while one in the southern hemisphere will always see a clockwise vortex. However there are two reasons for the apparent haphazard results of the everyday observations. The first source of confusion is that, even if many hours are allowed for settling the water in the vessel, air currents in the room or thermal currents associated with nonuniform temperatures around the vessel may easily cause water velocities larger than that due to the earth rotation. And the second source of confusion in the casual everyday observations is that the Coriolis force due to the earth rotation may in fact be much smaller than other forces normally present. For a water speed of 2.54 mm/s toward the drain, the Coriolis force at latitude $\varphi = 42°$ north is about 3×10^{-7} times the force of gravity. If the vessel is not perfectly symmetrical about the axis of the drain, the asymmetrical viscous forces at the side walls and at the bottom of the vessel may produce torques larger than that due to the Coriolis force; the direction of this asymmetrical torque would, of course, vary from vessel to vessel. On the other hand E. N. Andrade[6] gave a short comment to the Shapiro contribution.

Therefore in this section the author would like to explain physically these interesting observed phenomena by the Harada[7] idea and by the Sibulkin idea.

II. THEORETICAL CONSIDERATION

A. FUNDAMENTAL EQUATION

The equation of motion for an inviscid fluid may be represented as

$$\frac{D\mathbf{V}}{Dt} = -2(\zeta\,\mathbf{k} \times \mathbf{V}) - \frac{1}{\rho}\nabla p \qquad (13\text{–}1)$$

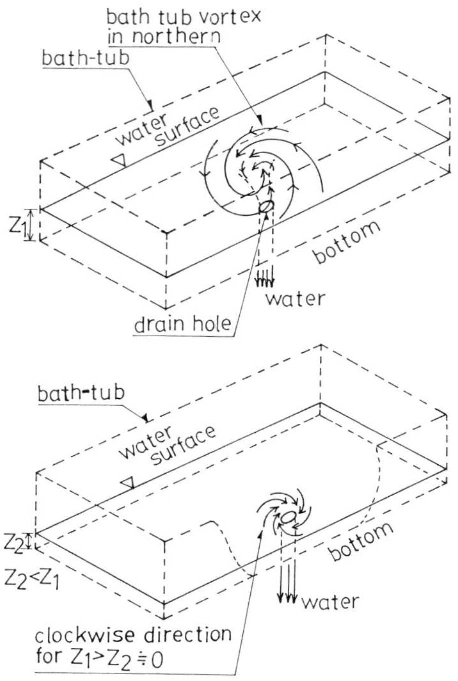

FIGURE 13–1. Schematic illustration of formation processes of bathtub vortex in the north hemisphere.

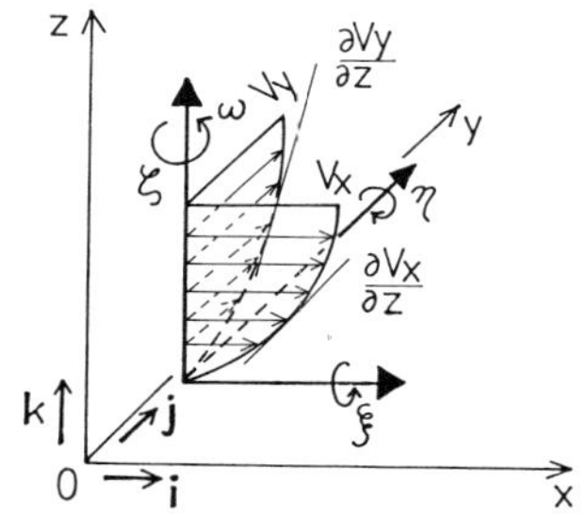

FIGURE 13–2. Illustration of three-dimensional flow.

where $2(\zeta\ \mathbf{k} \times \mathbf{V})$ is a Coriolis force acting on water as an external force with a reference coordinate system fixed to the earth.

When the primary vorticity component η as shown in Figure 13–2 rotates through a small angle θ, the secondary vorticity component ξ may be represented approximately by the following equation as

$$\xi = -2\eta \cdot \theta \tag{13–2}$$

where an axis of the rotation of angle θ is perpendicular to the vorticity vector η. The angular displacement θ is due to the earth's rotation. The angular velocity ω of the water surface at any point on the earth is expressed by

$$\theta = \int \omega\ dt = \int \omega_o \cdot \sin\varphi\ dt \tag{13–3}$$

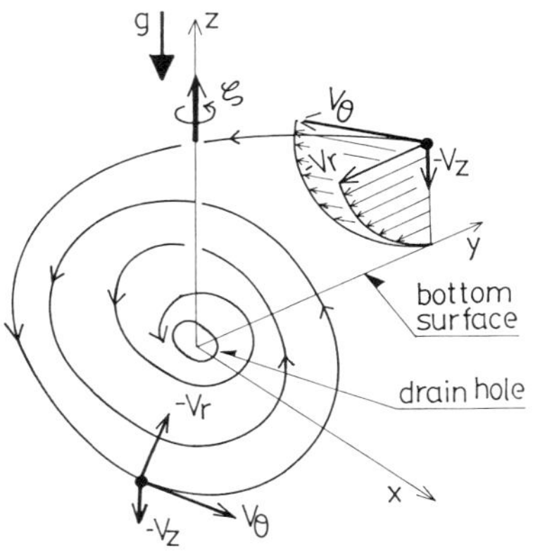

FIGURE 13–3. Axis-symmetrical flow of vortex motion near the bottom surface.

where ω_o is the angular velocity of the earth and φ is the latitude of the earth.

Then Equation 13–2 can be transformed to

$$\xi = -\frac{\partial V_y}{\partial z} = -2\eta \int \omega \, dt \tag{13–4}$$

Here the existence of the secondary flow can be obtained by Equations 13–3 and 13–4 as follows:

$$\frac{\partial V_y}{\partial t} = \int 2\omega \frac{\partial V_x}{\partial z} dz = 2\omega V_x \tag{13–5}$$

where we use the following relation

$$\frac{\partial}{\partial t}\left(\frac{\partial V_y}{\partial z}\right) = \frac{\partial}{\partial z}\left(\frac{\partial V_y}{\partial t}\right)$$

This equation shows the influence of secondary vorticity generation on the bathtub vortex.

B. AXIS-SYMMETRICAL FLOW

Assuming that the flow field is axis-symmetric and the axis of a drain hole is common to the symmetrical axis of this flow field, it is very useful to apply the cylindrical coordinates as shown in Figure 13–3. The equation of an axis-symmetric motion for the tangential direction can be written as follows:

$$\frac{\partial V_\theta}{\partial t} + V_r \frac{\partial V_\theta}{\partial r} + \frac{V_r V_\theta}{r} + V_z \frac{\partial V_\theta}{\partial z} = -2\omega V_r \tag{13–6}$$

where in this equation we may neglect a term $V_z(\partial V_\theta/\partial z)$ due to the small values of V_θ and V_z, and also a term $-2\omega V_r$ is the Coriolis force. An effect of secondary flow expressed by Equation 13–5 can be written as

$$\frac{\partial V_\theta}{\partial t} = 2\omega V_r \tag{13–7}$$

Here substituting Equation 13–7 into Equation 13–6 , Equation 13–6 can be written as

$$\frac{\partial V_\theta}{\partial r} + \frac{V_\theta}{r} = -4\omega \tag{13–8}$$

A solution of this equation can be expressed under the initial condition of $V_\theta = 0$ at $r = R$ as follows:

$$V_\theta = 2\omega\left(\frac{R^2}{r} - r\right) = 2\omega R\left(\frac{R}{r} - \frac{r}{R}\right) \tag{13–9}$$

The distribution of the tangential velocity of Equation 13–9 is satisfied in the region far from the drain hole ($r = 0$). However, near the drain hole or $r = 0$, the tangential velocity V_θ becomes large, so we must take into account the existence of the vertical velocity component V_z and the variation of V_θ for the z-direction. Then we must consider the following equation from the original equation of motion of Equation 13–6.

$$V_r \cdot \frac{\partial V_\theta}{\partial r} + \frac{V_r \cdot V_\theta}{r} + V_z \cdot \frac{\partial V_\theta}{\partial z} = -4\omega \cdot V_r \tag{13–10}$$

Now Harada assumed the velocity components to be

$$V_\theta = j \cdot z^m \tag{13–11}$$

$$V_r = k \cdot z^n \tag{13–12}$$

Here from the approximated continuity equation ($\partial V_r/\partial r \fallingdotseq 0$)

$$\frac{dV_z}{dz} + \frac{V_r}{r} = 0 \tag{13–13}$$

the relationship between V_r and V_z can be obtained

$$\frac{V_z}{V_r} = \frac{2r \cdot z}{(n + 1)(R_o^2 - r^2)} \tag{13–14}$$

where R_o is the radius of the vessel. Hereafter using Equation 13–11, the following equation can be derived

$$\frac{V_z}{V_r} \cdot \frac{dV_\theta}{dz} = \frac{2m \cdot r \cdot V_\theta}{(n + 1)(R_o^2 - r^2)} \tag{13–15}$$

Substituting Equation 13–15 into Equation 13–10, we obtain

$$\frac{dV_\theta}{dr} + \left[\frac{1}{r} + \frac{2m \cdot r}{(n + 1)(R_o^2 - r^2)}\right] V_\theta = -4\omega \tag{13–16}$$

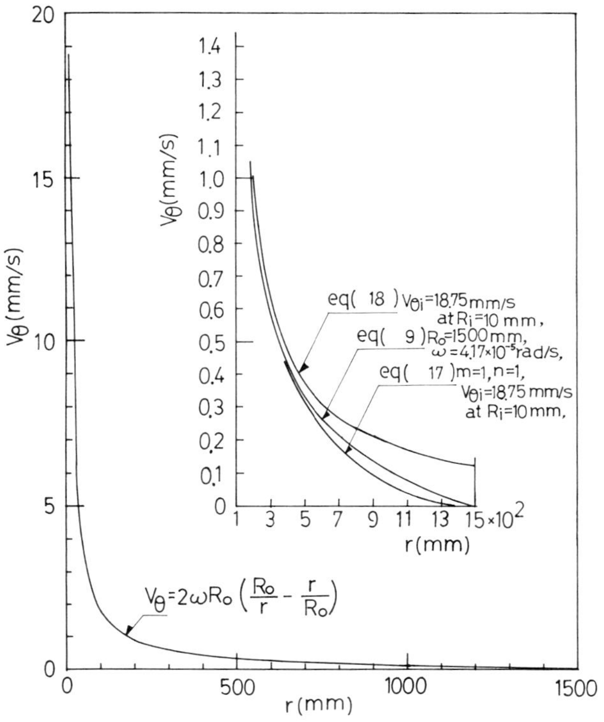

FIGURE 13–4. Tangential velocity distributions of Equations 9, 17, and 18 for $\varphi = 35°$, $R_o = 1500$ mm, $R_i = 10$ mm, $V_{\theta i} = 18.75$ mm/s.

Then denoting a boundary condition of $V_\theta = V_{\theta i}$ at $r = R_i$, so

$$V_\theta = \frac{V_{\theta i} \cdot R_i}{r} \cdot \left(\frac{R_o^2 - r^2}{R_o^2 - R_i^2}\right)^{[m/(n+1)]} + \frac{2\omega \cdot (1 + n)}{n + 1 - m}$$

$$\left[\left(\frac{R_o^2}{r} - r\right) - \left(\frac{R_o^2 - r^2}{R_o^2 - R_i^2}\right)^{[m/(n+1)]}\left(\frac{R_o^2 - R_i^2}{r}\right)\right] \qquad (13\text{–}17)$$

This equation is a general expression of the vortex motion in draining through a drain hole of a flat bottom vessel. If we assume that $V_{\theta i} = 0$ and m is nearly equal to zero, Equation 13–17 is the same as Equation 13–9. And further assuming that $m \fallingdotseq 0$ and the angular velocity component ω of the earth is negligible in comparison with that of $V_{\theta i}/R_i$, Equation 13–17 becomes

$$V_\theta = \frac{V_{\theta i} \cdot R_i}{r} \qquad (13\text{–}18)$$

The distribution of this tangential velocity is irrotational flow (free vortex).

Figure 13–4 shows the distributions of the tangential velocities V_θ of Equation 13–9, 13–17 and 13–18 for the latitude $\varphi = 35°$ north (Example: Tokyo latitude $\varphi = 35°$, $\omega = \Omega_o \cdot \sin\varphi = \Omega_o \cdot \sin 35° = 4.17 \times 10^{-5}$ rad/s and for the radius of vessel $R_o = 1500$ mm,

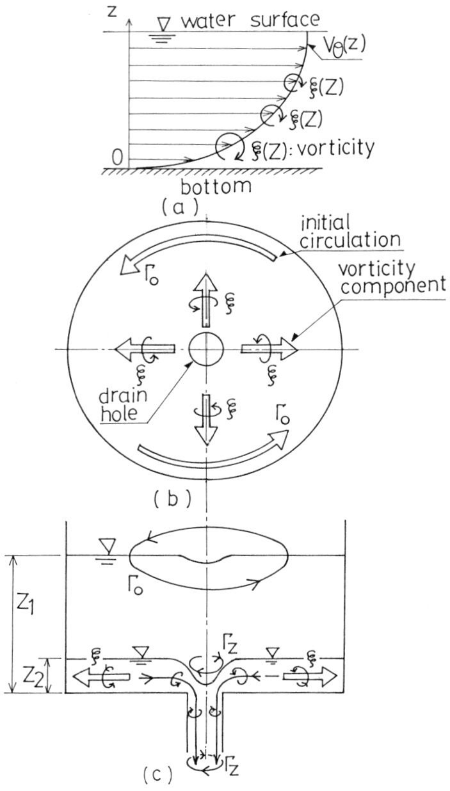

FIGURE 13–5. Schematic illustration of the physical explanation of reversed rotation by Sibulkin.

the radius R_i = 10 mm, and $V_{\theta i}$ = 18.75 mm/s. In this figure the values of V_θ of Equations 13–9, 13–17, and 13–18 are nearly equal in the radial region $r \leq 100$ mm. Therefore as a representative curve of V_θ, a curve of Equation 13–9 is drawn.

III. PHYSICAL EXPLANATION OF REVERSED ROTATION BY SIBULKIN

Already described in Section I, when the water depth in the bathtub decreases with lapse of time, so the bathtub vortex keeps the anticlockwise direction due to Coriolis force on the northern hemisphere. When the water surface approaches to the bottom surface of the bathtub, the bathtub vortex comes to a stop in a few minutes. Then after that the water surface near the drain hole rotates suddenly to the clockwise direction.

This observed phenomenon of the bathtub vortex was explained physically by Sibulkin. Therefore the author should like to introduce his idea. As shown in Figure 13–5, when water goes to flow down with rotation through a drain hole of a bathtub, the boundary layer with the velocity gradient develops near the bottom surface. Then as shown in Figure 13–5 (a), a vorticity component ξ having a rotation axis to the radial direction will be produced due to the velocity gradient dV_θ/dz of the tangential velocity V_θ. Assuming that the initial circulation Γ_o of the rotational flow on the water surface is in the anticlockwise direction and also that the vorticity component ξ (velocity gradient dV_θ/dz is weak due to the deep depth of water) having the axis to the radial direction is given as shown in Figure 13–5 (b) and, after that, the flowing fluid in the boundary layer approaches near the drain hole, the radial vorticity component ξ (velocity gradient dV_θ/dz is strong due to the shallower depth of water) gradually changes the direction of the vorticity axis to the direction of the z-axis due to the Poincare-Helmholtz theorem as shown in Figure 13–5 (c).

Consequently the total vorticity making the sum total of the vorticity ξ of the z-axis around the drain hole produces an induced circulation Γ_z to the clockwise direction. By this physical mechanism, the process of the change of rotation from the anticlockwise rotation in the initial state to the clockwise rotation in the very shallow depth of water in the bathtub can be explained. This interesting idea was made by Professor Sibulkin.

MAIN SYMBOLS

$\mathbf{k}(1)$;	unit vector
$m(1)$;	exponent
$n(1)$;	exponent
$p(Pa)$;	static pressure
$r(m)$;	radius
$\mathbf{V}(m/s)$;	velocity vector
$V_\theta(m/s)$;	tangential velocity
$V_r(m/s)$;	radial velocity
$V_z(m/s)$;	axial velocity
$Z(m)$;	axial distance
$\zeta(1/s)$;	vorticity component
$\eta(1/s)$;	vorticity component
$\theta(rad)$;	angle
$\xi(1/s)$;	vorticity component
$\rho(kg/m^3)$;	density of fluid
$\omega(rad/s)$;	angular velocity

REFERENCES

1. **Granger, R. A.,** Speed of a surge in a bath-tub vortex, *J. Fluid Mech.,* 34, 651, 1968.
2. **Kamel, M. Y. M.,** The effect of swirl on the flow of liquids through bottom outlets, Am. Soc. Mech. Eng. paper No. 64-WA/FE-37, 1965.
3. **Harthorne, W. R.,** Secondary circulation in fluid flow, *Proc. R. Soc. London Ser. A,* 206(A1084), 374, 1951.
4. **Sibulkin, M.,** A note on the bathtub-votex, *J. Fluid Mech.,* 14, 21, 1962.
5. **Shapiro, A. H.,** Bath-tub vortex, *Nature,* 146 (Dec.), 1080, 1962.
6. **Andrade, E. N.,** Bath-tub vortex, *Nature,* 197 (Feb.), 480, 1963.
7. **Harada, M.,** On the generalized bath-tub vortex, *J. Basic Eng.,* (Trans. Am. Soc. Mech. Eng.), Sept. 1967, 617.

Chapter 14

HURRICANE, TYPHOON, AND TORNADO

I. INTRODUCTION

On of the largest scale phenomena of vortex flows in nature are the typhoon, tornado, cyclone, and hurricane. The kinetic energy of these large scale vortex motions is about 10^{18} joule which is equivalent to one hundred hydrogen bombs. It is generally recognized that the source of the enormous energy is the latent heat of aqueous vapor being condensed.[1]

On the other hand it is very well-known that the U.S. is subject to frequent tornados. As a whole these tornados divide into three kinds: namely, tornado, funnel aloft, and water spout. Figure 14–1 shows an illustration of these tornados. The tornado frequently carries the huge funnel-shaped cloud which gives considerable damage to houses, trees, cars, and domestic animals due to the strong rotational vortex reaching the ground.[2] Here Table 14–1 shows the numerical values of the apparent eddy kinematic viscosity $\nu_t(m^2/s)$ which indicates the scale of the turbulent intensity of the vortex motion in the natural phenomena from the research of Logan.[3]

Figure 14–2 shows the schematic structure of a typhoon which rotates in an anticlockwise direction near the ground and near the center region at the altitude $Z = 15 \sim 16$ km, and rotates in a clockwise direction in the radial region $r = 200 \sim 400$ km at $Z = 15 \sim 16$ km. And also on the same figure the distributions of the tangential velocities at $Z = 2 \sim 4$ km, $Z = 12$ km, and $Z = 15 \sim 16$ km are shown. Therefore the tangential velocity V_θ of the typhoon depends not only on the radius r but also the altitude Z. This interesting research was done by Izawa.[4]

In addition, Figure 14–3 shows the distributions of the radial velocities at the radii $r = 26 \sim 200$ km and $r = 500 \sim 600$ km that were estimated by the analysis of tropical cyclones for the $0.25° \sim 2.0°$ latitude ring (solid curve) and the $5.0° \sim 6.0°$ latitude ring (dashed curve) by Palmén and Jordan.[5] From this figure it will be found that the radial velocity runs toward the inward direction near the ground level, but it runs toward the outward direction at the high altitude.

II. CYCLONIC AND ANTICYCLONIC MOTION OF AIR

Let us assume that when air moves by curvature motion the velocity of air does not change at a specific point. We may write the terms of the equation of motion as $\partial V_x/\partial t = dV_x/dt$, $\partial V_y/\partial t = dV_y/dt$. Then the equation of motion, neglecting the frictional force can be written as

$$\frac{dV_x}{dt} - f \cdot V_y = -\frac{1}{\rho}\frac{\partial p}{\partial x} \qquad (14\text{–}1)$$

$$\frac{dV_y}{dt} + f \cdot V_x = -\frac{1}{\rho}\frac{\partial p}{\partial y} \qquad (14\text{–}2)$$

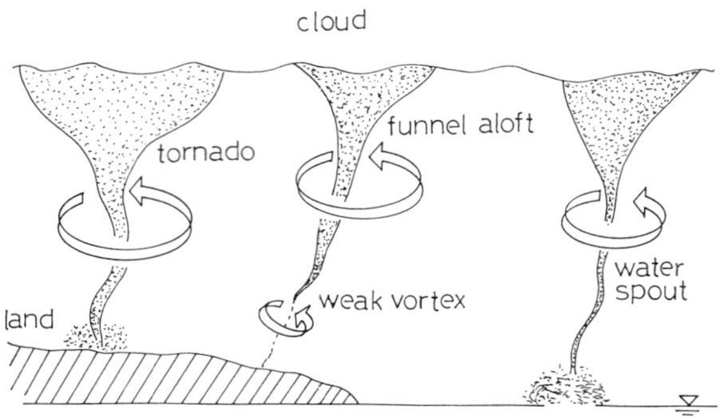

FIGURE 14–1. Schematic illustration of the three kinds of tornados.

TABLE 14–1
Estimated Values of the Eddy Kinematic Viscosity of the Vortex Motion in Nature (Kinematic Viscosity of Air $\nu = 1.5 \times 10^{-5}$ m^2/s)

Researcher	Year Published	Eddy Kinematic Viscosity $\nu_t(m^2/s)$	$\nu_t/\nu(1)$	Kinds of Vortices
Sinclair	1966	5 ~ 15	3.33 ~ 10 × 10^5	Dust devil
Kuo	1966	1	0.67 × 10^5	Dust devil
Kuo	1966	5	3.33 × 10^5	Tornado
Long	1958	1 ~ 10	0.67 ~ 6.7 × 10^5	Tornado
Rott & Lewellen	1966	10	6.7 × 10^5	Tornado
Gutman	1957	10	6.7 × 10^5	Water spout
Smith	1968	50	33.3 × 10^5	Hurricane

where f is a Coriolis parameter ($f = 2\omega \sin\varphi$; φ = latitude). Now as shown in Figure 14–4, changing from Cartesian coordinates to cylindrical coordinates, we obtain the following relations

$$x = r \cdot \cos\theta \quad , \quad y = r \cdot \sin\theta \tag{14–3}$$

$$V_x = \dot{r} \cdot \cos\theta - r \cdot \sin\theta \cdot \dot{\theta} \tag{14–4}$$

$$V_y = \dot{r} \cdot \sin\theta + r \cos\theta \cdot \dot{\theta} \tag{14–5}$$

Substituting these relations into Equations 14–1 and 14–2, Equations 14–1 and 14–2 become

$$\ddot{r} - r \cdot \dot{\theta}^2 - f \cdot r \cdot \dot{\theta} = -\frac{1}{\rho}\frac{\partial p}{\partial r} \tag{14–6}$$

$$r \cdot \ddot{\theta} + f \cdot \dot{r} + 2 \cdot \dot{r} \cdot \dot{\theta} = -\frac{1}{\rho r}\frac{\partial p}{\partial \theta} \tag{14–7}$$

In order to simplify the flow model of cyclonic and anticyclonic motions, we may assume that the pressure gradient only depends on the radius r and the flow is axis-symmetric, and further the velocity of air is independent of time. Equation 14–6 then becomes

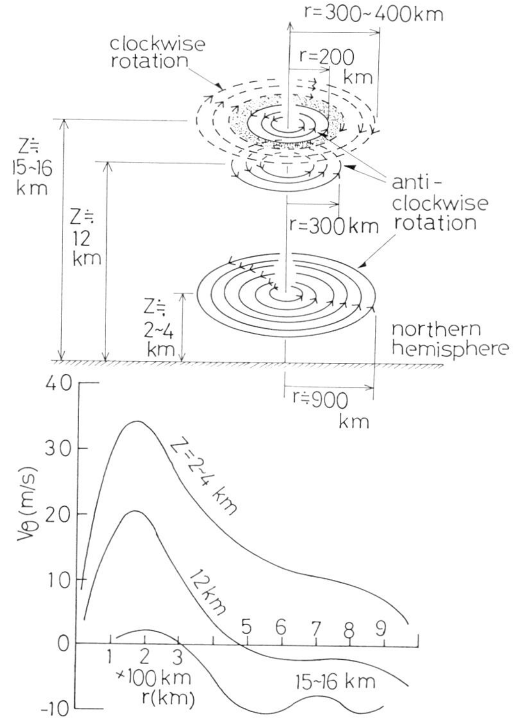

FIGURE 14–2. Estimated average tangential velocity distributions of a typhoon on the northern hemisphere by Izawa.[4]

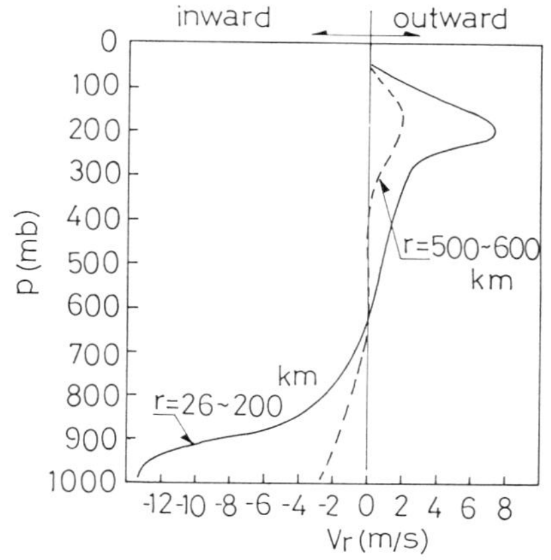

FIGURE 14–3. Estimated average radial velocity distributions by the analysis of the tropical cyclones by Palmén and Jordan.[5]

$$r \cdot \dot{\theta}^2 + f \cdot r \cdot \dot{\theta} = r \cdot \dot{\theta}^2 + 2\omega \sin\varphi \cdot r \cdot \dot{\theta} = \frac{1}{\rho}\frac{\partial p}{\partial r} \tag{14–8}$$

Solving Equation 14–8 for $\dot{\theta}$, the following equation is obtained

$$\dot{\theta} = -\omega \cdot \sin\varphi \pm \sqrt{\omega^2 \cdot \sin^2\varphi + \frac{1}{\rho r}\frac{\partial p}{\partial r}} \tag{14–9}$$

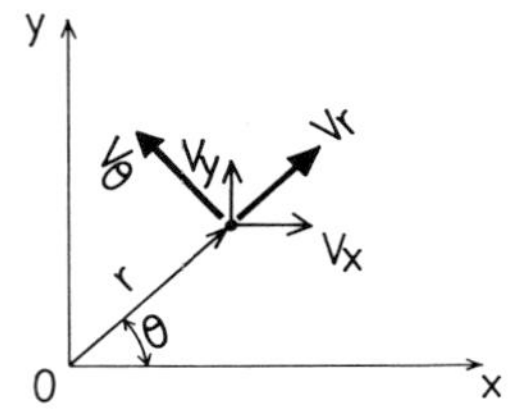

FIGURE 14–4. Illustration of the velocity components.

Therefore the tangential velocity ($V_\theta = r \cdot \dot{\theta}$) can be represented by

$$V_\theta = r \cdot \dot{\theta} = r\left(-\omega \cdot \sin\varphi \pm \sqrt{\omega^2 \cdot \sin^2\varphi + \frac{1}{\rho r}\frac{\partial p}{\partial r}}\right) \tag{14–10}$$

For the cyclonic motion, that is to say, $\partial p/\partial r > 0$ (the radial air flow may be directed to center), Equation 14–10 becomes

$$V_\theta = r\left(\sqrt{\omega^2 \cdot \sin^2\varphi + \frac{1}{\rho r}\left|\frac{\partial p}{\partial r}\right|} - \omega \cdot \sin\varphi\right) \tag{14–11}$$

On the other hand for the anticylonic motion, that is to say, $\partial p/\partial r < 0$ (the radial air flow may be directed outward), Equation 14–10 becomes

$$V_\theta = r\left(\omega \cdot \sin\varphi - \sqrt{\omega^2 \cdot \sin^2\varphi - \frac{1}{\rho r}\left|\frac{\partial p}{\partial r}\right|}\right) \tag{14–12}$$

Comparing Equation 14–11 with Equation 14–12, we can take the arbitrary value of the pressure gradient $|\partial p/\partial r|$ in the cylonic motion, but in the anticyclonic motion there is a defined maximum value of $|\partial p/\partial r|$, since a square root of Equation 14–12 must not be a negative value. So let us denote the maximum value of $|\partial p/\partial r|$ of Equation 14–12 as $|\partial p/\partial r| = P_m$, P_m being defined as

$$P_m = \omega^2 \cdot \sin^2\varphi \cdot r \cdot \rho \tag{14–13}$$

Then the maximum tangential velocity $V_{\theta m}$ of the anticyclonic motion is

$$V_{\theta m} = \omega \cdot r \cdot \sin\varphi \tag{14–14}$$

Further from another point of view, there is a critical radius r_c corresponding to the given pressure gradient $|\partial p/\partial r|$ of the anticyclonic motion

$$r_c = \frac{|\partial p/\partial r|}{\omega^2 \cdot \sin^2\varphi \cdot \rho} \tag{14–15}$$

and the tangential velocity $V_{\theta c}$ corresponding to r_c can be written as

$$V_{\theta c} = \frac{|\partial p/\partial r|}{\omega \cdot \sin\varphi \cdot \rho} \tag{14–16}$$

Next when we consider Equation 14–7 for the axis-symmetric motion, then

$$r \cdot \ddot{\theta} + f \cdot \dot{r} + 2 \cdot \dot{r} \cdot \ddot{\theta} = 0 \tag{14–17}$$

Transforming Equation 14–17

$$\frac{dV_\theta}{dt} + \left(f + \frac{V_\theta}{r}\right) \cdot V_r = 0 \tag{14–18}$$

and further retransforming Equation 14–18

$$\frac{dV_\theta}{dr} + \frac{V_\theta}{r} = -f \tag{14–19}$$

So we can obtain a general solution of Equation 14–19 as follows:

$$V_\theta = \frac{2C - f \cdot r^2}{2r} \tag{14–20}$$

Then assuming that the quiet lump of air ($V_\theta = 0$) at the radius r_o (distance from the center of cyclone) on the lower layer of atmosphere begins to blow into the center of cyclone, Equation 14–20 becomes

$$V_\theta = \frac{f \cdot (r_o^2 - r^2)}{2r} \tag{14–21}$$

This equation satisfies a conservation rule of an absolute angular momentum M = constant defined as

$$M = \frac{1}{2} f \cdot r^2 + r \cdot V_\theta \tag{14–22}$$

since, if we substitute Equation 14–21 into Equation 14–22, the absolute angular momentum M becomes

$$M = \frac{f \cdot r_o^2}{2} = \text{constant} \tag{14–23}$$

Therefore

$$\frac{dM}{dt} = 0 \tag{14–24}$$

Here, returning to Equation 14–21, when the radius r is small in comparison with the radius r_o, Equation 14–21 can be written as

$$V_\theta \fallingdotseq \frac{f \cdot r_o^2}{2r} \tag{14–25}$$

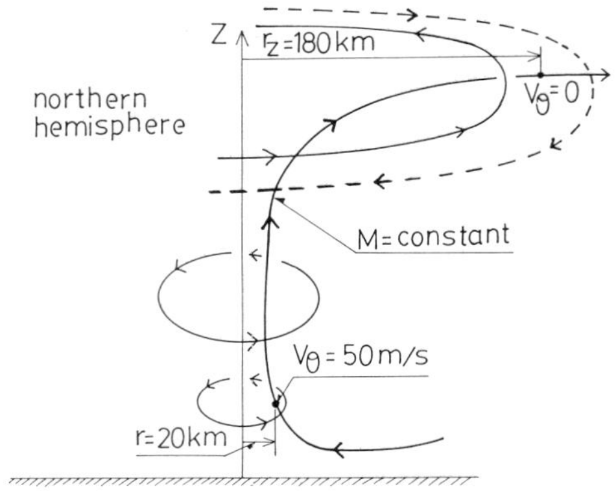

FIGURE 14–5. Numerical example of M = constant theorem at the latitude $\varphi = 25°$ on the northern hemisphere.

This equation shows the irrotational motion (free vortex) which means that when the radius r becomes very small, the tangential velocity increases to infinity. However due to the frictional force (generation of eddy viscosity due to the high Reynolds number and the frictional force between atmosphere and the surface of the earth), the maximum tangential velocity of hurricanes and of typhoons is restricted.

III. NUMERICAL EXAMPLE

When an ascending current of air near the center region of a hurricane blows up into the upper layer, the absolute angular momentum M may be conserved due to the small internal friction. At that time the tangential velocity V_θ at the radius r_z from the center in the upper layer goes to zero and then the tangential velocity becomes negative (clockwise rotation in the northern hemisphere). For example at the latitude 25° N ($f = 0.423 \times 2 \times 2 \times 3.14/24 \times 3600 = 6.16 \times 10^{-5}$), the lump of air in the lower layer having the tangential velocity $V_\theta = 50$ m/s at the radius r = 20 km from the center of hurricane ascends in accord with the conservation of the absolute angular momentum and begins to blow in the upper layer. So, the radius r_z can be estimated by Equation 14–22 as

$$r_z = 180 \text{ km}$$

This schematic illustration is shown in Figure 14–5.

IV. DALLAS TORNADO OF APRIL 2, 1957

Composite distributions of the tangential V_θ and upward V_z components of velocities of air flow (Dallas tornado of April 2, 1957) were estimated by tracing particles of debris and cloud tag movements in scaled movies of a tornado by W. H. Hoecker.[6] The solid debris particles which were mainly some uniformly sized rectangularly shaped thin sheets of material about 1.22 m × 2.44 m in size were tracked. Then the greatest tangential velocity measured was $V_{\theta Max} = 76$ m/s at a radius of r = 40 m and an elevation of z = 69 m, and also the greatest upward velocity was $V_{zMax} = 67$ m/s under the assumption of the radial symmetry.

Figure 14–6 shows the various types of elements that were used as air flow tracers in measuring the velocity distributions. In the upper portion the cloud tags and fragments were

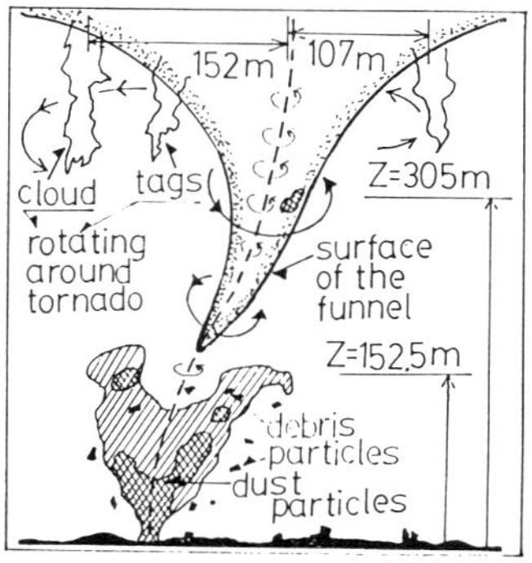

FIGURE 14–6. Schematic illustration of Dallas tornado of April 2, 1957 by Hoecker.[6]

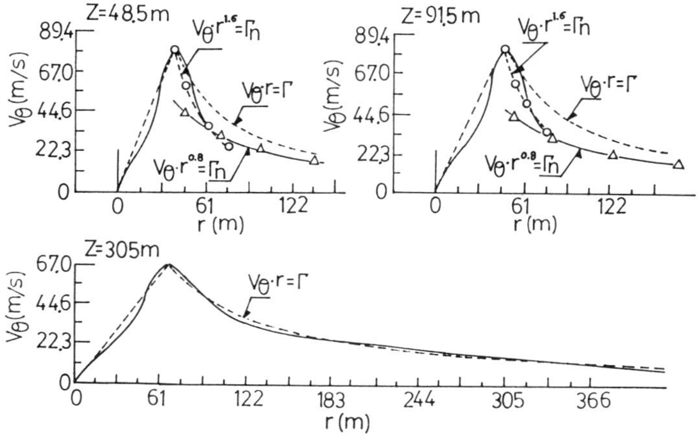

FIGURE 14–7. Distributions of the tangential velocities for elevations of Z = 48.5 m, 91.5 m, and 305 m by Hoecker.[6,7]

rotating around the tornado, while in the lower portion, that is, in and around the debris cloud, were the dust particles and pieces of debris.

Figure 14–7 shows the distributions of the tangential velocities at the elevation z = 45.8 m, 91.5 m, and 305 m, respectively. In this figure, the tangential velocity distribution of a Thomson-Rankine vortex is shown by the dotted line for comparing the distributions of the estimated tangential velocities. It must noted that there were different characteristics of the tangential velocities at the region of the boundary between the forced vortex and the free vortex for the actual velocity distribution and for the Thomson-Rankine vortex. That is to say, the distributions of vorticity show not only positive values but also negative values.

The static pressure field was derived by integrating the equation

$$\frac{dp}{dr} = \rho\frac{V_\theta^2}{r}$$

from a distribution of observed tangential velocity horizontally from the radius r = 460 m, stepwise, to near the center axis for each 30.5 m level.[7] For the boundary condition at the radius r = 460 m, a vertical lapse rate of pressure computed from the pressure and temperature

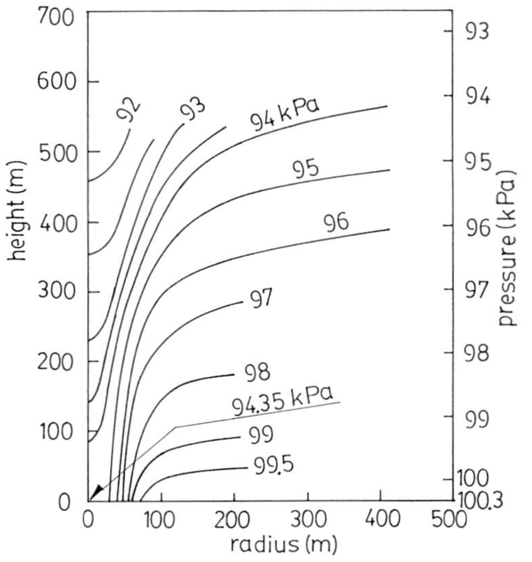

FIGURE 14–8. Estimated pressure field of Dallas tornado in the vertical plane by Hoecker.[6,7]

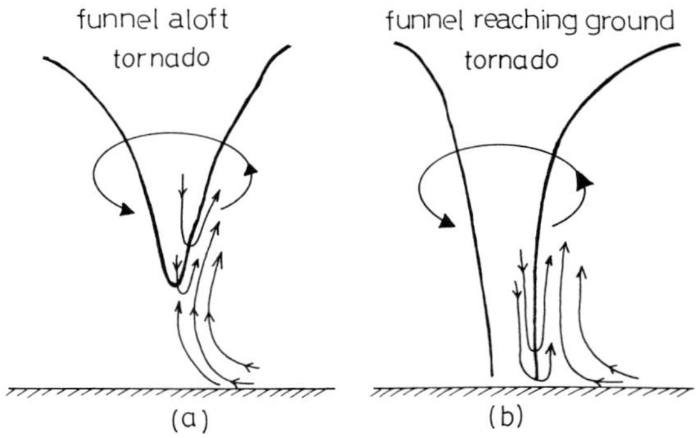

FIGURE 14–9. Schematic illustration of the two kinds of tornados.

conditions at Dallas for the time of the tornado was used. A total pressure drop of 6 kPa was computed at ground level at the axis. For estimating the static pressure field, the flow symmetry in the horizontal plane was assumed, giving pressure distribution in three dimensions. The resulting pressure distribution is shown in Figure 14–8.

Figure 14–9 shows the schematic illustration of the two kinds of air flow of the tornado near the ground in the radius-height plane as estimated by the numerical results of the static pressure distribution derived by the tangential wind velocities of the Dallas tornado (April 1957) and the study of movies and still photographs taken of tornados. Figure 14–9(a) shows a funnel aloft and Figure 14–9(b) shows a funnel reaching ground.

IV. ELLIPTICAL EYE OF TYPHOON

Mitsuta and Yoshizumi caught the rotation of an elliptical eye of the Miyako-jima typhoon.[8] The radar observations taken at Miyako-jima at 0453 JST on September 5, 1966 revealed that the tropical cyclone had not a circle but an elliptical eye rotating counterclockwise. They said that the second Miyako-jima typhoon had an elliptical eye while the pressure oscillation was occurring. And the major axis of the elliptical eye was rotating counterclockwise at an angular speed of half a rotation per about one hour. The recorded pressure oscillation with the period of about 50 min on September 5, 1966 extended over half a day and had a maximum double amplitude exceeding 10 mb.

In addition to this, they described the path of the second Miyako-jima typhoon and the orientation of the major axis of this elliptical eye in detail by the figures.

Concerning the fluid dynamical characteristics of hurricanes, namely, the tracks of hurricane Hilda, September 28 through October 5, 1964, the stream lines at 200 mb, the best single frame satellite photograph of hurricane Hilda, the relative wind velocity profiles, and also the mathematical expression of the mean flux of absolute angular momentum M were described in detail by Hawkins and Rubsam.[9,10]

MAIN SYMBOLS

f(rad/s);	Coriolis parameter
$M(m^2/s)$;	angular momentum
$P_m(Pa/m)$;	maximum pressure gradient
p(Pa);	static pressure
r(m);	radius
$r_c(m)$;	critical radius
$r_o(m)$;	radius of $V_\theta = 0$
$r_z(m)$;	radius on the upper layer for $V_\theta = 0$
$V_x(m/s)$;	fluid velocity for the x-axis
$V_y(m/s)$;	fluid velocity for the y-axis
$V_\theta(m/s)$;	tangential velocity
$V_{\theta M}(m/s)$;	maximum tangential velocity
$V_{\theta c}(m/s)$;	tangential velocity corresponding to $r = r_c$
$V_z(m/s)$;	axial velocity
θ(rad);	angle
$\nu_t(m^2/s)$;	eddy kinematic viscosity
$\rho(kg/m^3)$;	density of fluid
φ(deg);	latitude

REFERENCES

1. NHK-correspondence course of high school, physical geography-I, 1979.
2. **Fujita, T.,** Tornados (in Japanese), Kyoritzu-Pub. Co., Tokyo, 1974.
3. **Logan, S. E.,** An approach to the dust devil vortex, *A. I. A. A. J.,* 9(4), 660,
4. **Izawa, T.,** On the mean wind structure of typhoon, Typhoon Research Laboratory, Technical Note No. 2, Japan Meteorological Agency, 1964-March.
5. **Palmén, E. and Jordan, C. L.,** Note on the release of kinetic energy in tropical cyclones, *Tellus,* 7(2), 186, 1955.
6. **Hoecker, W. H.,** Wind speed and air patterns in the Dallas tornado of April 2, 1957, *Mon. Weather Rev.,* 88(5), 167, 1960.
7. **Hoecker, W. H.,** Three dimensional pressure pattern of the Dallas Tornado and some resultant implications, *Mon. Weather Rev.,* 89, 533, 1961.
8. **Mitsuta, Y. and Yoshizumi, S.,** Periodic vibrations of pressure, wind and rainfall observed at Miyako-jima during the Second Miyako-jima typhoon, *J. Meteorol. Soc. Jpn.,* 51(6), 475, 1973.

9. **Hawkins, H. F. and Rubsam, D. T.,** Hurricane Hilda, 1964, I. Genesis, as revealed by satellite photographs, conventional and aircraft data, *Mon. Weather Rev.,* 96(7), 428, 1968.
10. **Hawkins, H. F. and Rubsam, D. T.,** Hurricane Hilda, 1964, II. Structure and budgets of the hurricane on October 1, 1964, *Mon. Weather Rev.,* 96(7), 617, 1968.

Chapter 15

PHYSICS OF THERMAL BOUNDARY LAYER

I. INTRODUCTION

In this chapter a relationship between the velocity boundary layer thickness and the thermal boundary layer thickness which depends on the dimensionless quantities, that is to say, Reynolds number, Prandtl number, Eckert number, and Grashof number and also the exact solution of the temperature distribution for Couette flow between parallel plates and for the rotating flow in concentric cylinders, is described in detail.

II. THERMAL BOUNDARY LAYER

When a solid body at high temperature T_o is set in the high speed air flow of temperature T_∞ as shown in Figure 15–1, the heat transfer into the air flow is mainly limited to a thin layer on the surface of the solid body. This thin layer is called a thermal boundary layer. In this section the relationship between the velocity boundary layer and the thermal boundary layer will be briefly described.[1–4]

III. ENERGY EQUATION

In order to derive an energy equation for the incompressible fluid, as a simplification, an equilibrium equation of the thermal energy on the two-dimensional case is considered.

Let us denote the main symbols as follows:

Thermal conductivity	λ(W/mK),
Temperature of gas	T(K),
Specific heat of constant pressure	C_p(J/kg · K).

As shown in Figure 15–2, when we consider a small lump of fluid ABCD (the unit width), the heat flux per unit time coming from the outer surface into the inner region of this small fluid particle ABCD by heat conduction can be calculated. The heat flux flowing through the unit area of the surface AD is proportional to the negative temperature gradient $(-\partial T/\partial x)$ perpendicular to the surface AD; so, it can be represented as

$$Q_x = \lambda\left(-\frac{dT}{dx}\right) \qquad (15\text{–}1)$$

The heat flux flowing out from the unit area of the surface BC is

$$Q_x + \frac{\partial Q_x}{\partial x}dx = \lambda\left(-\frac{dT}{dx}\right) + \frac{\partial}{\partial x}\left[\lambda\left(-\frac{dT}{dx}\right)\right]dx \qquad (15\text{–}2)$$

Therefore the difference between Q_x and $Q_x + (\partial Q_x/\partial x)dx$ is the heat quantity remaining in this small body for the x-direction

$$\Delta Q_x(\mathrm{J/m^3 \cdot s}) = Q_x - \left(Q_x + \frac{\partial Q_x}{\partial x}dx\right) = \frac{\partial}{\partial x}\left(\lambda\frac{\partial T}{\partial x}\right) \qquad (15\text{–}3)$$

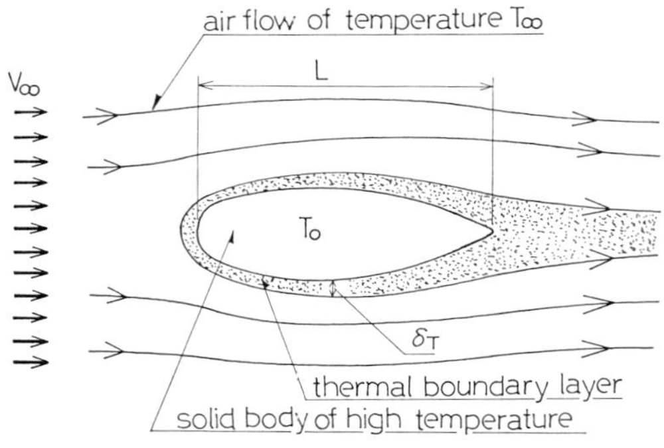

FIGURE 15–1. Schematic illustration of the thermal boundary layer.

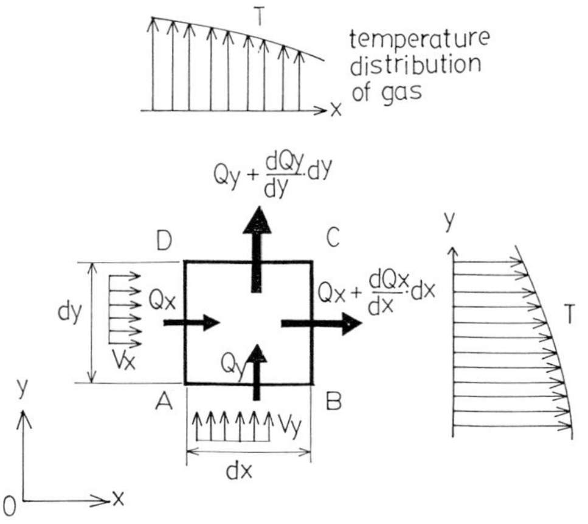

FIGURE 15–2. Schematic illustration of heat conductivity.

Similarly the heat quantity remaining in this small body for the y-direction can be represented as

$$\Delta Q_y(\mathrm{J/m^3 \cdot s}) = Q_y - \left(Q_y + \frac{\partial Q_y}{\partial y}\,dy\right) = \frac{\partial}{\partial y}\left(\lambda \frac{\partial T}{\partial y}\right) \tag{15–4}$$

At the same time the energy dissipation per unit volume of fluid and per unit time is

$$\Phi(\mathrm{J/m^3 \cdot s}) = \eta\left[2\left(\frac{\partial V_x}{\partial x}\right)^2 + 2\left(\frac{\partial V_y}{\partial y}\right)^2 + \left(\frac{\partial V_y}{\partial x} + \frac{\partial V_x}{\partial y}\right)^2\right] \tag{15–5}$$

Then the unit volume of this fluid particle ABCD shows the increment of the enthalpy $\rho C_p T$ for the steady flow condition

$$\Delta H(\mathrm{J/m^3 \cdot s}) = \rho C_p\left(V_x \frac{\partial T}{\partial x} + V_y \frac{\partial T}{\partial y}\right) \tag{15–6}$$

Consequently the steady equilibrium equation of the thermal energy is

$$\rho C_p\left(V_x\frac{\partial T}{\partial x} + V_y\frac{\partial T}{\partial y}\right) = \lambda\left(\frac{\partial^2 T}{\partial x^2} + \frac{\partial^2 T}{\partial y^2}\right) + \Phi \qquad (15\text{–}7)$$

The buoyant force affects the momentum equation by the thermal expansion. So the Navier-Stokes equation for the incompressible two-dimensional flow is

$$\rho\left(V_x\frac{\partial V_x}{\partial x} + V_y\frac{\partial V_y}{\partial y}\right) = -\frac{\partial p}{\partial x} + \rho g_x \beta \Delta T + \eta \nabla^2 V_x \qquad (15\text{–}8)$$

$$\rho\left(V_x\frac{\partial V_y}{\partial x} + V_y\frac{\partial V_y}{\partial y}\right) = -\frac{\partial p}{\partial y} + \rho g_y \beta \Delta T + \eta \nabla^2 V_y \qquad (15\text{–}9)$$

where the following defined symbols are used:

Temperature coefficient of expansion (for a gas, β is replaced by $\beta = 1/T$)	$\beta(1/K) = -\frac{1}{\rho}\left(\frac{\partial \rho}{\partial T}\right)p$
Temperature of unheated fluid	$T_\infty(K)$
Temperature difference	$\Delta T(K) = T - T_\infty$
Gravitational acceleration for x and y directions	$g_x(m/s^2)$ and $g_y(m/s^2)$.

And also, we need an equation of continuity as

$$\frac{\partial V_x}{\partial x} + \frac{\partial V_y}{\partial y} = 0 \qquad (15\text{–}10)$$

These four equations 15–7, 15–8, 15–9, and 15–10 are the fundamental differential equation for the unknown quantities V_x, V_y, p, and T. Let us denote the representative quantities of the fluid velocity, of the size, and of the temperature difference between the temperature T_W of the solid wall and the temperature T_∞ of the unheated fluid as V_∞, L, and $\Delta T_o = T_W - T_\infty$. So the following dimensionless quantities of the velocity, of the coordinate, of the temperature, and of the pressure are introduced.

$$\overline{V}_x = \frac{V_x}{V\infty} \quad , \quad \overline{x} = \frac{x}{L} \quad , \quad \overline{T} = \frac{T}{\Delta T_o} \quad \text{and} \quad \overline{p} = \frac{p}{\rho V_\infty^2} \qquad (15\text{–}11)$$

Then the Equation 15–7 can be transformed to

$$\begin{aligned}\rho C_p \frac{V\infty \Delta T_o}{L}\left(\overline{V}_x\frac{\partial \overline{T}}{\partial x} + \overline{V}_y\frac{\partial \overline{T}}{\partial y}\right) &= \frac{\lambda \Delta T_o}{L^2}\left(\frac{\partial^2 \overline{T}}{\partial \overline{x}^2} + \frac{\partial^2 \overline{T}}{\partial \overline{y}^2}\right) + \\ &+ \frac{\eta V_\infty^2}{L^2}\left\{2\left[\left(\frac{\partial \overline{V}_x}{\partial \overline{x}}\right)^2 + \left(\frac{\partial \overline{V}_y}{\partial \overline{y}}\right)^2\right] + \left(\frac{\partial \overline{V}_y}{\partial \overline{x}} + \frac{\partial \overline{V}_x}{\partial \overline{y}}\right)^2\right\}\end{aligned} \qquad (15\text{–}12)$$

Therefore introducing the above dimensionless quantities into Equation 15–12, Equation 15–12 can be represented as

$$\begin{aligned}\overline{V}_x\frac{\partial \overline{T}}{\partial \overline{x}} + \overline{V}_y\frac{\partial \overline{T}}{\partial \overline{y}} &= \frac{1}{P_r R_e}\left(\frac{\partial^2 \overline{T}}{\partial \overline{x}^2} + \frac{\partial^2 \overline{T}}{\partial \overline{y}^2}\right) + \frac{E}{R_e}\left\{2\left[\left(\frac{\partial \overline{V}_x}{\partial \overline{x}}\right)^2\right.\right. \\ &\left.\left. + \left(\frac{\partial \overline{V}_y}{\partial \overline{y}}\right)^2\right] + \left(\frac{\partial \overline{V}_y}{\partial x} + \frac{\partial \overline{V}_x}{\partial y}\right)^2\right\}\end{aligned} \qquad (15\text{–}13)$$

where the following dimensionless quantities are applied

$$\begin{aligned}
&\text{Reynolds number } R_e = V_\infty L/\nu \\
&\text{Prandtl number } P_r = \eta C_p/\lambda = \nu/D_t \\
&\text{Eckert number } E = V_\infty^2/C_p\Delta T,
\end{aligned}$$

in which a symbol $D_t(m^2/s) = \lambda/\rho C_p$ is called the thermal diffusivity.

Next similarly Equation 15–8 can be transformed to

$$\bar{V}_x\frac{\partial \bar{V}_x}{\partial x} + \bar{V}_y\frac{\partial \bar{V}_x}{\partial y} = -\frac{\partial \bar{p}}{\partial \bar{x}} + \frac{G_r}{R_e^2}\left(\bar{T} - \frac{T\infty}{\Delta T_o}\right) + \frac{1}{R_e}\left(\frac{\partial^2 \bar{V}_x}{\partial \bar{x}^2} + \frac{\partial \bar{V}_y}{\partial \bar{y}^2}\right) \tag{15–14}$$

where G_r is the Grashof number defined as Grashof number $G_r = g\beta\Delta T_o L^3/\nu^2$.

Here let us denote the thickness of the thermal boundary layer and the thickness of the velocity boundary layer as δ_T and δ_V, and also let us assume that L is larger than δ_T and δ_V; then the following rough estimate can be contemplated. When the viscous term $(1/R_e) \cdot (\partial^2\bar{V}_x/\partial\bar{y}^2) \sim (1/R_e) \cdot (L/\delta_v)^2$ of Equation 15–14 is nearly equivalent to the inertia term $\bar{V}_x(\partial\bar{V}_x/\partial\bar{x}) \sim 1$, the following relation can be obtained

$$R_e = (L/\delta_V)^2 \tag{15–15}$$

Further when the buoyant force term of Equation 15–15 is nearly equivalent to the inertia term, the following relation can be obtained

$$G_r/R_e^2 \sim 1 \quad \text{or} \quad G_r \sim R_e^2 \tag{15–16}$$

While when the convective term $\bar{V}_x(\partial\bar{T}/\partial\bar{x}) \sim 1$ of Equation 15–13 is nearly equal to the term $(1/P_rR_e) \cdot (\partial^2\bar{T}/\partial\bar{y}^2) \sim (1/P_rR_e) \cdot (L/\delta_T)^2$ of the thermal conductivity, the following relation is obtained

$$\frac{1}{P_rR_e} \sim \left(\frac{\delta_T}{L}\right)^2 \tag{15–17}$$

Also the energy dissipative term $(E/R_e) \cdot (\partial\bar{V}_x/\partial\bar{y})^2 \sim (E/R_e) \cdot (L/\delta_V)$ of Equation 15–13 is nearly equivalent to the convective term $\bar{V}_x(\partial\bar{T}/\partial\bar{x}) \sim 1$; so, the following relation can be derived

$$\frac{E}{R_e} \sim \left(\frac{\delta_V}{L}\right)^2 \tag{15–18}$$

Therefore, a relationship between δ_T and δ_V can be obtained by Equations 15–15 and 15–17 as follows:

$$\frac{\delta_T}{\delta_V} \sim \frac{1}{\sqrt{P_r}} \tag{15–19}$$

Consequently in case of gas flow, the Prandtl number P_r is nearly equal to $P_r \sim 1$. So the thickness δ_T of the thermal boundary layer is nearly equal to that δ_V of the velocity boundary layer; however, in case of liquid flow, P_r is nearly equal to $P_r = 10 \sim 1000$. So $\delta_V >> \delta_T$. When we combine Equations 15–15 and 15–18, the Eckert number E becomes

$$E = \frac{V_\infty^2}{C_p \Delta T} \sim 1 \tag{15–20}$$

IV. EXACT SOLUTION OF THE TEMPERATURE DISTRIBUTION ON THE TWO-DIMENSIONAL LAMINAR FLOW

A. COUETTE FLOW IN PARALLEL PLATES

Let us consider a Couette flow between two parallel flat plates in which a lower plate ($y = 0$) of temperature $T = T_1$ is fixed while an upper plate ($y = h$) of temperature $T = T_2$ is moved with velocity V_h to the positive x-direction.[5] So the boundary conditions are $V_x = V_h = 0$ at $y = 0$ and $V_x = V_h$, $V_y = 0$ at $y = h$.

The boundary conditions of temperatures are $T = T_1$ at $y = 0$ and $T = T_2$ at $y = h$, and the following condition is given as $dT/dy = 0$ at $y = 0$.

Then an exact solution of the velocity distribution which satisfies the boundary condition is

$$V_x = V_h \frac{y}{h} \quad , \quad V_y = 0 \quad , \quad p = \text{constant} \tag{15–21}$$

Next the fundamental differential equation of the temperature distribution can be obtained as

$$\lambda \frac{d^2T}{dy^2} = -\eta \left(\frac{dV_x}{dy} \right)^2 \tag{15–22}$$

Substituting Equation 15–21 into Equation 15–22, we obtain

$$\lambda \frac{d^2T}{dy^2} = -\eta \frac{V_h^2}{h^2} \tag{15–23}$$

Therefore the solution of the temperature distribution which satisfies the above-stated condition is

$$T(y) = T_1 + (T_2 - T_1) \frac{y}{h} + \frac{\eta V_h^2}{2\lambda h} \left(1 - \frac{y}{h} \right) \cdot y \tag{15–24}$$

Transforming the above equation

$$\frac{T(y) - T_1}{T_2 - T_1} = \frac{y}{h} + \frac{\eta V_h^2}{2\lambda (T_2 - T_1)} \left(1 - \frac{y}{h} \right) \cdot \frac{y}{h} \tag{15–25}$$

Consequently the temperature distribution is composed of the linear distribution for the rest fluid and of the parabolic distribution due to the exchange of the energy dissipation by viscosity. The temperature distributions for the various values of the dimensionless quantity D defined as

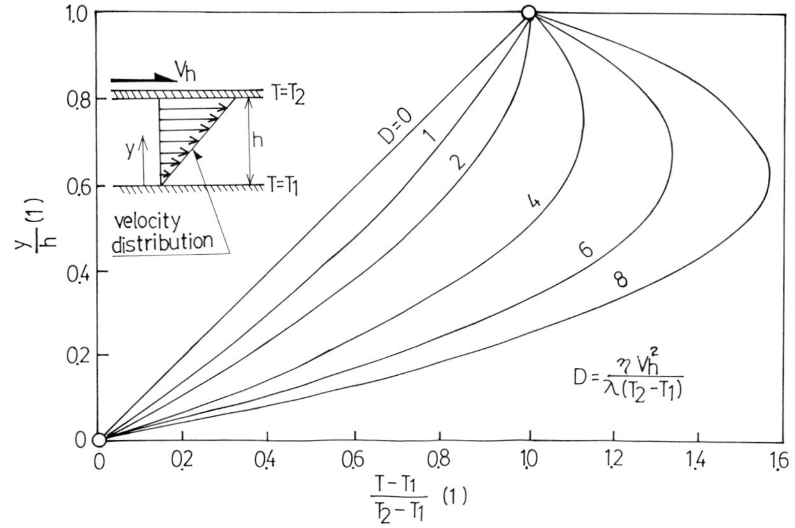

FIGURE 15–3. Temperature distributions between the two-dimensional flat plates.

$$D = \frac{\eta V_h^2}{\lambda(T_2 - T_1)} \tag{15–26}$$

are shown in Figure 15–3.

B. TWO-DIMENSIONAL FLOW IN THE CONCENTRIC CYLINDERS

Let us consider the exact solution for the rotational flow in the concentric cylinders, in which the outer cylinder of the radius R_1 is rotating with the angular velocity Ω_1, while the inner cylinder of the radius R_2 is rotating with the angular velocity Ω_2. So the tangential velocities on the surfaces of the two cylinders are

$$V_{\theta 1} = R_1\Omega_1 \quad , \quad V_{\theta 2} = R_2\Omega_2 \tag{15–27}$$

The momentum equation for the two-dimensional axial-symmetric flow on the incompressible fluid is

$$\rho\frac{V_\theta^2}{r} = \frac{dp}{dr} \tag{15–28}$$

$$\frac{d^2V_\theta}{dr^2} + \frac{d}{dr}\left(\frac{V_\theta}{r}\right) = 0 \tag{15–29}$$

and the fundamental equation of the temperature distribution on the steady condition is

$$\lambda\left(\frac{d^2T}{dr^2} + \frac{1}{r}\frac{dT}{dr}\right) + \eta\left(\frac{dV_\theta}{dr} - \frac{V_\theta}{r}\right)^2 = 0 \tag{15–30}$$

First of all, if we assume that the angular velocities of the outer and inner cylinders are Ω_1 and Ω_2, respectively, the distribution of the tangential velocity is

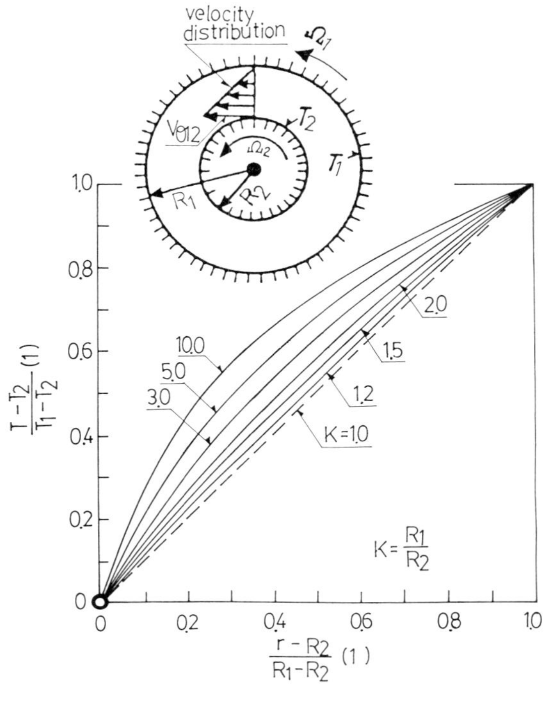

FIGURE 15–4. Temperature distributions between the two concentric cylinders.

$$V_\theta = A \cdot r + \frac{B}{r} \tag{15–31}$$

where A and B are

$$A = \frac{\Omega_1 R_1^2 - \Omega_2 R_2^2}{R_1^2 - R_2^2}, \quad B = \frac{(\Omega_2 - \Omega_1) \cdot R_1^2 \cdot R_2^2}{R_1^2 - R_2^2} \tag{15–32}$$

On the other hand, when we neglect the term of the thermal conductivity of Equation 15–30, the following equation results

$$\frac{T(r) - T_2}{T_1 - T_2} = \frac{\ln(r/R_2)}{\ln(R_1/R_2)} \tag{15–33}$$

The distributions of temperature for the parameter $K = R_1/R_2$ are given in Figure 15–4. For the very small span of $(R_1 - R_2)/R_2$ between the two concentric cylinders, the temperature distribution of Equation 15–33 for $K = R_1/R_2 \rightarrow 1$ gradually approaches the linear distribution of the Couette flow.

Next considering the term of the thermal conductivity of Equation 15–33, Equation 15–30 becomes

$$\frac{1}{r}\frac{d}{dr}\left(r\frac{dT}{dr}\right) = -4\frac{\eta}{\lambda} \cdot \frac{B^2}{r^4} \tag{15–34}$$

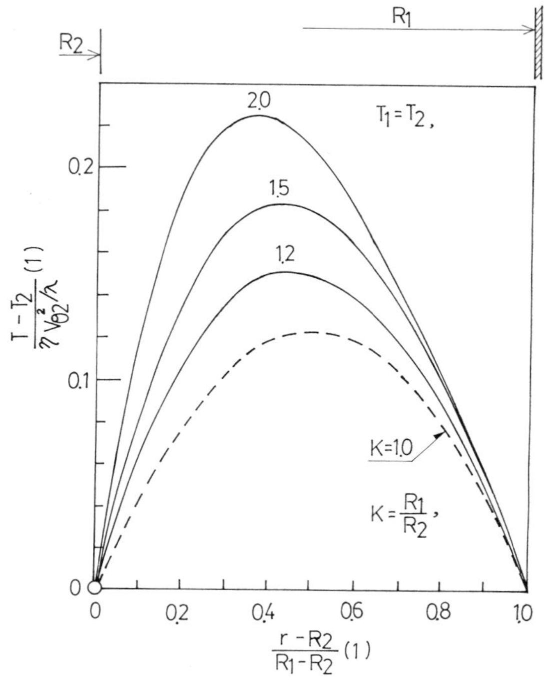

FIGURE 15–5. Temperature distributions between the two concentric cylinders for $T_1 = T_2$.

The solution which satisfies the boundary condition can be obtained as

$$T - T_2 = \frac{\eta}{\lambda} B^2 \left[\left(\frac{1}{R_1^2} - \frac{1}{R_2^2} \right) \frac{\ln(r/R_2)}{\ln(R_1/R_2)} + \frac{1}{R_2^2} - \frac{1}{R_1^2} \right] \tag{15–35}$$

Transforming the above equation with $V_{\theta 2} = (\Omega_2 - \Omega_1) \cdot R_2$, $K = R_1/R_2$, and $\xi = r/R_2$, the dimensionless form is

$$\frac{T - T_2}{(\eta/\lambda) V_{\theta 2}^2} = \frac{1}{\left(1 - \frac{1}{K^2}\right)^2} \left[\left(\frac{1}{K^2} - 1 \right) \frac{\ln \xi}{\ln K} + \left(1 - \frac{1}{\xi^2} \right) \right] \tag{15–36}$$

The calculated distributions of the temperature are given for the parameter $K = R_1/R_2$ in Figure 15–5. In addition to this, in case of $(R_1 - R_2)/R_2 \rightarrow 0$, that is to say, $K \rightarrow 1$, the temperature distribution for $T_1 = T_2$ becomes a parabolic curve.

MAIN SYMBOLS

C_p(J/kg K);	specific heat of constant pressure
D_t(m^2/s);	thermal diffusivity
E(1);	Eckert number
G_r(1);	Grashof number
P_r(1);	Prandtl number

$R_e(1)$;	Reynolds number
$T(K)$;	temperature
$V_\theta(m/s)$;	tangential velocity
$V_\infty(m/s)$;	velocity
$\beta(1/K)$;	temperature coefficient of expansion
$\delta_T(m)$;	thickness of the thermal boundary layer
$\delta_V(m)$;	thickness of the velocity boundary layer
$\eta(Pa \cdot s)$;	viscosity
$\lambda(W/mK)$;	thermal conductivity
$\Phi(J/m^3s)$;	energy dissipation per unit time and per unit volume
$\Omega_1(rad/s)$;	angular velocity of the outer cylinder
$\Omega_2(rad/s)$;	angular velocity of the inner cylinder

REFERENCES

1. **Bejan, A.,** *Entropy Generation through Heat and Fluid Flow,* John Wiley & Sons, New York, 1982.
2. **Furuya, Y.,** *Fluid Mechanics II* (in Japanese), Kyoritzu-Pub. Co., Tokyo, 1973.
3. **Hughes, W. F.,** *An Introduction to Viscous Flow,* Hemisphere-Pub., 1979.
4. **Schlichting, H.,** *Boundary layer theory,* McGraw-Hill, New York, 1960.
5. **Schlichting, H.,** Einige exakte Lösungen für die Temperaturverteilung in einer laminaren Strömung, *Z. A. M. M.,* 31(3), 78, 1951.

Chapter 16

VORTEX MOTION OF COMPRESSIBLE FLUID

I. INTRODUCTION

In this chapter the fundamental idea of the vortex motion of compressible fluid and the relationship between entropy and vorticity in the two-dimensional flow and also the Crocco equation, which can be applied to the fluid motion of gas with rotation, are described in detail.

II. FUNDAMENTAL EQUATION OF INVISCID AND NONCONDUCTING GAS

When a considered system is insulated for the flow of heat from the outside of this system, or when heat does not flow out from this system to the outside, the continuity equation, equation of motion, and energy equation can be written as

$$\frac{\partial \rho}{\partial t} + \mathrm{div}(\rho \mathbf{V}) = 0 \qquad (16\text{–}1)$$

$$\frac{\partial \mathbf{V}}{\partial t} - \mathbf{V} \times \mathbf{\Omega} + \frac{1}{2}\,\mathrm{grad}\,\mathbf{V}^2 = -\frac{1}{\rho}\,\mathrm{grad}\,p + \mathbf{F} \qquad (16\text{–}2)$$

$$\frac{DS}{Dt} = \frac{\partial S}{\partial t} + \mathbf{V} \cdot \mathrm{grad}\,S = 0 \qquad (16\text{–}3)$$

where ρ, $\mathbf{V}$, and S are density of fluid, velocity (vorticity, $\mathbf{\Omega} = \nabla \times \mathbf{V}$), and entropy, respectively. The body force $\mathbf{F}$ produced by the gravity force can be neglected in comparison with the body force of liquid, and further this term $\mathbf{F}$ is in very small quantity in comparison with the inertia and pressure terms in high speed fluid flow.[1] Let us assume that the fluid flow is very high speed and also the change of density of fluid follows the adiabatic change, which does not bring about the interchange of heat between this system and outside, and moreover the fluid flow is an isentropic flow of the reversible adiabatic change.

Here, denoting the velocity of sound by a, a relationship between pressure, density, and sound velocity can be represented by

$$a^2 = \frac{dp}{d\rho} = \frac{kp}{\rho} = kRT \qquad (16\text{–}4)$$

where k is the ratio of specific heat, R is the gas constant, and T is the absolute temperature. Substituting this equation into Equation 16–2 and after transforming this equation, we obtain

$$\frac{\partial \mathbf{V}}{\partial t} - \mathbf{V} \times \mathbf{\Omega} + \frac{1}{2}\,\mathrm{grad}\,\mathbf{V}^2 = -\frac{a^2}{\rho}\,\mathrm{grad}\,\rho \qquad (16\text{–}5)$$

Table 16–1 shows the physical contents of the main materials on the conditions of $p = 101.33$ kPa and $T = 293°K$ ($T = 20°C$).

TABLE 16–1
Main Physical Constants on 101.325 kPa and 20°C

		M (—)	R $\left(\frac{J}{kg \cdot K}\right)$	ρ $\left(\frac{kg}{m^3}\right)$	c_p $\left(\frac{J}{kg \cdot K}\right)$	c_v $\left(\frac{J}{kg \cdot K}\right)$	k (—)	η × 10^6 (Pa · s)	ν × 10^6 $\left(\frac{m^2}{s}\right)$	λ × 10^3 $\left(\frac{W}{m \cdot K}\right)$	D × 10^6 $\left(\frac{m^2}{s}\right)$	Pr (—)
Air	—	28.967	287.03	1.204	1,007	720	1.40	18.1	15.0	25.7	21.2	0.71
Carbon dioxide	CO_2	44.0098	188.92	1.839	847	658	1.29	14.8	8.05	16.3	10.4	0.77
Helium	He	4.0026	2,077.2	0.1664	5197	3120	1.67	19.5	117	148	171	0.68
Hydrogen	H_2	2.0158	4,124.6	0.0838	14,288	10,162	1.41	8.80	105	178	149	0.71
Nitrogen	N_2	28.0134	296.80	1.165	1,041	743	1.40	17.5	15.0	25.5	21.0	0.71
Oxygen	O_2	31.9988	259.83	1.331	919	658	1.40	20.3	15.2	26.2	21.4	0.71
Methane	CH_4	16.0426	518.27	0.6682	2,226	1,702	1.31	10.9	16.3	33.3	22.4	0.73
Water	H_2O	18.0152	461.52	998.2	4,182	4,154	1.01	1,002.6	1.004	598	0.143	7.02

Note: M(1), molecular weight; R(J/kg · K), gas constant; ρ(kg/m^3), density; c_p(J/kg · K), specific heat of constant pressure; c_v(J/kg · K), specific heat of constant volume; k(1), ratio of specific heat; η(Pa · s), viscosity; ν(m^2/s), kinematic viscosity; λ(W/m · K), thermal conductivity; D(m^2/s), thermal diffusivity; Pr(1), Prandtl number.

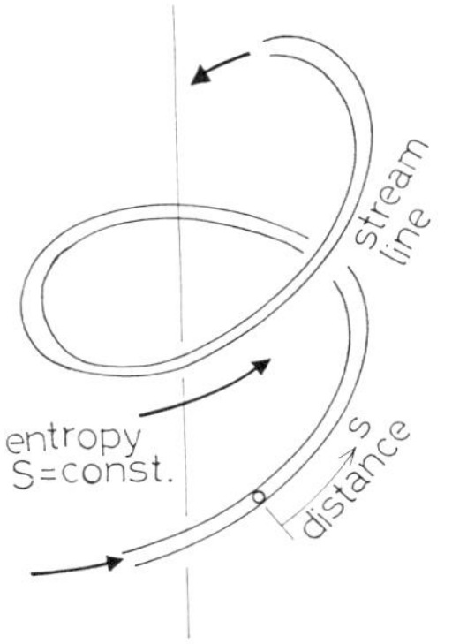

FIGURE 16–1. Entropy along a stream line.

III. STEADY MOTION OF INVISCID AND NONCONDUCTING GAS

Let us neglect the effects of the viscosity and the thermal conductivity of the gas; so, the equations of continuity, equation of motion, and energy equation on the steady state condition ($\partial/\partial t = 0$) can be written as

$$\mathrm{div}(\rho \mathbf{V}) = 0 \tag{16–6}$$

$$\mathbf{V} \times \boldsymbol{\Omega} = \frac{1}{\rho}\,\mathrm{grad}\ p + \frac{1}{2}\,\mathrm{grad}\ \mathbf{V}^2 + \mathrm{grad}\ \Phi \tag{16–7}$$

$$\mathbf{V} \cdot \mathrm{grad}\ S = 0 \tag{16–8}$$

where an external force **F** is assumed to be derived by Potential Φ. Then taking a distance s along a stream line as shown in Figure 16–1, a term dS/ds from Equation 16–8 goes to zero. Therefore the entropy S is constant along the stream line. And further an enthalpy I can be represented by the definition of thermodynamics as follows:

$$\mathrm{grad}\ I = T\ \mathrm{grad}\ S + \frac{1}{\rho}\,\mathrm{grad}\ p \tag{16–9}$$

Substituting Equation 16–9 into Equation 16–7, Equation 16–7 can be transformed

$$\mathbf{V} \times \boldsymbol{\Omega} = \mathrm{grad}\ I - T\ \mathrm{grad}\ S + \frac{1}{2}\,\mathrm{grad}\ \mathbf{V}^2 + \mathrm{grad}\ \Phi \tag{16–10}$$

Moreover taking a scalar product of **V** for both sides of Equation 16–10, it becomes

$$\mathbf{V} \cdot (\mathbf{V} \times \boldsymbol{\Omega}) = \mathbf{V} \cdot \mathrm{grad}\ I - \mathbf{V} \cdot T\ \mathrm{grad}\ S + \mathbf{V}\frac{1}{2}\,\mathrm{grad}\ \mathbf{V}^2 + \mathbf{V} \cdot \mathrm{grad}\ \Phi \tag{16–11}$$

As mentioned above, $\mathbf{V} \cdot \mathrm{grad}\ S = 0$ along a stream line; so, Equation 16–11 can be transformed by the method of vector analysis as follows:

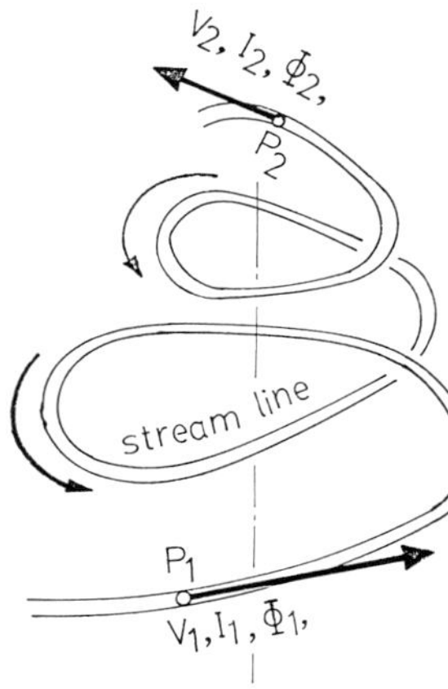

FIGURE 16–2. Illlustration of a Bernoulli theorem along a stream line for a perfect gas.

$$\mathbf{V} \cdot \text{grad}\left(I + \frac{1}{2}\mathbf{V}^2 + \Phi\right) = 0 \tag{16–12}$$

Consequently changing a magnitude of a velocity vector **V** to V, we can obtain the following equation from Equation 16–12

$$I + \frac{1}{2}V^2 + \Phi = K_s = \text{constant} \tag{16–13}$$

This equation is equivalent to Bernoulli's equation, in which the value of K_s is constant along a stream line. Hence as shown in Figure 16–2, denoting that the physical quantities I_1, V_1, and Φ_1 at a point P_1 and I_2, V_2, and Φ_2 at a point P_2 along a spiral stream line, Equation 16–13 can be written as

$$I_1 + \frac{1}{2}V_1^2 + \Phi_1 = I_2 + \frac{1}{2}V_2^2 + \Phi_2 = K_s \tag{16–14}$$

Next since the entropy along a stream line keeps constant value (dS = 0), the following equation can be obtained

$$dI = \frac{dp}{\rho} \tag{16–15}$$

Substituting Equation 16–15 into Equation 16–14, the equation along a stream line becomes

$$\frac{1}{2}V^2 + \int\frac{dp}{\rho} + \Phi = K_s = \text{constant} \tag{16–16}$$

Then a relationship between a sound velocity a, pressure p, and the ratio of the specific heat k becomes

$$a^2 = \frac{dp}{d\rho} = \frac{kp}{\rho} \tag{16–17}$$

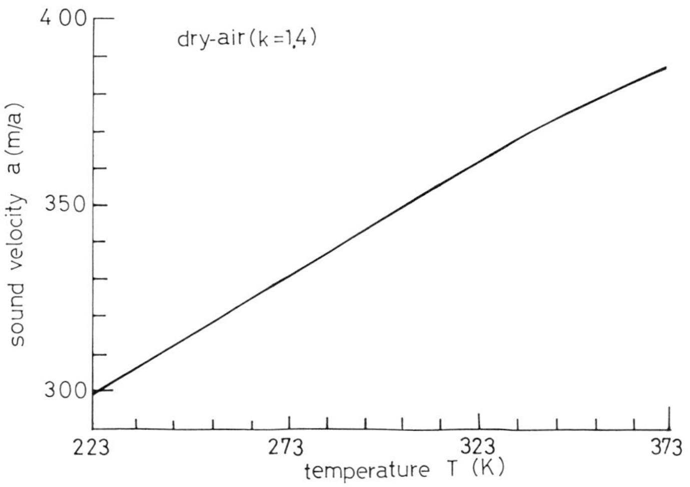

FIGURE 16–3. Relationship between the sound velocity and the gas temperature for k = 1.4.

Now transforming Equation 16–16 with Equation 16–17, the following equation is obtained

$$\frac{1}{2}V^2 + \int a^2 \cdot \frac{d\rho}{\rho} + \Phi = K_s = \text{constant} \tag{16–18}$$

For a perfect gas of a constant specific heat we can transform Equation 16–13 to the following form

$$\begin{aligned} \frac{1}{2}V^2 + C_pT + \Phi &= \frac{1}{2}V^2 + \frac{k}{k-1} \cdot \frac{p}{\rho} + \Phi \\ &= \frac{1}{2}V^2 + \frac{1}{k-1} \cdot a^2 + \Phi = K_s = \text{constant} \end{aligned} \tag{16–19}$$

For reference, Figure 16–3 shows the sound velocity a of dry air (k = 1.4) as a function of temperature T(K).

In case of a motion of gas, we can neglect an external force in comparison with an external force for a motion of liquid. So neglecting a potential Φ, the following equation can be obtained from Equation 16–19

$$\frac{1}{2}V^2 + \frac{1}{k-1} \cdot a^2 = \frac{1}{2}V_{max}^2 = \frac{1}{k-1} \cdot a_o^2 = K_s \tag{16–20}$$

where a_o is a sound velocity at a stationary condition and V_{max} is a value corresponding to a sound velocity a zero (a = 0); that is to say, as shown in Figure 16–4, the theoretically obtained maximum gas velocity spouts through an orifice into a vacuum.

Therefore when we consider the motion of a perfect gas spouting through an orifice of the large chamber into a vacuum (a = 0), the maximum spouting velocity V_{max} can be obtained from Equation 16–20

$$V_{max} = a_o\sqrt{\frac{2}{k-1}} \tag{16–21}$$

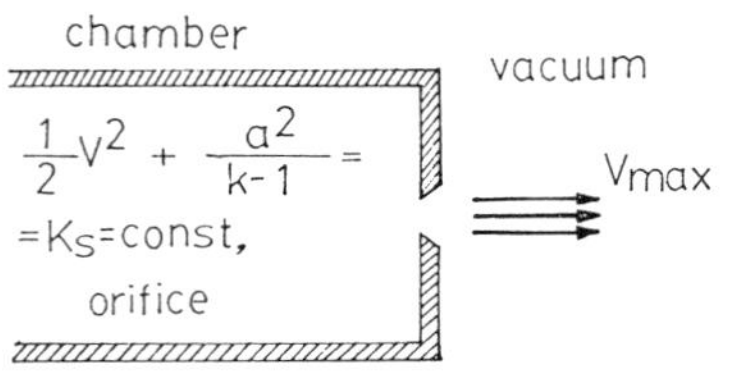

FIGURE 16–4. Relationship between a velocity of gas and a sonic velocity of gas spouting from a chamber into a vacuum.

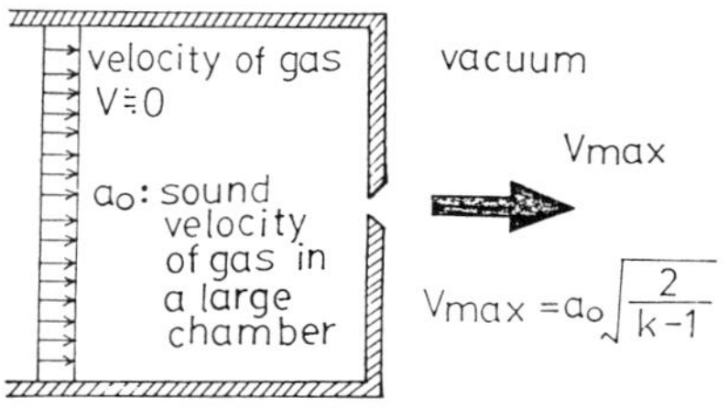

FIGURE 16–5. Maximum spouting velocity of gas ejecting into a vacuum.

where a_o means a sound velocity in this large chamber, since the velocity V of gas may be regarded as $V \fallingdotseq 0$ due to the large chamber as shown in Figure 16–5. Consequently in case of air of the ratio of the specific heat $k = 1.4$ which spouts from this large chamber, the following equation can be obtained from Equation 16–21

$$\frac{V_{max}}{a_o} = \sqrt{\frac{2}{k-1}} = 2.236 \tag{16–22}$$

Hence, the maximum spouting velocity of air reaches about 2.24 times of the sound velocity. In addition to this, when a local fluid velocity V of flow at a point P is equal to a local velocity a^*, that is to say, under a condition of $V = a^*$, we can obtain the following equation from Equation 16–20

$$\frac{1}{2}a^{*2} + \frac{1}{k-1}\cdot a^{*2} = \frac{1}{2}\left(\frac{k-1}{k+1}\right)\cdot a^{*2} = \frac{1}{2}V_{max}^2 \tag{16–23}$$

Then a relationship between a^*, a_o, and V_{max} along a stream tube of an isentropic flow as shown in Figure 16–6 can be written as

$$a^* = V_{max}\cdot\sqrt{\frac{k-1}{k+1}} = a_o\cdot\sqrt{\frac{2}{k+1}} \tag{16–24}$$

IV. VORTEX MOTION

Here let us consider a vortex motion of the simplest two-dimensional fluid flow. Assume that each stream line is a steady rotational motion around a center axis, so that the tangential velocity of fluid flow is a constant value along a stream line of the circular locus of an arbitrary radius r.

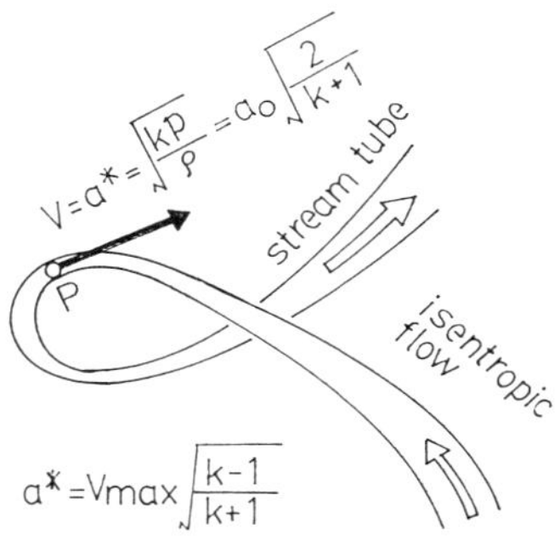

FIGURE 16–6. Illustration of a relation between a local gas velocity and a local sound velocity at a point p on an isentropic flow.

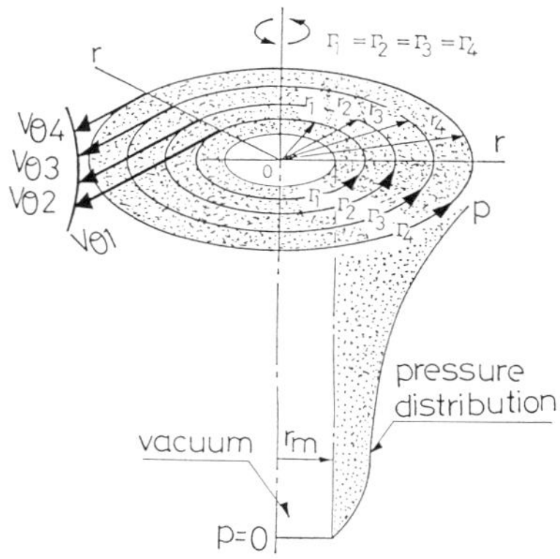

FIGURE 16–7. Illustration of rotational flow of irrotational motion on adiabatic process.

Denoting V_θ as a tangential velocity on a circular locus of a radius r and further assuming that rotational flow is an irrotational motion which has the same circulation Γ along the stream lines as shown in Figure 16–7, the following relation can be written as

$$2\pi r \cdot V_\theta = \Gamma = \text{constant} \tag{16–25}$$

Therefore, from Equation 16–20 of a Bernoulli equation, we obtain the following relation

$$\frac{1}{2}V_\theta^2 + \frac{a^2}{k-1} = \frac{1}{2}V_{max}^2 \tag{16–26}$$

So transforming Equation 16–26,

$$\frac{1}{2}V_\theta^2 = \frac{V_{max}^2}{2}\left(1 - \frac{2}{k-1}\cdot\frac{a^2}{V_{max}^2}\right) = \frac{V_{max}^2}{2}\left(1 - \frac{\rho_o}{\rho}\frac{p}{p_o}\right) \tag{16–27}$$

Here, in the case of the adiabatic process of fluid flow, a relationship between density ρ and pressure p is obtained

$$\frac{p}{p_o} = \left(\frac{\rho}{\rho_o}\right)^k \tag{16–28}$$

So, substituting Equation 16–28 into Equation 16–27, we obtain

$$V_\theta^2 = V_{max}^2\left[1 - \left(\frac{\rho}{\rho_o}\right)^{k-1}\right] = a_o^2\left(\frac{2}{k-1}\right)\left[1 - \left(\frac{\rho}{\rho_o}\right)^{k-1}\right] \quad (16\text{–}29)$$

Consequently, substituting Equation 16–26 into Equation 16–29, a relationship between an arbitrary radius r and density ρ can be represented as

$$r = \frac{\Gamma}{2\pi V_{max}} \cdot \frac{1}{\sqrt{1 - (\rho/\rho_o)^{k-1}}} \quad (16\text{–}30)$$

In general, the following relation for gas must be satisfied as

$$0 < \rho < \rho_o$$

so there exists a minimum radius r_m which keeps a rotational condition from Equation 16–30. Accordingly a relationship between the minimum radius r_m and V_{max} of the maximum tangential velocity becomes

$$2\pi r_m \cdot V_{max} = \Gamma \quad (16\text{–}31)$$

Then combining Equations 16–30 and 16–31, as shown in Figure 16–7, the following relation is obtained

$$\frac{r}{r_m} = \frac{1}{\sqrt{1 - (\rho/\rho_o)^{k-1}}} \quad (16\text{–}32)$$

Therefore, a relationship between density ρ and radius r is

$$\frac{\rho}{\rho_o} = \left[1 - \left(\frac{r_m}{r}\right)^2\right]^{1/(k-1)} \quad (16\text{–}33)$$

Further, a relationship between pressure p and radius r is

$$\frac{p}{p_o} = \left[1 - \left(\frac{r_m}{r}\right)^2\right]^{k/(k-1)} \quad (16\text{–}34)$$

and moreover, a relationship between the tangential velocity V_θ and the radius r is

$$\frac{V_\theta}{V_{max}} = \frac{r_m}{r} = \sqrt{1 - \left(\frac{\rho}{\rho_o}\right)^{k-1}} \quad (16\text{–}35)$$

As a result of this gas motion, the density ρ and the pressure p go to zero at a radius r_m, but the density ρ, the pressure p, and the tangential velocity V_θ become to ρ_o, p_o, and $V_\theta = 0$ at $r = \infty$ (infinity, stagnation point).[2,3]

Moreover let us derive a relationship between the Mach number and radius r. So transform Equation 16–26 with Equation 16–25 as follows:

$$r = \frac{\Gamma}{2\pi a_o} \cdot \sqrt{\left(\frac{k-1}{2}\right) + \frac{1}{M^2}} \quad (16\text{–}36)$$

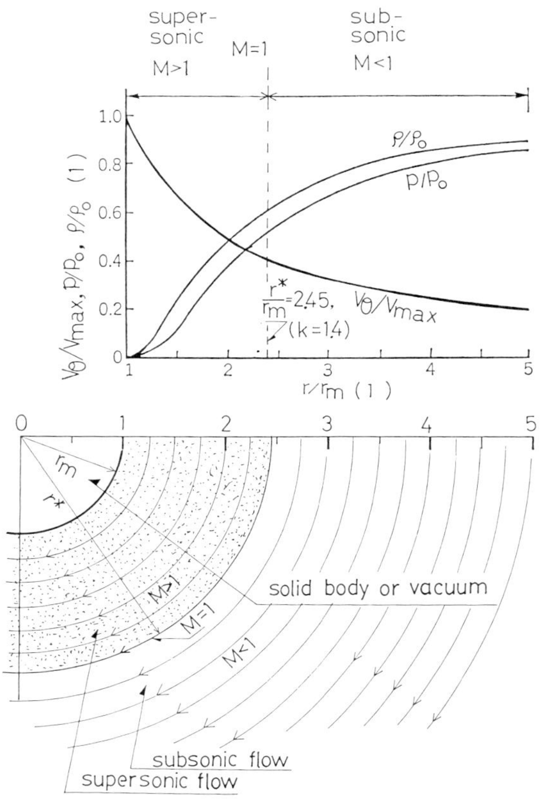

FIGURE 16–8. Distribution of V_θ/V_{max}, p/p_o, and ρ/ρ_o for r/r_m on the adiabatic processes of $k = 1.4$.

where a symbol $M = V_\theta/a$ is called the Mach number. When M goes to infinity, there is a minimum radius $r = r_{min}$ which can be written as

$$r_{min} = \frac{\Gamma}{2\pi a_o} \cdot \sqrt{\frac{k-1}{2}} \tag{16–37}$$

And when M goes to unity, that is to say, when the tangential velocity V_θ is equal to the sound velocity a^*, the radius r^* corresponding to $M = 1$ can be represented by Equation 16–36 as follows:

$$r^* = \frac{\Gamma}{2\pi a_o} \cdot \sqrt{1 + \frac{k-1}{2}} \tag{16–38}$$

Consequently the ratio of r^* to r_{min} is

$$\frac{r}{r_{min}} = \frac{\sqrt{1 + (k-1)/2}}{\sqrt{(k-1)/2}} = \sqrt{\frac{k+1}{k-1}} \tag{16–39}$$

Here Figure 16–8 shows the dimensionless distributions of ρ/ρ_o, p/p_o, and V_θ/V_{max} to the dimensionless radius r/r_m for a gas (air) of the ratio for the specific heat ($k = 1.4$).

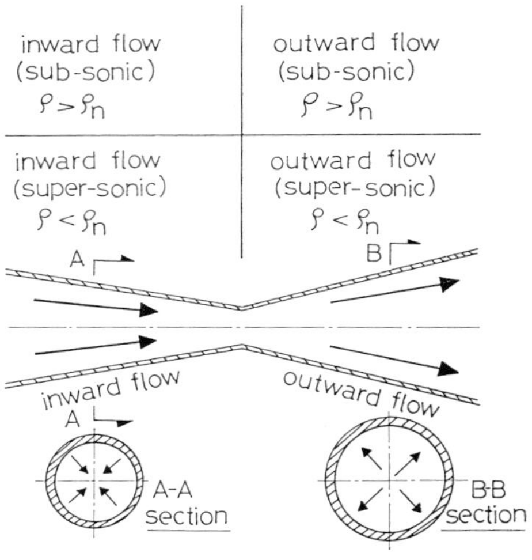

FIGURE 16–9. Illustration of the four kinds of flow regimes.

V. RADIAL FLOW

Next let us consider the two-dimensional motion of radial flow. As shown in Figure 16–9, a symbol V_r is the radial velocity at the radius r; so, the continuity equation is

$$2\pi\rho r V_r = m = \lambda\rho_o \qquad (16\text{–}40)$$

where a symbol m means a strength of source per unit length. From Equation 16–40 we can obtain the following equation

$$V_r = \frac{\lambda}{2\pi r(\rho/\rho_o)} \qquad (16\text{–}41)$$

Then combining Equation 16–41 with the Bernoulli equation of Equation 16–20, the following relation can be written

$$V_r^2 = V_{max}^2\left[1 - \left(\frac{\rho}{\rho_o}\right)^{k-1}\right] \qquad (16\text{–}42)$$

Substituting Equation 16–41 into Equation 16–42, the following equation can be written

$$\left(\frac{\rho}{\rho_o}\right)^2\left[1 - \left(\frac{\rho}{\rho_o}\right)^{k-1}\right] = \frac{\lambda^2}{4\pi^2 r^2\, V_{max}^2} \qquad (16\text{–}43)$$

From this equation we can guess the maximum value of ρ between $\rho = 0$ and $\rho = \rho_o$, since Equation 16–43 goes to zero at $\rho = 0$ and $\rho = \rho_o$. The maximum value of ρ can be easily derived by differentiating Equation 16–43 as follows

$$\left(\frac{\rho}{\rho_o}\right)^{k-1} = \frac{2}{k + 1} \qquad (16\text{–}44)$$

Moreover applying Equation 16–24 to Equation 16–44, Equation 16–44 can be transformed

$$\frac{2}{k+1} = \left(\frac{a^*}{a_o}\right)^2 = \frac{T}{T_o} = \frac{\rho_o}{\rho} \cdot \frac{p}{p_o} = \left(\frac{\rho}{\rho_o}\right)^{k-1} \tag{16–45}$$

Here denoting a suffix n as the maximum value of ρ, so $\rho = \rho_n$. Therefore denoting a symbol $r = r_n$ corresponding to $\rho = \rho_n$, Equation 16–43 becomes

$$r_n^2 = \frac{k+1}{k-1} \cdot \frac{\lambda^2}{4\pi^2 V_{max}^2} \left(\frac{\rho}{\rho_n}\right)^2 \tag{16–46}$$

On the other hand, from Equation 16–42 we can obtain a relationship between V_{max} and V_{rn} as

$$V_{rn} = \sqrt{\frac{k-1}{k+1}} \cdot V_{max} \tag{16–47}$$

Substituting Equation 16–47 into Equation 16–46, we obtain

$$r_n = \frac{\lambda}{2\pi V_{rn}} \left(\frac{\rho_o}{\rho_n}\right) \tag{16–48}$$

and further substituting Equation 16–48 into Equation 16–43, the following equation is obtained

$$\left(\frac{\rho}{\rho_o}\right)^2 \left[1 - \left(\frac{\rho}{\rho_o}\right)^{k-1}\right] = \frac{k-1}{k+1} \cdot \left(\frac{2}{k+1}\right)^{2/(k-1)} \left(\frac{r_n}{r}\right)^2 \tag{16–49}$$

Then a relationship between p and r can be derived with Equations 16–28 and 16–48 as

$$\frac{V_r}{V_{rn}} = \frac{\rho_n}{\rho_o} \frac{\rho_o}{\rho} \frac{r_n}{r} = \left(\frac{2}{k+1}\right)^{1/(k-1)} \cdot \frac{\rho_o}{\rho} \cdot \frac{r_n}{r} \tag{16–50}$$

Consequently the flow regimes for radial flow of gas are divided into four modes, that is to say, as shown in Figure 16–9, the direction of radial flow is outward or inward, and also the density ρ of gas is larger or smaller than the density ρ_n. When the density ρ of gas is larger than ρ_n, the flow of gas is subsonic, and when ρ is smaller than ρ_n, the flow of gas is everywhere supersonic. Here the distributions of ρ/ρ_o, p/p_o, and V_r/V_{max} for r/r_n are shown in Figure 16–10, where a solid line corresponds to subsonic flow and a dotted line corresponds to supersonic flow.

VI. SPIRAL FLOW

Here we consider the characteristics of the spiral vortex flow. Then the continuity equation, the circulation, and the equation of energy can be represented by the following equations

$$2\pi\rho r V_r = m = \lambda\rho_o = \text{constant} \tag{16–51}$$

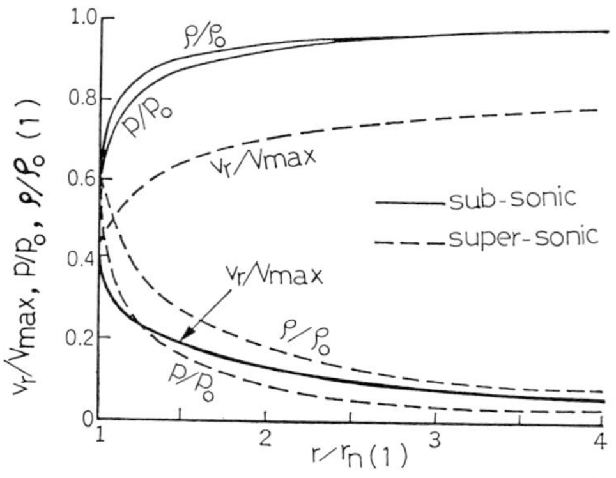

FIGURE 16–10. Illustration of radial flow for subsonic ($M < 1$) and supersonic ($M > 1$) flows.

$$2\pi r V_\theta = \Gamma = \text{constant} \tag{16–52}$$

$$V_r^2 + V_\theta^2 = V_{max}^2\left[1 - \left(\frac{\rho}{\rho_o}\right)^{k-1}\right] \tag{16–53}$$

Combining these three equations, we obtain

$$r^2 = \frac{1}{4\pi^2 V_{max}^2}\,\frac{\Gamma^2(\rho/\rho_o)^2 + \lambda^2}{(\rho/\rho_o)^2[1 - (\rho/\rho_o)^{k-1}]} \tag{16–54}$$

This equation shows that, when the density of fluid goes to zero or approaches $\rho = \rho_o$, the radius r approaches infinity. Consequently there is a miminum value of the radius r for a given density ρ between $\rho = 0$ and $\rho = \rho_o$. This relation can be determined by the practice of the logarithmic differentiation as

$$\frac{\lambda^2}{\Gamma^2(\rho/\rho_o)^2 + \lambda^2} = \frac{1}{2}(k - 1)\cdot\left[\frac{(\rho/\rho_o)^{k-1}}{1 - (\rho/\rho_o)^{k-1}}\right] \tag{16–55}$$

The value of ρ/ρ_o corresponding to the minimum value of the radius for the given values of Γ and λ can be estimated by the trial and error method. Then calculating the radial velocity V_r for the given value of λ^2, at the position of the minimum radius r_s, V_r can be represented as

$$V_r^2 = \frac{\lambda^2}{4\pi^2 r^2(\rho/\rho_o)^2} = \frac{1}{2}\cdot(k - 1)\cdot V_{max}^2\left(\frac{\rho}{\rho_o}\right)^{k-1} = a_o^2\cdot\left(\frac{\rho}{\rho_o}\right)^{k-1} = a^2 \tag{16–56}$$

This equation shows that the radial velocity V_r approaches the sound velocity at $r = r_s$. This physical phenomenon is similar to the relation for the sound velocity (Mach number $M = 1$) at the throat of the minimum cross-section in the Laval nozzle.

The stream line can be described by using the stream function Ψ which is defined as

$$\frac{1}{r}\frac{\partial\Psi}{\partial\theta} = \frac{\rho}{\rho_o}V_r = \frac{\lambda}{2\pi r} \tag{16–57}$$

$$\frac{\partial \Psi}{\partial r} = -\frac{\rho}{\rho_o} V_\theta = -\frac{\Gamma}{2\pi} \cdot \frac{(\rho/\rho_o)}{r} \tag{16–58}$$

Here the relationship between the radius r and the density ratio ρ/ρ_o is given by Equation 16–54, and also Equations 16–57 and 16–58 satisfy the two-dimensional continuity equation of

$$\frac{1}{r}\frac{\partial(\rho r V_r)}{\partial r} + \frac{1}{r}\frac{\partial(\rho V_\theta)}{\partial \theta} = 0 \tag{16–59}$$

Now let us proceed in the calculation for air of the ratio of the specific heat $k = 1.4$ as a fluid. The relationship between ρ/ρ_o and $A = a/a_o$ is

$$\frac{\rho}{\rho_o} = \left(\frac{a}{a_o}\right)^{2/(k-1)} = \left(\frac{a}{a_o}\right)^5 = A^5 \tag{16–60}$$

So Equation 16–54 is reduced to

$$r^2 = \frac{1}{4\pi^2 V_{max}^2} \cdot \frac{\Gamma^2 A^{10} + \lambda^2}{A^{10}(1 - A^2)} \tag{16–61}$$

Differentiating the above equation and transforming, we obtain

$$\frac{dr}{r} = \left(\frac{A^2}{1 - A^2} - \frac{5\lambda^2}{\Gamma^2 A^{10} + \lambda^2}\right)\frac{dA}{A} \tag{16–62}$$

Further, substituting Equation 16–62 into Equation 16–58, the following equation is obtained

$$d\Psi = -\frac{\Gamma}{2\pi}\left(\frac{\rho}{\rho_o}\right)\frac{dr}{r} = -\frac{\Gamma}{2\pi}\left(\frac{A^6}{1 - A^2} - \frac{5\lambda^2 A^4}{\Gamma^2 A^{10} + \lambda^2}\right) dA \tag{16–63}$$

Integrating the above equation, we obtain

$$\Psi = \frac{\Gamma}{2\pi}\left(\frac{\lambda}{\Gamma}\tan^{-1}\frac{\Gamma A^5}{\lambda} - \frac{1}{2}\ln\frac{1 + A}{1 - A} + A + \frac{1}{3}A^3 + \frac{1}{5}A^5\right) + f(\theta) \tag{16–64}$$

From Equation 16–57 with Equation 16–64, the following equation can be obtained as

$$\frac{\partial \Psi}{\partial \theta} = \frac{\partial f}{\partial \theta} = \frac{\lambda}{2\pi} \tag{16–65}$$

The undetermined function $f(\theta)$ can be calculated as

$$f(\theta) = \frac{\lambda}{2\pi} \cdot \theta \tag{16–66}$$

Therefore the stream function is

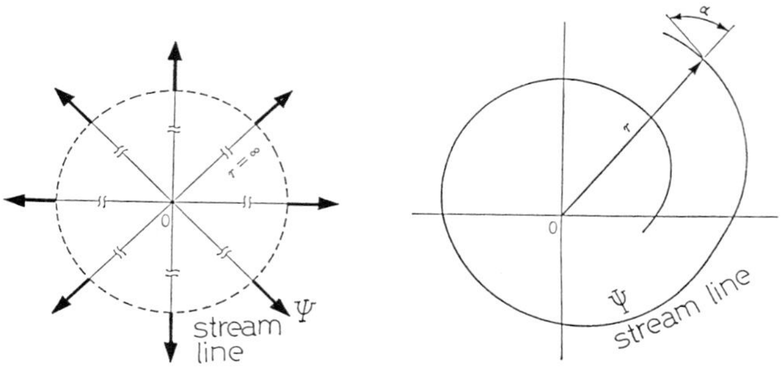

Stream line for A → 0.

Stream line for A=1 (equi-angular spiral)

FIGURE 16–11. Stream lines for the two kinds of value of A.

$$2\pi\Psi = \lambda\left(\theta + \tan^{-1}\frac{\Gamma A^5}{\lambda}\right) - \Gamma\left(\frac{1}{2}\ln\frac{1 + A}{1 - A} - A - \frac{1}{3}A^3 - \frac{1}{5}A^5\right) \quad (16\text{–}67)$$

Now let us consider the limiting value of $A = (a/a_o) \to 0$, so the values of Ψ and r become

$$\Psi = \frac{\lambda\theta}{2\pi} \quad , \quad r \to \infty \quad (16\text{–}68)$$

Consequently the stream line gradually approaches the radial flow as shown in Figure 16–11. On the other hand when the value of A approaches $A \to 1$, Equation 16–54 reduces to

$$2 \cdot \ln r + \ln(1 - A) = \text{constant} \quad (16\text{–}69)$$

Therefore Equation 16–67 becomes

$$2\pi\Psi \to \lambda\theta - \Gamma\ln r + \text{constant} \quad (16\text{–}70)$$

This equation shows the stream line gradually approaching to the equiangular spiral as

$$r \propto \exp(\lambda\theta/\Gamma) \quad (16\text{–}71)$$

As shown in Figure 16–12, this relation can be expressed as

$$\cot\alpha = \frac{1}{r}\cdot\frac{dr}{d\theta} = \frac{1}{\exp(\lambda\theta/\Gamma)}\cdot\frac{\lambda}{\Gamma}\exp\left(\frac{\lambda\theta}{\Gamma}\right) = \frac{\lambda}{\Gamma} = \text{constant} \quad (16\text{–}72)$$

This relation shows the angle α which is independent of r to be the constant value, that is to say, the equiangular spiral (logarithmic spiral) gives the angle between the tangent of this stream line and the direction perpendicular to this stream line to be a constant value. This is a feature of this equiangular spiral.

Figure 16–12 shows the stream lines which have a point of intersection of the subsonic and supersonic lines on the critical radius of $r/r_s = 1$ for $\lambda = \Gamma$.

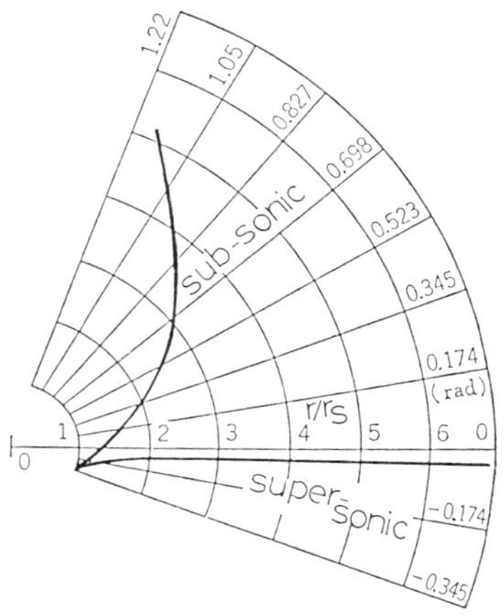

FIGURE 16–12. Stream lines for $\lambda = \Gamma$. (From Howarth, L., *Modern Developments in Fluid Dynamics, High Speed Flow,* Vols. I and II, Oxford University Press, 1956. With permission.)

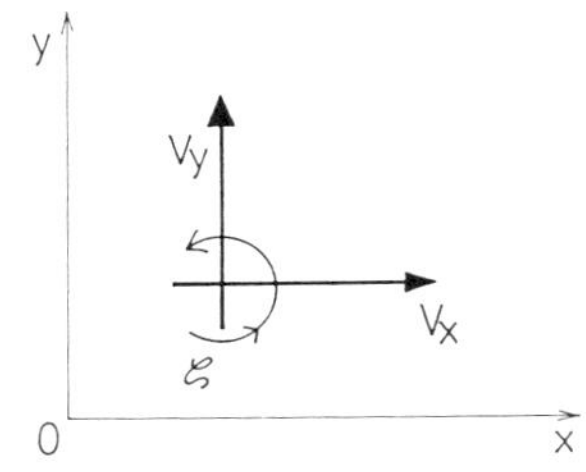

FIGURE 16–13. Vorticity and velocity components on the two-dimensional flow.

VII. RELATIONSHIP BETWEEN ENTROPY AND VORTICITY IN THE TWO-DIMENSIONAL FLOW

A. EQUATION OF VORTICITY

A vorticity equation for two-dimensional flow can be represented as shown in Figure 16–13 as

$$\zeta = \frac{\partial V_y}{\partial x} - \frac{\partial V_x}{\partial y} \tag{16–73}$$

Then the equation of Euler and the continuity equation can be represented as

$$\frac{\partial V_x}{\partial t} + V_x \frac{\partial V_x}{\partial x} + V_y \frac{\partial V_x}{\partial y} = -\frac{1}{\rho}\frac{\partial p}{\partial x} \tag{16–74}$$

$$\frac{\partial V_y}{\partial t} + V_x \frac{\partial V_y}{\partial x} + V_y \frac{\partial V_y}{\partial y} = -\frac{1}{\rho}\frac{\partial p}{\partial y} \tag{16–75}$$

$$\frac{\partial \rho}{\partial t} + \frac{\partial(\rho V_x)}{\partial x} + \frac{\partial(\rho V_y)}{\partial y} = 0 \qquad 16–76)$$

Here eliminating pressure p with combining Equations 16–74 and 16–75 and after that substituting Equation 16–73, the following equation is obtained

$$\frac{\partial \zeta}{\partial t} + V_x \frac{\partial \zeta}{\partial z} + V_y \frac{\partial \zeta}{\partial y} = -\left(\frac{\partial V_x}{\partial x} + \frac{\partial V_y}{\partial y}\right) \zeta \qquad (16\text{–}77)$$

Next, transforming Equation 16–77 by combining with Equation 16–76, we obtain

$$\frac{\partial \zeta}{\partial t} + V_x \frac{\partial \zeta}{\partial x} + V_y \frac{\partial \zeta}{\partial y} = \frac{1}{\rho}\left(\frac{\partial \rho}{\partial t} + V_x \frac{\partial \rho}{\partial x} + V_y \frac{\partial \rho}{\partial y}\right) \qquad (16\text{–}78)$$

or by the simplified form

$$\frac{D\zeta}{Dt} = \frac{\zeta}{\rho} \frac{D\rho}{Dt} \qquad (16\text{–}79)$$

In the case of the incompressible fluid, the density of fluid does not change; so, the right-hand side of Equation 16–79 always goes to zero. But in the case of the compressible fluid, the density of fluid is not always constant. So the change of density contributes to the change of the vorticity.

Moreover, in two-dimensional steady flow ($\partial\rho/\partial t = 0$), let us assume a function $\Psi(x,y)$ which satisfies the continuity equation as

$$\frac{\partial(\rho V_x)}{\partial x} + \frac{\partial(\rho V_y)}{\partial y} = 0 \qquad (16\text{–}80)$$

So, the velocity components V_x and V_y can be represented by

$$V_x = \frac{\rho_o}{\rho} \frac{\partial \Psi}{\partial y} \quad , \quad V_y = -\frac{\rho_o}{\rho} \frac{\partial \Psi}{\partial x} \qquad (16\text{–}81)$$

Therefore, the function $\Psi(x,y)$ is not inconsistent with the rule of the conservation of mass. Further let us assume that

$$\Psi(x,y) = \text{constant} \qquad (16\text{–}82)$$

or

$$d\Psi = \frac{\partial \Psi}{\partial x} dx + \frac{\partial \Psi}{\partial y} dy = 0 \qquad (16\text{–}83)$$

so we can obtain the following equation by combining with Equation 16–81

$$-\rho\, V_y\, dx + \rho\, V_x\, dy = 0 \qquad (16\text{–}84)$$

or we can transform as

$$\frac{dx}{V_x} = \frac{dy}{V_y} \qquad (16\text{–}85)$$

This equation gives the equation of the stream line in the two-dimensional flow; consequently, the function $\Psi(x,y)$ is equivalent to the stream function, while the equation of the vorticity can be represented by combining with Equation 16–81 as follows:

$$-\zeta = \frac{\partial}{\partial x}\left(\frac{\rho_o}{\rho}\frac{\partial \Psi}{\partial x}\right) + \frac{\partial}{\partial y}\left(\frac{\rho_o}{\rho}\cdot\frac{\partial \Psi}{\partial y}\right) \tag{16–86}$$

B. CROCCO EQUATION

First of all let us transform Equations 16–74 and 16–75 as follows:

$$\frac{\partial V_x}{\partial t} + \frac{\partial}{\partial x}\left[\frac{1}{2}(V_x^2 + V_y^2)\right] - V_y\zeta = -\frac{1}{\rho}\frac{\partial p}{\partial x} \tag{16–87}$$

$$\frac{\partial V_y}{\partial t} + \frac{\partial}{\partial y}\left[\frac{1}{2}(V_x^2 + V_y^2)\right] + V_x\zeta = -\frac{1}{\rho}\frac{\partial p}{\partial y} \tag{16–88}$$

Then when we define the enthalpy h as $h = e + p/\rho$ and the heat quantity q per unit mass as $q = de + P\,d(1/\rho)$, where e is the internal energy, we can obtain the following equation

$$dq = d\left(h - \frac{p}{\rho}\right) + p\,d\left(\frac{1}{\rho}\right) = dh - \frac{dp}{\rho} = dh - \left(\frac{1}{\rho}\frac{\partial p}{\partial x}dx + \frac{1}{\rho}\frac{\partial p}{\partial y}dy\right)$$
$$= dh + \left(V_x\frac{\partial V_x}{\partial x} + V_y\frac{\partial V_x}{\partial y}\right)dx + \left(V_x\frac{\partial V_y}{\partial x} + V_y\frac{\partial V_y}{\partial y}\right)dy$$

However, on the stream line the above equation can be transformed with $V_x dy = V_y dx$ as $dq = dh + d[{}^1\!/\!_2(V_x^2 + V_y^2)]$. Therefore in the flow process of the adiabatic change $dq = 0$, we can obtain the following equation, $h_o = h + {}^1\!/\!_2(V_x^2 + V_y^2)$.

Here let us define the energy equation of the adiabatic flow as

$$h + \frac{1}{2}(V_x^2 + V_y^2) = h_o \tag{16–89}$$

where the symbols h(J/kg) and h_o(J/kg) denote the enthalpy and the stagnation enthalpy per unit mass, and then substituting Equation 16–89 into Equations 16–87 and 16–88, the above equations can be transformed

$$\frac{\partial V_x}{\partial t} + \frac{\partial h_o}{\partial x} - V_y\zeta = \frac{\partial h}{\partial x} - \frac{1}{\rho}\frac{\partial p}{\partial x} \tag{16–90}$$

$$\frac{\partial V_y}{\partial t} + \frac{\partial h_o}{\partial y} + V_x\zeta = \frac{\partial h}{\partial y} - \frac{1}{\rho}\frac{\partial p}{\partial y} \tag{16–91}$$

The relationship between enthalpy h and entropy s(J/kg K) can be defined as

$$ds = \frac{dh}{T} - \frac{dp}{T\rho} \tag{16–92}$$

So, Equations 16–90 and 16–91 can be transformed to

$$\frac{\partial V_x}{\partial t} + \frac{\partial h_o}{\partial x} - V_y\zeta = T\frac{\partial s}{\partial x} \tag{16–93}$$

$$\frac{\partial V_y}{\partial t} + \frac{\partial h_o}{\partial y} + V_x\zeta = T\frac{\partial s}{\partial y} \tag{16–94}$$

These equations give the relationship between the change of the entropy and the vorticity, and also these equations are generally called the Crocco equation.[4,5] Further the more general result was derived by Vazsonyi.[6] Especially for steady flow the following two cases are important.

1. In Case of the Stagnation Enthalpy h_o Being a Constant Value All Over the Steady Flow Field

When the stagnation enthalpy h_o is a constant value all over the steady flow field, Equations 16–93 and 16–94 can be reduced to

$$-V_y\zeta = T\frac{\partial s}{\partial x} \quad , \quad V_x\zeta = T\frac{\partial s}{\partial y} \tag{16–95}$$

Therefore, in the case of the stagnation enthalpy h_o = constant, the change of the entropy is accompanied by the production of the vorticity. For example on the surface of the shock wave there is an increment of the entropy. So in the downstream region from this shock wave, the vorticity is produced by Equation 16–95. Therefore in the downstream region from the shock wave, the fluid flow changes to flow with the vorticity.

2. In the Case of the Steady Flow with h_o = constant and $\zeta = 0$

When the stagnation enthalpy is constant and at the same time the vorticity is zero, Equations 16–93 and 16–94 can be reduced to

$$T\frac{ds}{dx} = 0 \quad , \quad T\frac{ds}{dy} = 0 \tag{16–96}$$

Therefore when the entropy s(x,y) becomes constant, the potential flow is equivalent to equientropy flow.

C. POTENTIAL FLOW

If we assume that the flow has no vorticity ($\zeta = 0$), there is a potential flow; accordingly the velocity components V_x and V_y can be represented by a potential $\Phi(x,y)$ as follows:

$$V_x = \frac{\partial \Phi}{\partial x} \quad , \quad V_y = \frac{\partial \Phi}{\partial y} \tag{16–97}$$

The continuity equation in the steady flow condition can be written as

$$V_x\frac{\partial \rho}{\partial x} + \rho\frac{\partial V_x}{\partial x} + V_y\frac{\partial \rho}{\partial y} + \rho\frac{\partial V_y}{\partial y} = 0 \tag{16–98}$$

in which the density gradients $\partial\rho/\partial x$ and $\partial\rho/\partial y$ can be transformed with the equation of motion as follows:

$$\frac{\partial \rho}{\partial x} = \frac{\partial \rho}{\partial p}\frac{\partial p}{\partial x} = -\rho\frac{\partial \rho}{\partial p}\left(V_x\frac{\partial V_x}{\partial x} + V_y\frac{\partial V_x}{\partial y}\right) = -\frac{\rho}{a^2}\left(V_x\frac{\partial V_x}{\partial x} + V_y\frac{\partial V_x}{\partial y}\right) \quad (16\text{–}99)$$

$$\frac{\partial \rho}{\partial y} = \frac{\partial \rho}{\partial p}\frac{\partial p}{\partial y} = -\rho\frac{\partial \rho}{\partial p}\left(V_x\frac{\partial V_y}{\partial x} + V_y\frac{\partial V_y}{\partial y}\right) = -\frac{\rho}{a^2}\left(V_x\frac{\partial V_y}{\partial x} + V_y\frac{\partial v_y}{\partial y}\right) \quad (16\text{–}100)$$

Then Equation 16–98 can be transformed to

$$\left(1 - \frac{V_x^2}{a^2}\right)\frac{\partial V_x}{\partial x} + \left(1 - \frac{V_y^2}{a^2}\right)\frac{\partial V_y}{\partial y} - \frac{V_x V_y}{a^2}\left(\frac{\partial V_x}{\partial y} + \frac{\partial V_y}{\partial x}\right) = 0 \quad (16\text{–}101)$$

Here substituting Equation 16–97 into Equation 16–101, Equation 16–101 can be represented as

$$\frac{\partial^2 \Phi}{\partial x^2} + \frac{\partial^2 \Phi}{\partial y^2} = \frac{1}{a^2}\left[\left(\frac{\partial \Phi}{\partial x}\right)^2 \frac{\partial^2 \Phi}{\partial x^2} + \left(\frac{\partial^2 \Phi}{\partial y^2}\right)\frac{\partial^2 \Phi}{\partial y^2} + 2\frac{\partial \Phi}{\partial x}\frac{\partial \Phi}{\partial y}\frac{\partial^2 \Phi}{\partial x \partial y}\right] \quad (16\text{–}102)$$

This equation is the fundamental equation which determines the velocity potential of the irrotational steady flow of the two-dimensional compressible fluid.

MAIN SYMBOLS

a_o(m/s);	sound velocity
a^*(m/s);	local velocity of sound
e(J/kg);	internal energy per unit mass
C_p(J/kg · K);	specific heat at constant pressure
C_v(J/kg · K);	specific heat at constant volume
h(J/kg);	enthalpy per unit mass
h_o(J/kg);	stagnation enthalpy
k(1);	ratio of specific heat
M(1);	Mach number
p(pa);	static pressure
R(J/kg · K);	gas constant
r_m(m);	minimum radius keeping a rotating condition
r_s(m);	critical radius
s(J/kg · K);	entropy per unit mass
T(K);	temperature
V(m/s);	velocity vector
V_{max}(m/s);	maximum spouting velocity
Γ(m^2/s);	circulation
ζ(1/s);	vorticity
ρ(kg/m^3);	density
Φ(m^2/s);	potential
Ψ(m^2/s);	stream function
$\mathfrak{Q}$(1/s);	vorticity vector

REFERENCES

1. **Howarth, L.,** *Modern Developments in Fluid Dynamics,* High Speed Flow, Vols. I and II, Oxford University Press, New York, 1956.
2. **Piesche, M.,** Über die laminare Wirbelsenkenströmung kompressibler Mediem, *Acta Mech.,* 43, 177, 1982.
3. **Milne-Thomson.,** *Theoretical Hydrodynamics,* 5th ed., Macmillan, New York.
4. **Crocco, L.,** Eine neue Stromfunktion für die Erforschung der Bewegung der Gase mit Rotation, *Z. A. M. M.,* 17(1), 1, 1937.
5. **Crocco, L.,** *Rend. Lincei,* 23(6), 115, 1936.
6. **Vazsonyi, A.,** *Q. Appl. Math.,* Vol. 3, 29, 1945.

Chapter 17

PHYSICS OF THE THERMAL CONDUCTION IN TURBULENT ROTATIONAL FLOW OF COMPRESSIBLE FLUID

I. INTRODUCTION

In this chapter, the fundamental physical idea for the thermal conduction in turbulent rotational flow of compressible fluid by Deissler and Perlmutter and also by Hinze is explained in detail.

This chapter is very important in order to understand the physical meaning of the temperature separation such that the center region of the vortex tube is cool and the outer region is hot.

II. FUNDAMENTAL EQUATION

Let us assume that the rotational flow has axis-symmetry and a steady flow condition, so the Navier-Stokes equation of the incompressible and viscous fluid for the tangential direction can be represented as

$$\rho\left(V_r \frac{dV_\theta}{dr} + \frac{V_r V_\theta}{r}\right) = \eta\left(\frac{d^2V_\theta}{dr^2} + \frac{1}{r}\cdot\frac{dV_\theta}{dr} - \frac{V_\theta}{r^2}\right) \tag{17–1}$$

The shear stress $\tau_{r\theta}$ acting on the cylindrical surface for the tangential direction can be written as

$$\tau_{r\theta} = \eta\left(\frac{dV_\theta}{dr} - \frac{V_\theta}{r}\right) \tag{17–2}$$

Substituting Equation 17–2 into Equation 17–1, we obtain

$$\rho\left(V_r \frac{dV_\theta}{dr} + \frac{V_r V_\theta}{r}\right) = \frac{d\tau_{r\theta}}{dr} + 2\frac{\tau_{r\theta}}{r} = \frac{1}{r^2}\cdot\frac{d}{dr}(r^2\tau_{r\theta}) \tag{17–3}$$

III. REYNOLDS STRESS AND OGAWA COMBINED VORTEX MODEL

Let us denote that the time mean value and the fluctuating value of the velocity in the turbulent rotational flow can be written as $\overline{V}$ and v, respectively, so the velocity components as shown in Figure 17–1 can be represented as

$$\text{radial velocity} \quad V_r = \overline{V}_r + v_r \tag{17–4}$$

$$\text{tangential velocity} \quad V_\theta = \overline{V}_\theta + v_\theta \tag{17–5}$$

$$\text{axial velocity} \quad V_z = \overline{V}_z + v_z \tag{17–6}$$

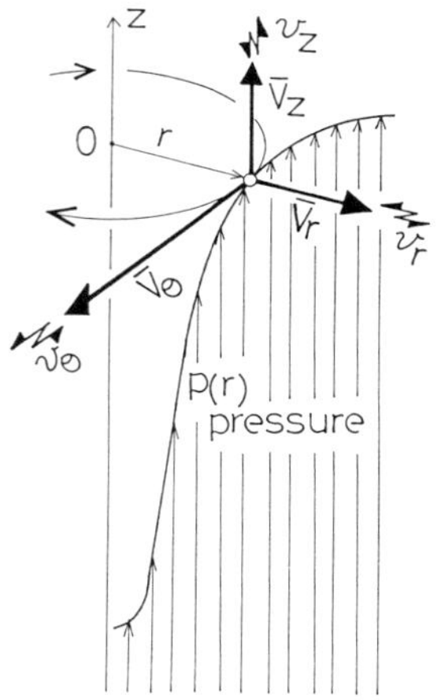

FIGURE 17–1. Velocity components and static pressure.

Then substituting Equations 17–4 and 17–5 into the left-hand side of Equation 17–1 and taking the time mean averaged treatment, each term can be transformed

$$\overline{V_r \frac{dV_\theta}{dr}} = \overline{(\bar{V}_r + v_r)\frac{d}{dr}(\bar{V}_\theta + v_\theta)} = \bar{V}_r \cdot \frac{d\bar{V}_\theta}{dr} + \overline{v_r \frac{dv_\theta}{dr}} \tag{17–7}$$

$$\frac{V_r V_\theta}{r} = \frac{\overline{(\bar{V}_r + v_r)(\bar{V}_\theta + v_\theta)}}{r} = \frac{\bar{V}_r \bar{V}_\theta}{r} + \frac{\overline{v_r v_\theta}}{r} \tag{17–8}$$

Further, substituting Equations 17–7 and 17–8 into Equation 17–1 and assuming that the viscous terms can be neglected in comparison with the Reynolds stress terms in the rotational flow of high turbulence, we obtain

$$\rho \bar{V}_r \frac{d\bar{V}_\theta}{dr} + \rho \frac{\bar{V}_r \bar{V}_\theta}{r} + \rho \overline{v_r \frac{dv_\theta}{dr}} + \rho \frac{\overline{v_r v_\theta}}{r} = 0 \tag{17–9}$$

Next the continuity equation for the axis-symmetric flow can be written as

$$\frac{dV_r}{dr} + \frac{V_r}{r} + \frac{dV_z}{dz} = 0 \tag{17–10}$$

Neglecting the axial gradient of V_z as a small quantity and substituting Equation 17–4 into Equation 17–10, the following equation can be represented as

$$\frac{d\bar{V}_r}{dr} + \frac{\bar{V}_r}{r} = -\frac{dv_r}{dr} - \frac{v_r}{r} = 0 \tag{17–11}$$

Now, multiplying a term ρv_θ on the right-hand side of Equation 17–11 and taking the time averaged treatment, the following equation can be obtained as

$$\rho \overline{v_\theta \frac{dv_r}{dr}} + \rho \frac{\overline{v_r v_\theta}}{r} = 0 \tag{17–12}$$

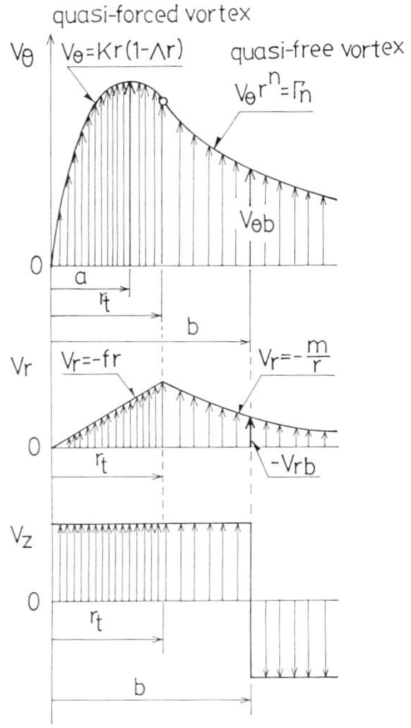

FIGURE 17–2. Schematic illustration of the Ogawa combined vortex model.

Adding Equation 17–12 to the left-hand side of Equation 17–9 and transforming it, we obtain

$$\rho \overline{V}_r \frac{d\overline{V}_\theta}{dr} = -\rho \frac{\overline{V}_r \overline{V}_\theta}{r} - \rho \frac{d}{dr} (\overline{v_r v_\theta}) - 2\rho \frac{(\overline{v_r v_\theta})}{r} \tag{17–13}$$

In order to simplify the representation of the turbulent shear stress τ_t due to the Reynolds stress, we may assume that the apparent eddy kinematic viscosity ν_t keeps a constant value for the radial direction; so, the turbulent shear stress can be represented as

$$\tau_t = -\rho \overline{v_r v_\theta} = \rho \nu_t \left(\frac{d\overline{V}_\theta}{dr} - \frac{\overline{V}_\theta}{r} \right) \tag{17–14}$$

Substituting Equation 17–14 into Equation 17–13, we obtain the following expression

$$\underbrace{\rho \overline{V}_r \frac{d\overline{V}_\theta}{dr} \doteqdot \rho \frac{d\overline{V}_\theta}{dt}}_{\text{acceleration term}} = \underbrace{-\rho \frac{\overline{V}_r \overline{V}_\theta}{r}}_{\text{inertia term}} + \underbrace{\frac{1}{r^2} \cdot \frac{d}{dr} (r^2 \tau_t)}_{\text{turbulent viscous term}} \tag{17–15}$$

First of all we consider concretely how the inertia and turbulent viscous terms contribute to an acceleration term of Equation 17–15. For the distributions of the tangential and radial velocities, the Ogawa combined vortex model may be applied as shown in Figure 17–2.

Let us denote the tangential velocity $V_\theta = V_{\theta b}$ at the radius $r = b$, corresponding to an axial velocity of zero and the exponent $n = 0.5$ of the quasi-free vortex, so the equation of the tangential velocity on the quasi-free vortex region can be represented as

$$V_\theta \cdot r^{0.5} = V_{\theta b} \cdot b^{0.5} \tag{17–16}$$

We obtain the following equation for the boundary condition of the tangential velocity of the quasi-free and quasi-forced vortices at the radius $r = r_t$

$$K \cdot r_t \cdot \left(1 - \frac{r_t}{2a}\right) = V_{\theta b}\left(\frac{b}{r_t}\right)^{0.5} \tag{17–17}$$

Therefore, a relationship between K and $V_{\theta b}$ can be represented as

$$K = \frac{V_{\theta b}(b/r_t)^{0.5}}{r_t[1 - (r_t/2a)]} \tag{17–18}$$

and also from the relationship $r_t \cdot \wedge = (1 + n)/(2 + n)$ we can obtain the following equation for $n = 0.5$

$$r_t \cdot \wedge = 0.6 \tag{17–19}$$

Further, let us denote the inward radial velocity $V_r = -V_{rb}$ at the radius $r = b$, so we can derive the relationship between V_{rb} and m on the quasi-free vortex region and between V_{rb} and f on the quasi-forced vortex region as

$$m = b \cdot V_{rb} \tag{17–20}$$

$$f = b \cdot V_{rb}/r_t^2 \tag{17–21}$$

We will examine how the inertia and viscous terms influence the accelerative term of Equation 17–15 for the quasi-forced and quasi-free vortex regions, respectively.

IV. QUASI-FORCED VORTEX

In the quasi-forced vortex region $(0 \leq r \leq r_t)$, substituting $V_\theta = K \cdot r(1 - \wedge r)$ and $V_r = -f \cdot r$ into Equation 17–15, the inertia term becomes

$$-\rho\frac{\overline{V}_r\overline{V}_\theta}{r} = \rho f\frac{V_{\theta b} \cdot (b/r_r)^{0.5}}{r_t[1 - (r_t/2a)]} \cdot r \cdot \left(1 - \frac{r}{2a}\right) \tag{17–22}$$

and the turbulent shear stress τ_t becomes

$$\tau_t = \rho\nu_t\left(\frac{d\overline{V}_\theta}{dr} - \frac{\overline{V}_\theta}{r}\right) = -\rho\nu_t \cdot \frac{V_{\theta b} \cdot (b/r_t)^{0.5}}{2ar_t[1 - (r_t/2a)]} \cdot r \tag{17–23}$$

Therefore, the turbulent viscous term becomes

$$\frac{1}{r^2} \cdot \frac{d}{dr}(r^2 \cdot \tau_t) = -\frac{3}{2}\rho\nu_t \cdot \frac{V_{\theta b} \cdot (b/r_t)^{0.5}}{ar_t[1 - (r_t/2a)]} \tag{17–24}$$

Consequently, the dimensionless representation of Equation 17–15 can be represented as

$$\underbrace{\frac{d\overline{V}_\theta/dt}{\overline{V}_{\theta b}\cdot \overline{V}_{rb}/b}}_{\text{accelerative term}} = \frac{(b/r_t)^{7/2}}{[1-(r_t/2a)]}\cdot\left[\underbrace{\left(\frac{r}{b}\right)\cdot\left(1-\frac{b}{2a}\cdot\frac{r}{b}\right)}_{\text{inertia term}} - \underbrace{\frac{3}{2}\cdot\frac{\nu_t}{a\cdot V_{rb}}\left(\frac{r_t}{b}\right)^2}_{\text{turbulent viscous term}}\right] \quad (17\text{–}25)$$

From this equation you will find that the turbulent viscous term is independent of the radius r but depends on the values of $\nu_t \cdot (r_t/b)^2/a \cdot V_{rb}$.

V. QUASI-FREE VORTEX (IN CASE OF n = 0.5)

On that region ($r \geqslant r_t$), the tangential and radial velocities can be represented as

$$V_\theta = V_{\theta b}\cdot\sqrt{b/r} \quad , \quad V_r = -V_{rb}\cdot(b/r) \quad (17\text{–}26)$$

Substituting the equations of V_θ and V_r into Equation 17–15, each term can be described

$$-\rho\frac{\overline{V}_r\overline{V}_\theta}{r} = \rho\frac{\overline{V}_{rb}\cdot\overline{V}_{\theta b}}{b\cdot(r/b)^2\sqrt{r/b}} \quad (17\text{–}27)$$

$$\tau_t = \rho\nu_t\cdot\left(\frac{dV_\theta}{dr}-\frac{V_\theta}{r}\right) = -\frac{3}{2}\rho\nu_t\cdot\sqrt{b}\cdot\overline{V}_{\theta b}\cdot r^{-1.5} \quad (17\text{–}28)$$

Therefore, the turbulent shear stress term can be written as

$$\frac{1}{r^2}\cdot\frac{d}{dr}(r^2\cdot\tau_t) = -\frac{3}{4}\rho\nu_t\cdot\sqrt{b}\cdot\overline{V}_{\theta b}\cdot r^{-2.5} \quad (17\text{–}29)$$

So Equation 17–15 can be described by a dimensionless form

$$\frac{d\overline{V}_\theta/dt}{\overline{V}_{rb}\cdot\overline{V}_{\theta b}/b} = \left(\frac{b}{r}\right)^{2.5} - \frac{3}{4}\cdot\frac{\nu_t}{b\cdot\overline{V}_{rb}}\cdot\left(\frac{b}{r}\right)^{2.5} \quad (17\text{–}30)$$

The turbulent viscous term depends on $\nu_t \cdot (b/r)^{2.5}/b \cdot \overline{V}_{rb}$. So the effect of this term on the accelerative term becomes small, since the eddy kinematic viscosity ν_t in the quasi-free vortex is much lower than the eddy kinematic viscosity on the quasi-forced vortex region.

VI. NUMERICAL EXAMPLE

In order to estimate the effects of the inertia and turbulent terms on the accelerative term, the following numerical values are given:

The radius a corresponding to the maximum tangential velocity, a = 20 mm
The radius b corresponding to the axial velocity to be zero, b = 50 mm
The boundary radius r_t between the quasi-forced and quasi-free vortices,
r_t = 24 mm

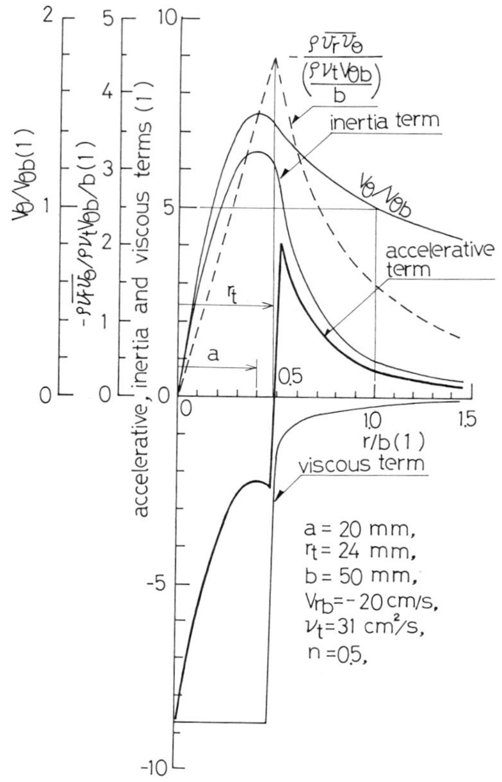

FIGURE 17–3. Numerical example of the accelerative, inertia, and viscous terms on the Ogawa combined vortex model.

The radial inward velocity at r = b, $V_{rb} = -20$ cm/s
The velocity exponent of the quasi-free vortex, n = 0.5
The kinematic eddy viscosity, $\nu_t = 31\ cm^2/s$

The calculated results of the dimensionless forms are shown in Figure 17–3. When the tangential accelerative term is a positive value in the radial region of $r > r_t$, the tangential velocity will be accelerated. While in the region of $0 < r < r_t$, the tangential accelerative term has a negative value; so, the tangential velocity will be decelerated. Since as shown in this figure of calculated results of the turbulent viscous term, the tangentially rotating fluid is strongly retarded by the Reynolds stress in the quasi-forced vortex region, so most of the energy dissipation due to the Reynolds stress is generated in this region. Therefore the temperature of the incompressible rotating fluid in this region increases due to the exchange of the energy dissipation into heat as described in the next section.

For a compressible fluid, as will be described in a later section, the work done by the high Reynolds stress in the quasi-forced vortex region will be transferred into the quasi-free vortex region, so the gas near the center region of the rotating fluid becomes cooled. This physical phenomenon is one of the most important actions for heat separation in the vortex tube.

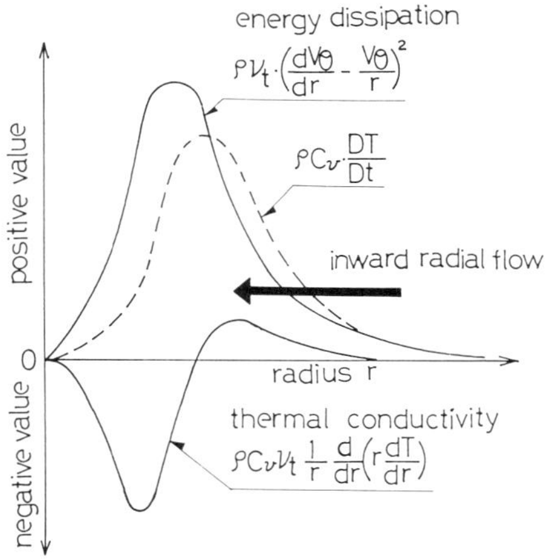

FIGURE 17–4. Schematic illustration of the thermal transfer to the lump of fluid.

VII. TEMPERATURE DISTRIBUTION OF INCOMPRESSIBLE FLUID

Assuming that all of the physical quantities do not change in the z-direction, the energy equation for incompressible fluid is

$$\rho C_v \frac{DT}{Dt} = \lambda \frac{1}{r}\frac{d}{dr}\left(r\frac{dT}{dr}\right) + \eta\left(\frac{dV_\theta}{dr} - \frac{V_\theta}{r}\right)^2 \tag{17–31}$$

where λ(W/m · K) is the thermal conductivity and η(Pa · s) is the viscosity of gas.[1] The fluid flow is considered in this section to be a turbulent condition. So let us exchange the symbols λ and η as follows:

$$\lambda \rightarrow \rho C_v \nu_k \sim \rho C_v \nu_t \quad \text{and} \quad \eta \rightarrow \rho \nu_t \tag{17–32}$$

where a symbol ν_k(m²/s) is the thermal diffusivity, and in case of the simplest assumption the Prandtl number $P_r = \nu/\nu_k$ is to be one.

Substituting Equation 17–32 into Equation 17–31, Equation 17–31 becomes

$$\rho C_v \frac{DT}{Dt} = \rho C_v \nu_t \frac{1}{r} \cdot \frac{d}{dr}\left(r\frac{dT}{dr}\right) + \rho\nu_t\left(\frac{dV_\theta}{dr} - \frac{V_\theta}{r}\right)^2 \tag{17–33}$$

This equation means that the change of the static temperature of an incompressible fluid is affected by heat quantities added to the lump of fluid by thermal conductivity and by energy dissipation due to the eddy viscosity.

Here let us explain the physical meaning of Equation 17–33 as the left-hand side means the rate of change of the internal energy per unit volume of the lump of fluid, the first term of the right-hand side means the heat quantity transferred to the lump of fluid due to the turbulent thermal conductivity, and the second term means the energy dissipation due to the eddy viscosity.

Figure 17–4 shows the schematic illustration of the thermal transfer. This figure is quoted from a paper of Deissler and Perlmutter.[2] In the case of the homogeneous mass flow to the z-axis, when the lump of fluid flows to the center point with the spiral current, so the energy dissipation term works upon an increment of the temperature of the lump of fluid. Then the thermal conductivity term in the central region of the high temperature becomes negative due to losing heat by the thermal conductivity; therefore the lump of fluid receives this transmitted heat in the outside region. The net effect of the rate of change of the static temperature for the lump of fluid per unit time increases when the lump of fluid flows into the center region with the spiral motion.

Consequently, in the incompressible fluid the total temperature at the arbitrary radius is always higher than the static temperature. So it may be impossible to bring about the energy separation in the central region. In view of these facts, the necessary condition which brings on the possibility of the energy separation in the vortex chamber is compressibility of fluid.

However, expansion of the incompressible fluid does not always cause cooling of fluid. So in order to cool down fluid, for example, for the fluid flow in the turbine, the fluid must work in the external region during the expansion of fluid or heat must be transported from the fluid to the external region.

Let us estimate the distribution of the static temperature in the region of the vortex core by the heat conductivity due to the energy dissipation based upon the turbulent flow. The left-hand side of Equation 17–33 can be assumed to be negligible. Then Equation 17–33 becomes

$$\rho C_v \nu_t \frac{1}{r}\frac{d}{dr}\left(r\frac{dT}{dr}\right) + \rho\nu_t\left(\frac{dV_\theta}{dr} - \frac{V_\theta}{r}\right)^2 = 0 \qquad (17\text{–}34)$$

Substituting the equation of the tangential velocity on the quasi-forced vortex flow, the following equation is obtained

$$\frac{d}{dr}\left(r\frac{dT}{dr}\right) = -\frac{K^2\wedge^2}{C_v}r^3 = -\Psi r^3 \qquad (17\text{–}35)$$

The solution of this equation is

$$T = -\frac{\Psi}{16}\cdot r^4 + C_1 \ln r + C_2 \qquad (17\text{–}36)$$

where C_1 and C_2 are the constants of integration. Now let us denote a boundary condition $T = T_o$ at $r = 0$ (center), and also it is necessary to give $C_1 = 0$ for Equation 17–36 to be a finite value. Therefore Equation 17–36 becomes

$$T = -\frac{\Psi}{16}\cdot r^4 + T_o \qquad (17\text{–}37)$$

where the relationship between $\wedge$ and K is

$$\wedge = \frac{1}{2a} \quad , \quad K = \frac{5}{2}\cdot V_{\theta b}\cdot\left(\frac{b}{r_t}\right)^{0.5}\cdot\frac{1}{r_t} \qquad (17\text{–}38)$$

Figure 17–5 shows the distribution of the static pressure of air flow in the region of the quasi-forced vortex for the given conditions as:

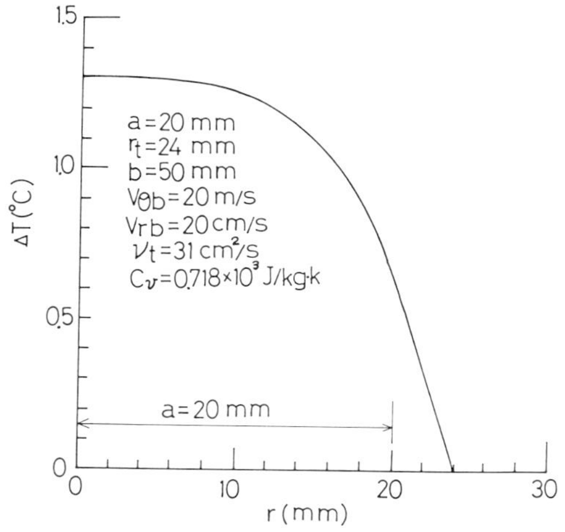

FIGURE 17–5. Distributions of static temperature, in the case of neglecting the energy transfer due to the radial air flow of incompressible rotating fluid.

specific heat of the constant volume, $C_v = 0.718$ kJ/kg · K,
tangential velocity at $r = b = 50$ mm, $v_{\theta b} = 20$ m/s,
$T - T_o = 0$ at $r = r_t = 24$ mm, $T_o = 274.3$ K.

From this figure you will find that the static temperature of air at the vortex center increases until T = 1.3°C, when the energy dissipation of the incompressible fluid is transformed to heat by the turbulence.

VIII. THERMAL CONDUCTIVITY IN THE STABLE ATMOSPHERIC LAYER

Here a physical concept of the turbulent mixing processes in the stable atmospheric layer by F. S. Grunow (1963) will be explained. Then it may be possible to assume that the turbulent mixing motion of fluid is sufficiently fast, so the heat exchange is not produced during the movement of the lump of fluid. As shown in Figure 17–6, when a lump of fluid of pressure p_1 and temperature T_1 at the altitude z_1 at a point 1 is moved to the upward direction, this lump of fluid is assumed to expand adiabatically. Next this lump of fluid, arriving at point 2′ at altitude Z_2, becomes to the state of pressure p_2' and temperature T_2'; therefore this lump of fluid is very chilled. On the other hand it may be possible to assume that during holding of this lump of fluid at the altitude Z_2, this lump heats from temperature T_2' to temperature T_2 of the atmosphere at the altitude Z_2 of a point 2. And after that during a descent motion from the altitude Z_2 of a point 2 to the altitude Z_1 of a point 1′, this lump of fluid experienced adiabatic compression. Finally, before moving again to the upward motion from the altitude Z_1 to the altitude Z_2, this lump of fluid of the temperature T_1' at a point 1′ heated by the adiabatic compression gradually attains the temperature T_1 of the atmosphere of altitude Z_1 at a point 1 due to the surrounding cooled atmosphere.

Therefore, the action of the thermal conductivity can be produced in the opposite direction against the heat flow of the Fourier law by the turbulent mixing processes. So heat is not exchanged by the temperature differences of the two altitudes Z_1 and Z_2 of the lumps of atmospheric fluid, but heat is exchanged by the differences of the newly produced additive

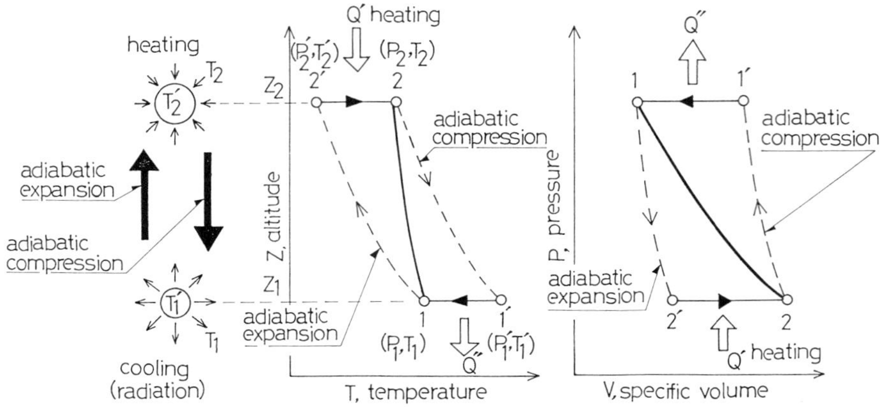

FIGURE 17–6. Schematic illustration of the circulatory processes of the turbulent thermal conductivity on the stable atmospheric layer.

temperatures with the adiabatic expansion process and with the adiabatic compression process. As a result of these processes the transportation of heat produced in the higher layer of the atmosphere moves toward the lower layer. This concept brings a possible explanation of the physical process of how to decrease the temperature and the pressure of the surrounding fluid near the center axis and of how to increase the temperature and the pressure of the surrounding fluid near the outer region far away from the outer axis in the vortex chamber.

IX. PHYSICAL EXPLANATION OF THE SEPARATION OF THE TOTAL TEMPERATURE FOR THE COMPRESSIBLE FLUID

A. THERMAL CONDUCTIVITY DUE TO THE PRESSURE GRADIENT WITH TURBULENT MIXING

As shown in Figure 17–7, it may be assumed that the lump of fluid can be moved from the radius r to the radius $r + \Delta r$ in the radial direction by the turbulent mixing. We regard that Δr is equivalent to the mixing length. We can assume that the lump of fluid moves to another radial position by the development of the instability of the rotational flow at the radius r or at the radius $r + \Delta r$, and after that these moved lumps of fluids will be mixed with the surrounding fluid. Then let us consider the lump of fluid at the radius r, where the mean temperature and the pressure at the radius r is T_1 and p_1, respectively. When this lump of fluid in these conditions traverses to the region of the high pressure p_2 at the radius $r + \Delta r$, this lump of fluid can be regarded to be in a compressed isentropic state. Therefore when this lump of fluid is moved from r to $r + \Delta r$, at the same time according to the rule of the mass conservation the same mass of fluid must be moved out from the radius $r + \Delta r$ to r. Consequently the total heat quantity which is transferred from r to $r + \Delta r$ is

$$C_p(T_2' - T_2) \tag{17–39}$$

where a symbol T_2' is the temperature of the compressed state moved from r to $r + \Delta r$. In consequence the turbulent heat transfer which is transported per unit time and per unit area from r to $r + \Delta r$ is

$$Q_t = C_p \rho f v_r (T_2' - T_2) \tag{17–40}$$

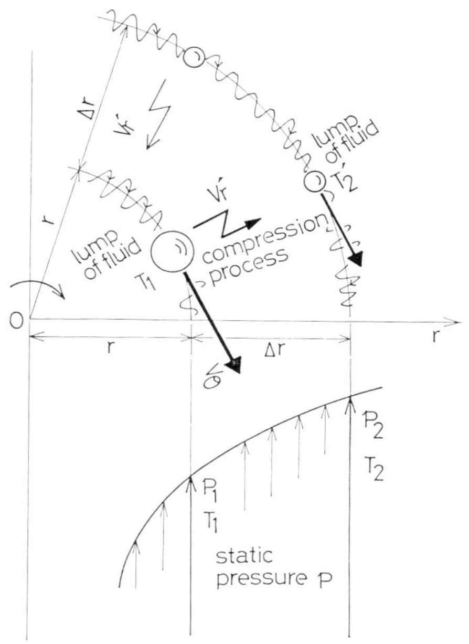

FIGURE 17–7. Schematic illustration of the thermal conductivity due to the pressure gradient.

where a symbol v_r is the mean radial velocity and a symbol f is the rate of the effective area, when the lump of fluid is reached from r to r + Δr. Therefore the above equation may be transformed to

$$Q_t = -C_p\rho f v_r \Delta r\left(\frac{\Delta T_f}{\Delta r} - \frac{\Delta T_e}{\Delta r}\right) \tag{17–41}$$

where $\Delta T_f = T_2 - T_1$ shows the difference of the mean fluid temperatures between r and r + Δr and also $\Delta T_e' = T_2' - T_1$ shows the difference of the temperature change, when this lump of fluid is moving from r to r + Δr.

Therefore substituting $dT = d(T_2 - T_1)$ into the above equation, we obtain

$$Q_t = -C_p\rho f v_r \Delta r\left(\frac{dT}{dr} - \frac{dT_e}{dr}\right) \tag{17–42}$$

Accordingly in the isentropic compressed process, the above equation may be transformed as follows. When temperature, pressure, and density of the lump of fluid in the standard state is T_1, p_1, and ρ_1, the relationship betwen p and T in the isentropic compressed process can be written as

$$T = T_1\left(\frac{p}{p_1}\right)^{(k-1)/k} \tag{17–43}$$

Differentiating by the radius r and transforming, the following equation is obtained

$$\frac{dT_e}{dr} = \frac{1}{\rho C_p} \cdot \frac{dp}{dr} \tag{17–44}$$

This equation shows the relationship between the temperature gradient and the pressure gradient for the radial direction of the lump of fluid. Substituting the above equation in Equation 17–42, we obtain

$$Q_t = -C_p \rho f v_r \Delta r\left(\frac{dT}{dr} - \frac{1}{\rho C_p} \cdot \frac{dp}{dr}\right) \tag{17–45}$$

In this equation when we assume that a term $fv_r\Delta r$ is equivalent to the coefficient of the turbulent diffusivity ν_t, Equation 17–45 can be transformed to

$$Q_t = -C_p \rho \nu_t\left(\frac{dT}{dr} - \frac{1}{\rho C_p} \cdot \frac{dp}{dr}\right) \tag{17–46}$$

This equation is equivalent to the Knoernschild effect which was derived theoretically by Hinze[3] as follows. When there is a pressure difference between the two different altitudes, the transport of heat by turbulent motion is shown by the turbulent mixing of the layer of the atmosphere for the processes of compression and expansion. Readers interested in this effect should refer to the paper by J. O. Hinze.

B. ENERGY EQUATION FOR COMPRESSIBLE FLUID

The energy equation for axis-symmetric steady flow and for the compressible perfect gas can be represented as

$$\rho C_p\left(V_r\frac{\partial T}{\partial r} + V_z\frac{\partial T}{\partial z}\right) - V_r\frac{\partial p}{\partial r} - V_z\frac{\partial p}{\partial z} = \frac{1}{r}\frac{\partial}{\partial r}\left(\lambda r\frac{\partial T}{\partial r}\right) +$$

$$+ \frac{\partial}{\partial z}\left(\lambda\frac{\partial T}{\partial z}\right) + \eta\left\{2\left[\left(\frac{\partial V_r}{\partial r}\right)^2 + \left(\frac{V_r}{r}\right)^2 + \left(\frac{\partial V_z}{\partial z}\right)^2\right] + \left(\frac{\partial V_r}{\partial z} + \frac{\partial V_z}{\partial r}\right)^2 +\right.$$

$$\left. + \left(\frac{\partial V_\theta}{\partial r} - \frac{V_\theta}{r}\right)^2\right\} - \frac{2}{3}\eta\left(\frac{\partial V_r}{\partial r} + \frac{V_r}{r} + \frac{\partial V_z}{\partial z}\right)^2 \tag{17–47}$$

Now in order to facilitate the explanation, the temperature T of gas, the tangential velocity V_θ, and the radial velocity V_r are assumed to be independent of z, and further the terms V_r and V_z are negligible in comparison with a term V_θ. In addition to this, the considered flow condition is to be turbulent flow; so, we can exchange dynamic viscosity for $\rho\nu_t$ and thermal conductivity λ to $\rho C_p \nu_k \doteqdot \rho C_p \nu_t$. Then the above equation is transformed to

$$\rho C_p V_r\frac{dT}{dr} - V_r\frac{dp}{dr} = \frac{1}{r} \cdot \frac{d}{dr}\left(\rho C_p \nu_t r\frac{dT}{dr}\right) + \rho\nu_t\left(\frac{dV_\theta}{dr} - \frac{V_\theta}{r}\right)^2 \tag{17–48}$$

Here ν_k is an eddy diffusivity of heat; we may assume approximately $\nu_k \doteqdot \nu_t$. On the other hand multiplying a term V_θ in the equation of the turbulent rotational flow for the tangential direction, we obtain

$$\rho V_r V_\theta\frac{dV_\theta}{dr} + \frac{\rho V_r V_\theta^2}{r} = \rho\nu_t V_\theta\left(\frac{d^2V_\theta}{dr^2} + \frac{1}{r} \cdot \frac{dV_\theta}{dr} - \frac{V_\theta}{r^2}\right) \tag{17–49}$$

And also the equation of motion for the radial direction can be represented as

$$\frac{dp}{dr} \doteqdot \frac{\rho V_\theta^2}{r}$$

Then substituting the above equation into Equation 17–49, the following equation is obtained

$$\rho V_r V_\theta \frac{dV_\theta}{dr} + V_r \frac{dp}{dr} = \rho \nu_t V_\theta \left(\frac{d^2 V_\theta}{dr^2} + \frac{1}{r} \cdot \frac{dV_\theta}{dr} - \frac{V_\theta}{r^2} \right) \qquad (17\text{–}50)$$

Next substituting an equation of the turbulent shear stress τ_t

$$\tau_t = \rho \nu_t \left(\frac{dV_\theta}{dr} - \frac{V_\theta}{r} \right) \qquad (17\text{–}51)$$

into the above equation, Equation 17–51 is transformed to

$$\frac{1}{r} \cdot \frac{d}{dr}(rV_\theta \tau_t) - \rho \nu_t \left(\frac{dV_\theta}{dr} - \frac{V_\theta}{r} \right)^2 = \rho \nu_t V_\theta \left(\frac{d^2 V_\theta}{dr^2} + \frac{1}{r} \cdot \frac{dV_\theta}{dr} - \frac{V_\theta}{r^2} \right) \qquad (17\text{–}52)$$

Here substituting the above equation into Equation 17–48, the following equation can be represented as

$$\rho C_p r V_r \frac{dT}{dr} + \frac{\rho r V_r}{2} \cdot \frac{dV_\theta^2}{dr} = \frac{d}{dr}\left(\rho C_p \nu_t r \frac{dT}{dr} \right) + \frac{d}{dr}(rV_\theta \tau_t) \qquad (17\text{–}53)$$

Further transforming the above equation, we obtain

$$\rho r V_r \frac{d}{dr}\left(C_p T + \frac{V_\theta^2}{2} \right) = \frac{d}{dr}\left(\rho C_p \nu_t r \frac{dT}{dr} \right) + \rho \nu_t \frac{d}{dr}\left[rV_\theta \left(\frac{dV_\theta}{dr} - \frac{V_\theta}{r} \right) \right] \qquad (17\text{–}54)$$

Therefore from the above equation we may conjecture the following important result that when the radial flow velocity V_r is very strong, or the value of the eddy viscosity ν_t becomes very low, the following equation can be obtained

$$C_p T + \frac{V_\theta^2}{2} = \text{constant} \qquad (17\text{–}55)$$

Namely, total temperature shows the constant value throughout every part of flow.

Further taking into consideration the effect of the radial pressure gradient on the turbulent thermal conductivity for the first term $\rho C_p \nu_t (dT/dr)$ of the right-hand side of Equation 17–54 and then substituting Equation 17–45 into Equation 17–54, Equation 17–54 becomes

$$\rho r V_r \frac{d}{dr}\left(C_p T + \frac{V_\theta^2}{2} \right) = \frac{d}{dr}\left[\rho r C_p \nu_t \left(\frac{dT}{dr} - \frac{1}{C_p} \cdot \frac{dp}{dr} \right) \right] + \\ + \frac{d}{dr}\left[\rho \nu_t r V_\theta \left(\frac{dV_\theta}{dr} - \frac{V_\theta}{r} \right) \right] \qquad (17\text{–}56)$$

And transforming Equation 17–56, we can obtain the following equation as

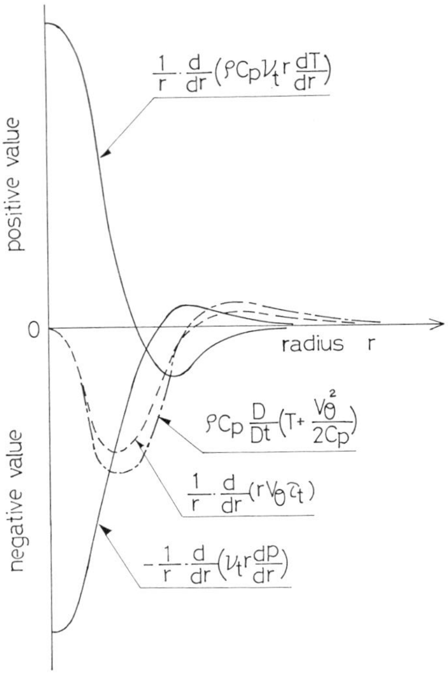

FIGURE 17–8. Schematic illustration of how to separate total temperature of the compressible fluid. (From Deissler, R. G. and Perlmutter, M., *Inst. J. Heat Mass Transfer,* 1, 173, 1960.)

$$\rho C_p \frac{D}{Dt}\left(T + \frac{V_\theta^2}{2C_p}\right) = \frac{1}{r}\cdot\frac{d}{dr}\left(\rho r C_p \nu_t \frac{dT}{dr}\right) - \frac{1}{r}\cdot\frac{d}{dr}\left(\nu_t r \frac{dp}{dr}\right) + \frac{1}{r}\cdot\frac{d}{dr}(rV_\theta\tau_t) \quad (17\text{–}57)$$

Consequently when the lump of fluid with a spiral flow motion wanders into the central axis of vortex motion, the rate of change of total temperature of the lump of fluid of Equation 17–57 will be affected by the following physical effects

$\frac{1}{r}\cdot\frac{d}{dr}\left(\rho C_p \nu_t r \frac{dT}{dr}\right)$;	flux of turbulent thermal conductivity transported to the lump of fluid by the temperature gradient
$-\frac{1}{r}\cdot\frac{d}{dr}\left(\nu_t r \frac{dp}{dr}\right)$;	flux of turbulent thermal conductivity transported to the lump of fluid by the pressure gradient during the processes of compression or expansion of the lump of fluid
$\frac{1}{r}\cdot\frac{d}{dr}(rV_\theta\tau_t)$;	work done by the turbulent shear stress to the lump of fluid.

Here in order to explain more clearly the physical meaning of the rate of change of total temperature in case of the homogeneous axial flow, the author should like to utilize a reference of Deissler and Perlmutter.

Figure 17–8 shows the schematic illustration of how to separate total temperature of the compressible fluid by Deissler and Perlmutter. From this figure you will find that, even if the quantities of the flux of the turbulent thermal conductivity by the temperature gradient (positive value) and by the pressure gradient (negative value) keep high values near the vortex core region, these two terms have positive and negative values and cancel each other out. This means that flux of heat conductivity to the region of the vortex core by the temperature gradient is equivalent to heat flux towards the outward direction by the expansion or by the compression of the lump of fluid with the pressure gradient.

Consequently most of the turbulent shear work contributes to the energy separation. For that reason the value of the turbulent shear work comes up to the value of the rate of change of the total temperature.

And already as shown in Figure 17–3, the value of the negative shear stress increases very abruptly, when the inward spiral flow comes from the region of the quasi-free vortex into the region of the quasi-forced vortex. And further the work done on the fluid element by this shear stress leads to the lowering of the total temperature and also leads to the reduction of the tangential velocity by the eddy viscosity in the quasi-forced vortex region. Accordingly the energy separation strongly depends on the characteristics of the tangential velocity in the region of the quasi-forced vortex. Therefore as a total effect, while the lump of fluid in the high turbulent rotational flow in the region of the quasi-forced vortex flows with a spiral motion into the vortex core region of the rapidly decreasing static pressure, the lump of fluid carries out the turbulent shear work to the outer region during the expansion process. In view of these facts, the energy with the separation of the total temperature is transported from the region of the vortex core to the outer region.

MAIN SYMBOLS

C_p(J/kg · K);	specific heat of the constant pressure
C_v(J/kg · K);	specific heat of the constant volume
p(Pa);	static pressure
P_r(1);	Prandtl number
Q_t(J/m^2 · s);	turbulent heat transfer
r(m);	radius
r_t(m);	boundary radius between quasi-free and quasi-forced vortices
V_r(m/s);	radial velocity
V_z(m/s);	axial velocity
V_θ(m/s);	tangential velocity
η(Pa · s);	viscosity
λ(W/m · K);	thermal conductivity
ν(m^2/s);	kinematic viscosity
ν_t(m^2/s);	eddy kinematic viscosity
ν_k(m^2/s);	thermal diffusivity
ρ(kg/m^3);	density
$\tau_{r\theta}$(Pa);	shear stress

REFERENCES

1. **Schlichting, H.,** Einige exakte Lösungen für die Temperaturverteilung in einer laminaren Strömung, *Z. A. M. M.*, 31, 78, 1951.
2. **Deissler, R. G. and Perlmutter, M.,** Analysis of the flow and energy separation in a turbulent vortex, *Inst. J. Heat Mass Transfer,* 1, 173, 1960.
3. **Hinze, J. O.,** The effect of compressibility on the turbulent transport on heat in a stable stratified atmosphere, Symposium on Atmospheric Diffusion and Air Pollution, Oxford, August 1958, 229.

Chapter 18

VORTEX TUBE

I. INTRODUCTION

In this chapter the physical mechanisms of the vortex tube by the paper of A. J. Reynolds are explained in detail. The Knoernschild effect and the Archimedes effect especially are described. Also, it is shown that the contribution of the work flux due to the Reynolds stress is the most important factor for the energy separation in the turbulent rotational air flow in the vortex tube.[1–8]

II. ENERGY EQUATION IN HIGH TURBULENT FLOW FIELD

A. ENTHALPY IN THE HIGH TURBULENT FLOW

In this section, the author should like to explain physically how the turbulent shear work of compressible fluid is affected by the various distributions of the radial and tangential velocities of gas as discussed in the papers of A. J. Reynolds.[1]

Here let us describe the energy equation of the compressible fluid in the high turbulent flow condition. The velocity vector **V** which is composed of the radial V_r, tangential V_θ, and axial V_z velocities can be written as

$$\mathbf{V}^2 = V_r^2 + V_\theta^2 + V_z^2 \tag{18–1}$$

An enthalpy i(J/kg) can be defined as

$$i = C_p \cdot T \tag{18–2}$$

Therefore, the total enthalpy i_o(J/kg) per unit mass of fluid is composed of the kinetic energy and enthalpy

$$i_o = i + \frac{\mathbf{V}^2}{2} = C_p \cdot T + \frac{\mathbf{V}^2}{2} \tag{18–3}$$

In the high turbulent flow condition we can neglect the viscous and body forces; so, the generalized energy equation can be written

$$\frac{\partial}{\partial t}(\rho i_o) + \nabla \cdot (\rho i_o \mathbf{V}) = \frac{\partial p}{\partial t} \tag{18–4}$$

Now taking a mean value of time for the above equation, and after that assuming the mean motion to be a steady state, the following equation is obtained

$$\nabla \cdot (\overline{\rho i_o \mathbf{V}}) = 0 \tag{18–5}$$

Consequently, in case of the axially symmetrical flow, Equation 18–5 can be transformed to

$$\frac{1}{r} \cdot \frac{\partial}{\partial r} (\overline{r\, \rho V_r i_o}) + \frac{\partial}{\partial z} (\overline{\rho V_z i_o}) = 0 \tag{18–6}$$

Therefore one term ($\rho V_r i_o$) can be represented

$$\overline{\rho V_r i_o} = \overline{(\rho_o + \rho')(\overline{V}_r + v_r)(i_o + i_o')} \tag{18–7}$$

where the symbols ρ', v_r, and i_o' mean the fluctuating components of ρ, V_r, and i_o, respectively. The enthalpies i_o and i can be represented as

$$i_o = I_o + i_o' \quad , \quad i = I + i' \tag{18–8}$$

where a symbol i' means a fluctuating component of i. Accordingly the terms I_o and i_o' can be represented as follows:

$$I_o = I + \frac{1}{2}(V_\theta^2 + V_r^2 + V_z^2 + \overline{v_\theta^2} + \overline{v_r^2} + \overline{v_z^2}) \tag{18–9}$$

$$i_o' = i' + v_r\overline{V}_r + v_\theta\overline{V}_\theta + v_z\overline{V}_z \tag{18–10}$$

B. ORDER ESTIMATION

For the analysis of vortex flow in the vortex tube, we should apply the above-stated energy equation. Let us assume the orders of the velocity components of the turbulent rotational flow in the vortex tube as follows:

Tangential velocity	$V_\theta \sim O(1)$	
Radial velocity	$V_r \sim O(\epsilon)$,	where $\epsilon < 1$
Radius	$r \sim O(1)$	
Radial gradient	$\partial/\partial r \sim O(1)$	
Axial gradient	$\partial/\partial z \sim O(\delta)$,	where $\delta < 1$
Components of the squared turbulent velocities	$\overline{v_\theta^2}$, $\overline{v_r^2}$ and $\overline{v_z^2} \sim O(\beta)$,	where $\beta < 1$
Density fluctuation	$\rho' \sim O(\eta)$,	where $\eta < 1$
Enthalpy fluctuation	$i' \sim O(\alpha)$,	where $\alpha < 1$

Then from a rule of the mass conservation, we can obtain the following equation as

$$\rho_o r V_r = C + O(\delta) = \text{Constant} \tag{18–11}$$

Therefore, the order of the density becomes $\rho_o \sim O(1)$, since the order of $O(\delta)$ is smaller than one. Further let us assign that the order of the total enthalpy is $\Delta I_o \sim O(1)$. So, the energy separation is no more than several times of the dynamic pressure. Certainly all of the quantities α, β, δ, ϵ, η are smaller in comparison with the order O(1); so, the higher orders of the products of these small quantities naturally can be neglected.

Next, concerning the order of the axial velocity V_z, the continuity equation of the compressible fluid and also that of the order estimation can be represented as

$$\frac{1}{r} \cdot \frac{\partial}{\partial r}(\rho r V_r) + \frac{\partial}{\partial z}(\rho V_z) = 0 \tag{18–12}$$

$$\frac{1}{O(1)} \cdot O(1)[O(1) \cdot O(1) \cdot O(\epsilon)] + O(\delta)[O(1) \cdot O(V_z)] \tag{18–13}$$

Therefore, the equation of the order of V_z is

$$O(\epsilon) + O(\delta) \cdot O(V_z) = 0 \tag{18–14}$$

Then from the above equation the order of V_z can be estimated

$$V_z \sim O(\epsilon/\delta) \tag{18–15}$$

Consequently, the orders of the continuity equation which include the turbulent components are

$$\underbrace{\frac{1}{r}\cdot\frac{\partial}{\partial r}(\rho_o r V_r)}_{O(\epsilon)} + \underbrace{\frac{\partial}{\partial z}(\rho_o V_z)}_{O(\epsilon)} + \underbrace{\frac{1}{r}\cdot\frac{\partial}{\partial r}(r\overline{\rho' v_r})}_{O(\beta^{1/2}\eta)} + \underbrace{\frac{\partial}{\partial z}(\overline{\rho' v_z})}_{O(\beta^{1/2}\eta\delta)} = 0 \tag{18–16}$$

From the above equation, neglecting the minimum order $O(\beta^{1/2}\eta\delta)$, we may write

$$\frac{1}{r}\cdot\frac{\partial}{\partial r}(\rho_o r V_r) + \frac{\partial}{\partial z}(\rho_o V_z) + \frac{1}{r}\frac{\partial}{\partial r}(r\overline{\rho' v_r}) \doteqdot 0 = O(\beta^{1/2}\delta\eta) \tag{18–17}$$

In addition to this, the orders of the energy equation can be estimated approximately as

$$\underbrace{\frac{1}{r}\cdot\frac{\partial}{\partial r}(\rho_o r V_r I_o)}_{O(\epsilon)} + \underbrace{\frac{\partial}{\partial z}(\rho_o V_z I_o)}_{O(\epsilon)} + \underbrace{\frac{1}{r}\cdot\frac{\partial}{\partial r}(rI_o\overline{\rho' v_r})}_{O(\beta^{1/2}\eta)} + \underbrace{\frac{1}{r}\frac{\partial}{\partial r}(r\rho_o\overline{v_r i'})}_{O(\alpha\beta^{1/2})}$$
$$+ \underbrace{\frac{1}{r}\frac{\partial}{\partial r}(r\rho_o V_\theta\overline{v_r v_\theta})}_{O(\beta)} + \underbrace{\frac{1}{r}\frac{\partial}{\partial r}(r\rho_o V_z\overline{v_r v_z})}_{O(\beta\epsilon/\delta)} = O(\beta\delta, \beta\epsilon, \alpha\beta^{1/2}\delta) \tag{18–18}$$

Substituting a term of the continuity equation 18–17

$$\frac{1}{r}\frac{\partial}{\partial r}(\rho_o r V_r) = -\frac{\partial}{\partial z}(\rho_o V_z) - \frac{1}{r}\frac{\partial}{\partial r}(r\overline{\rho' v_r})$$

into Equation 18–18, we obtain

$$\rho_o\left(V_r\frac{\partial I_o}{\partial r} + V_z\frac{\partial I_o}{\partial z}\right) = -\overline{\rho' v_r}\frac{\partial I_o}{\partial r} - \frac{1}{r}\frac{\partial}{\partial r}(\rho_o r\overline{v_r i'}) -$$
$$-\frac{1}{r}\frac{\partial}{\partial r}(\rho_o r V_\theta\overline{v_r v_\theta}) - \frac{1}{r}\frac{\partial}{\partial r}(\rho_o r V_z\overline{v_r v_z}) \tag{18–19}$$

Denoting the sum of the two terms of the left-hand side as $\rho_o DI_o/Dt$, Equation 18–19 becomes

$$\rho_o \frac{DI_o}{Dt} = -\frac{1}{r}\frac{\partial}{\partial r}(\rho_o r\overline{v_r i'}) - \overline{\rho' v_r}\frac{\partial I_o}{\partial r} - \frac{1}{r}\frac{\partial}{\partial r}(\rho_o r V_\theta \overline{v_r v_\theta}) - \frac{1}{r}\frac{\partial}{\partial r}(\rho_o r V_z \overline{v_r v_z}) \quad (18\text{–}20)$$

In this equation, the third term of the right-hand side represents the work done by the turbulent shear stress (Reynolds stress) of the turbulent rotational air flow which affect one of the most important factors in the vortex tube.

C. SURVEY OF THE FUNDAMENTAL ENERGY FLUX

From the simplified equation of energy, the four kinds of energy flux indicate the fundamental energy fluxes in the high turbulent rotational air flow.

1. Heat Flux ($\rho_o \overline{v_r i'}$)

Heat flux ($\rho_o \overline{v_r i'}$) is produced by the turbulent mixing based upon the radial gradient of the temperature and the pressure. This quantity was pointed out by Knoernschild.

2. Total Energy Flux ($\overline{\rho' v_r} I_o$)

Concerning the total energy flux ($\overline{\rho' v_r} I_o$), as mentioned by Reynolds in his paper "On the dynamics of turbulent vortical flow", the correlation between ρ' and v_r has a very low value. The lump of the lower density fluid in the turbulent rotational air flow is inclined to move toward the inward direction from the high pressure region into the lower pressure region due to the effect of the Archimedes bouyancy force. Therefore the time mean value of the product of the two terms ρ' and v_r has a tendency:

$$\overline{\rho' v_r} > 0 \quad (18\text{–}21)$$

3. The Work Flux by the Two Kinds of Reynolds Stresses ($\rho_o V_\theta \overline{v_r v_\theta}$ and $\rho_o V_z \overline{v_r v_z}$)

These two kinds of Reynolds stresses are the most important quantities for the energy separation in the vortex tube. The Reynolds stress ($-\rho_o \overline{v_r v_\theta}$) especially shows very high values in the turbulent rotational flow at high Reynolds numbers. These effects will be described in a later section in detail.

However it is necessary to estimate the orders of each term of Equation 18–18. Then from the experimental results of the turbulent rotational air flow in the vortex tube, the following approximate orders are given as

$$\delta \sim 1/500 \quad , \quad \epsilon \sim 1/200 \quad , \quad \beta \sim 1/100 \quad , \quad \eta \sim 1/100$$

Substituting the above orders into Equation 18–18, the orders of each term can be estimated as follows:

$$\underbrace{\frac{1}{r}\frac{\partial}{\partial r}(r\rho_o V_r I_o)}_{O(1/200)} + \underbrace{\frac{\partial}{\partial z}(\rho_o V_z I_o)}_{O(1/200)} + \underbrace{\frac{1}{r}\frac{\partial}{\partial r}(rI_o\overline{\rho' v_r})}_{O(1/1000)} + \underbrace{\frac{1}{r}\frac{\partial}{\partial r}(r\rho_o\overline{v_r i'})}_{O(1/10)}$$

$$+ \underbrace{\frac{1}{r}\frac{\partial}{\partial r}(r\rho_o V_\theta \overline{v_r v_\theta})}_{O(1/100)} + \underbrace{\frac{1}{r}\frac{\partial}{\partial r}(r\rho_o V_z \overline{v_r v_z})}_{O(1/400)} = O(1/5000,\ 2/500) \quad (18\text{–}22)$$

We must consider the physical processes of each term of the above equation in the next section.

III. KNOERNSCHILD EFFECT

Already mentioned in Chapter 17, in transforming Equation 17–46, we obtain

$$\overline{v_r i'} = -\nu_t\left(\frac{\partial I}{\partial r} - \frac{k-1}{k}\cdot\frac{I}{p}\cdot\frac{\partial p}{\partial r}\right) \tag{18–23}$$

Then using the radial equilibrium condition of rotational flow as

$$\frac{\partial p}{\partial r} = \rho_o\frac{V_\theta^2}{r} + O(\beta) \tag{18–24}$$

Equation 18–23 becomes

$$\overline{v_r i'} = -\nu_t\left(\frac{\partial I}{\partial r} - \frac{V_\theta^2}{r}\right) \tag{18–25}$$

This equation suggests that, when the radial gradient of the temperature

$$\frac{dI}{dr} = C_p\frac{dT}{dr} \tag{18–26}$$

is larger than that of the temperature, which can be defined by a rule of the adiabatic process, the heat flux due to the Knoernschild effect moves toward the outward direction (positive value) everywhere. And further, we can imagine that this term approaches its highest value at the vortex core where the centrifugal acceleration gives the maximum value.

IV. ARCHIMEDES EFFECT

The time mean value $(\overline{\rho' v_r})$ of the product of the fluctuating density ρ' and of the fluctuating radial velocity v_r is not always due to an Archimedes effect, since the lighter lump of fluid than the surrounding fluid covering the fixed lump of fluid can move toward the inner center axis by an effect of buoyancy. However, as shown in Figure 18–1, we may explain the above phenomenon as follows. The density difference $(\rho_1 - \rho_k < 0)$ between both lumps of fluid will produce, when the surrounding pressure P_k of the covering of the fixed lump of fluid is higher than the pressure p_1 of the fixed lump of fluid, or when the surrounding temperature T_1 of the covering of the fixed lump of fluid is lower than the temperature T_k of the fixed lump of fluid.

The latter effect is directly connected to the quantity of $\overline{v_r i'}$. Therefore the thermal conduction due to the Archimedes effect may be considered to be included among the Knoernschild effect. Accordingly, a term $\overline{\rho' v_r}$ in connection with a term $\overline{v_r i'}$ has a tendency of $\overline{\rho' v_r} > 0$ in the compressible turbulent rotational air flow; so, the behavior of the term $\overline{\rho' v_r}$ will show the movement toward the outward direction. The paper of Deissler and Perlmutter showed the correlation of $\overline{\rho' v_r}$ in the compressible rotating air flow, which is ascribed a positive value. So the energy flux $\overline{\rho' v_r} I_o$ always moves toward the outward

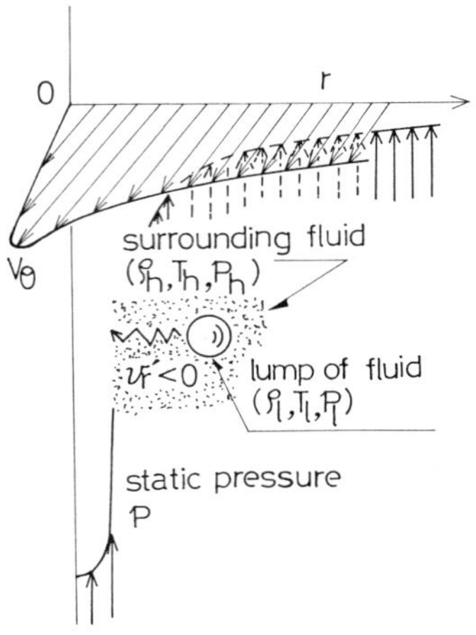

FIGURE 18–1. Schematic illustration of the Archimedes effect.

direction. In addition to this as shown by a term $\overline{\rho' v_r}(\partial I_o/\partial r)$ of Equation 18–20, the total effect of this outward heat flux contributes to the cooling effect of air, since the value of dI_o/dr (except in the vicinity of the solid wall surface in the vortex tube) has a positive value. Therefore, this term

$$\underset{\text{positive}}{-\overline{\rho' v_r}} \quad \underset{\text{positive}}{\frac{\partial I_o}{\partial r}} \quad < \quad \underset{\text{negative}}{0} \tag{18–27}$$

becomes negative; consequently, the contribution to DI_o/Dt becomes negative. The above description can be respresented as follows

$$\frac{\partial}{\partial r}(r\rho V_r) = \frac{\partial}{\partial r}[r(\rho_o + \rho')(V_r + v_r)] = 0 \tag{18–28}$$

Therefore, the above equation can be transformed to

$$\rho_o V_r = -\overline{\rho' v_r} \tag{18–29}$$

As mentioned above, the term $\overline{\rho' v_r}$ can be regarded as a positive value. So from Equation 18–29, the mean radial velocity can be represented as

$$v_r = -\frac{\overline{\rho' v_r}}{\rho_o} < 0 \tag{18–30}$$

Accordingly the radial velocity component shows the inward direction due to the negative value of V_r.

Then we can write the equivalent equation as

$$\rho_o V_r \frac{dI_o}{dr} = \rho_o \frac{dI_o}{dt} = -\overline{\rho' v_r}\frac{dI_o}{dr} \tag{18–31}$$

So the term $-\overline{\rho' v_r}(\partial I_o/\partial r)$ transfers to the inward direction, since the contribution of $\overline{\rho' v_r}$ to dI_o/dr becomes negative.

Let us consider a term $\overline{\rho' v_r}$. Assuming that the turbulent fluctuating velocity of the lump of fluid at an arbitrary point can be written as

$$v_r = v_{r1} + v_{r2} \tag{18–32}$$

where v_{r1} is the radial velocity produced by the Archimedes buoyancy $(\rho' V_\theta^2/r)$ per unit volume of fluid in the centrifugal field, and also v_{r2} is the radial velocity produced by the local flow due to the turbulent flow around the fixed lump of fluid. Therefore the following relation can be obtained

$$\overline{\rho' v_r} = \overline{\rho'(v_{r1} + v_{r2})} \fallingdotseq \overline{\rho' v_{r1}} \tag{18–33}$$

In addition to this, consider the motion of the lighter density fluid which moves in the inward direction. Then assuming that this lighter lump of fluid shows a behavior like a Brownian movement and also this moving lump of fluid experiences the Stokes drag force, the relationship between the fluctuating density ρ' and the fluctuating velocity v_{r1} can be stated approximately

$$\rho' \frac{V_\theta^2}{r} \propto \rho_o \nu_t v_{r1} \tag{18–34}$$

where a constant of proportionality is related to the scale of turbulence. Next assume that the relationship between the fluctuating pressure p' and the fluctuating velocity v_r is

$$p' \propto \rho_o \overline{v_r^2} \tag{18–35}$$

The ratio of ρ' to p' defines

$$K = \rho'/p' \tag{18–36}$$

where K means the compressibility of fluid. Consequently the term $\overline{\rho' v_r}$ can be described as

$$\overline{\rho' v_r} \propto \rho' \frac{\rho'}{\rho_o \nu_t} \cdot \frac{V_\theta^2}{r} = \frac{p'^2}{\rho_o \nu_t}\left(\frac{K^2 V_\theta^2}{r}\right) = \wedge_T \rho_o K^2 \frac{V_\theta^2}{r} \tag{18–37}$$

where a symbol $\wedge_T$ is equivalent to $p'^2/\rho_o^2 \nu_t$. Further transforming Equation 18–37

$$\overline{\rho' v_r} = \wedge_T K^2 \frac{dp}{dr} \tag{18–38}$$

This equation corresponds to the Archimedes correlation $\overline{\rho' v_r}$ as mentioned previously.

V. WORK FLUX DUE TO THE REYNOLDS STRESS FOR THE CIRCUMFERENTIAL DIRECTION

The momentum equation for the circumferential direction under the assumptions of the axial symmetry and independent of the axial direction $(\partial/\partial z = 0)$ in the turbulent rotational air flow can be described as follows:

$$\rho_o V_r\left(\frac{dV_\theta}{dr} + \frac{V_\theta}{r}\right) = -\frac{1}{r^2}\frac{d}{dr}(r^2\rho_o\overline{v_r v_\theta}) \tag{18–39}$$

Transform the above equation as follows:

$$\rho_o V_r r\frac{d}{dr}(rV_\theta) = -\frac{d}{dr}(r^2\rho_o\overline{v_r v_\theta}) \tag{18–40}$$

Here let us try to explain the above equation for the most simple example in detail.
In case of the constant radial flow, the continuity equation can be written as

$$\rho_o r V_r = \text{constant} = C \tag{18–41}$$

Substituting the above equation into Equation 18–40, the following equation is obtained

$$C\frac{d}{dr}(rV_\theta) = -\frac{d}{dr}(r^2\rho_o\overline{v_r v_\theta}) \tag{18–42}$$

Integrating the above equation once and multiplying both sides by V_θ, the following equation is obtained

$$r\,\rho_o V_\theta\overline{v_r v_\theta} = -CV_\theta^2 \tag{18–43}$$

Here this term $r\,\rho_o V_\theta\overline{v_r v_\theta}$ means the circumferential work flux due to the Reynolds stress acting on the unit surface, and also when the value of a constant C is a positive value, the fluid flow goes toward the outward direction. Therefore the flow of the work flux is toward the inward direction due to the negative value of Equation 18–43. When the constant C becomes negative, the flow of the work flux is toward the outward direction. In other words the flow direction of the mass flow to the radial direction is in the opposite direction of the flow of the work flux. From this physical meaning of the transport processes between the mass flow and the flow of the work flux, the contribution of this work flux term of Equation 18–20 to $\rho_o DI_o/Dt$ becomes more clear. Consequently, from Equation 18–20, the following relationship can be written as

$$\rho_o\frac{DI_o}{Dt} \propto -\frac{1}{r}\frac{d}{dr}(r\,\rho_o V_\theta\overline{v_r v_\theta}) = \frac{C}{r}\frac{dV_\theta^2}{dr} \tag{18–44}$$

Now when the fluid flow is the inward direction ($C < 0$) and also when the distribution of the tangential velocity is described by the quasi-free vortex as

$$V_\theta r^n = \Gamma_n = \text{constant} \tag{18–45}$$

Equation 18–44 can be transformed to

$$\rho_o\frac{DI_o}{Dt} \propto -2n\Gamma_n^2 r^{-2n-2}C \tag{18–46}$$

Therefore the contribution to $\rho_o DI_o/Dt$ becomes a positive value. This means that in such a flow condition, the fluid is heated locally.

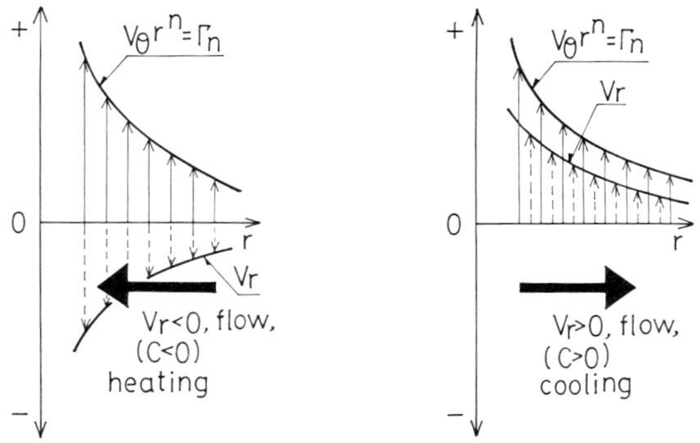

FIGURE 18–2. Contributions of flow directions to $\rho_o DI_o/Dt$ on the quasi-free vortex flow.

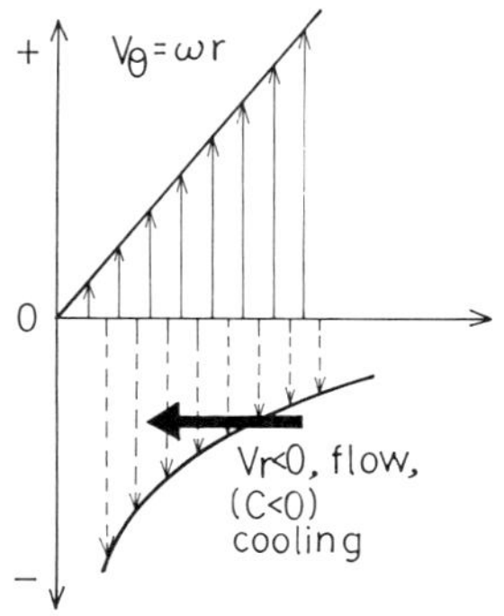

FIGURE 18–3. Contribution of flow direction to $\rho_o DI_o/Dt$ on the forced vortex flow.

When the fluid flow is the inward direction (C < 0) and also when the distribution of the tangential velocity is described by the forced vortex

$$V_\theta = \omega r \tag{18–47}$$

Equation 18–44 can be transformed to

$$\rho_o \frac{DI_o}{Dt} \propto 2C\omega^2 \tag{18–48}$$

Therefore, the contribution to $\rho_o DI_o/Dt$ becomes negative. Accordingly, the fluid is cooled locally. From these theoretical considerations, it is very important to emphasize that the positive and negative values of $\rho_o DI_o/Dt$ are corresponding to the increase and the decrease per unit time of the total enthalpy of fluid. These schematic illustrations are shown in Figures 18–2 and 18–3. Furthermore, let us consider the contributions of the various kinds of flow models to the work flux for the circumferential direction while assuming that the distribution of the tangential velocity of the quasi-free vortex can be written as

$$V_\theta \cdot r^{0.5} = \Gamma = \text{constant} \tag{18–49}$$

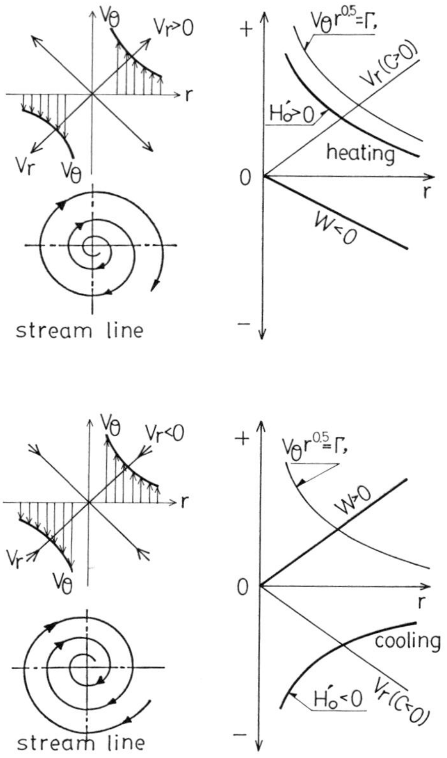

FIGURE 18–4. Schematic illustrations for $r\rho_o V_r = Cr^2$.

A. IN THE CASE OF $r\rho_o V_r = Cr^2 (V_r = Cr/\rho_o)$

From Equation 18–40, we can obtain the following equation

$$Cr^2 \frac{d}{dr}(rV_\theta) = -\frac{d}{dr}(r^2 \rho_o \overline{v_r v_\theta}) \tag{18–50}$$

Integrating the above equation once, the work flux W is

$$W = rV_\theta \rho_o \overline{v_r v_\theta} = -\frac{C\Gamma^2}{5} r \tag{18–51}$$

Therefore, the contribution to the total enthalpy is

$$H_o' = \rho_o \frac{DI_o}{Dt} \propto -\frac{1}{r}\frac{d}{dr}(rV_\theta \rho_o \overline{v_r v_\theta}) = \frac{C\Gamma^2}{5r} \tag{18–52}$$

Then when the fluid flow goes toward the outward direction ($C > 0$), from Equations 18–51 and 18–52, W becomes negative ($W < 0$) and also H_o' becomes positive ($H_o' > 0$). Consequently when the fluid is flowing from the center toward the outward, this fluid is heated.

When the fluid flow goes toward the center ($C < 0$), W becomes positive ($W > 0$) and also H_o' becomes negative ($H_o' < 0$). Then the flowing fluid is cooled during the flowing process. These physical processes are illustrated in Figure 18–4.

B. IN THE CASE OF $r\rho_o V_r = Cr(V_r = C/\rho_o = \text{constant})$

In case of the constant radial velocity (V_r = constant), from Equation 18–40 we obtain the following equation

$$Cr\frac{d}{dr}(rV_\theta) = -\frac{d}{dr}(r^2\rho_o\overline{v_r v_\theta}) \tag{18–53}$$

Integrating the above equation once, the following equation is obtained

$$\frac{1}{3}C\Gamma r^{3/2} = -r^2\rho_o\overline{v_r v_\theta} \tag{18–54}$$

Therefore, W and H_o' can be expressed as

$$W = rV_\theta\rho_o\overline{v_r v_\theta} = -\frac{C\Gamma^2}{3} \tag{18–55}$$

$$H_o' = \rho_o\frac{DI_o}{Dt} \propto -\frac{1}{r}\frac{d}{dr}(rV_\theta\rho_o\overline{v_r v_\theta}) = 0 \tag{18–56}$$

Therefore, when the fluid flow goes toward the outward direction ($C > 0$), W becomes negative ($W < 0$) and $H_o' = 0$. When the fluid flow goes toward the inward direction ($C < 0$), W becomes positive and $H_o' = 0$. Consequently in both cases the flowing fluid is not heated or cooled, since $H_o' = 0$. These schematic illustrations are shown in Figure 18–5.

C. IN THE CASE OF $r\rho_o V_r = C(V_r = C/\rho_o r)$

When the radial velocity is in inverse ratio to the radius r, from Equation 18–40 we can derive the following equation

$$C\frac{d}{dr}(rV_\theta) = -\frac{d}{dr}(r^2\rho_o\overline{v_r v_\theta}) \tag{18–57}$$

Integrating the above equation once, the following equation is obtained

$$CrV_\theta = -r^2\rho_o\overline{v_r v_\theta} \tag{18–58}$$

Therefore, W and H_o' can be expressed as

$$W = rV_\theta\rho_o\overline{v_r v_\theta} = -CV_\theta^2 = -\frac{C\Gamma^2}{r} \tag{18–59}$$

$$H_o' = \rho_o\frac{DI_o}{Dt} \propto -\frac{C\Gamma^2}{r^3} \tag{18–60}$$

Consequently, when the radial flow goes toward the outward direction ($C > 0$), so $H_o' < 0$ (cooling). When the radial flow goes toward the inward direction ($C < 0$), so $H_o' > 0$ (heating). These physical explanations are shown in Figure 18–6.

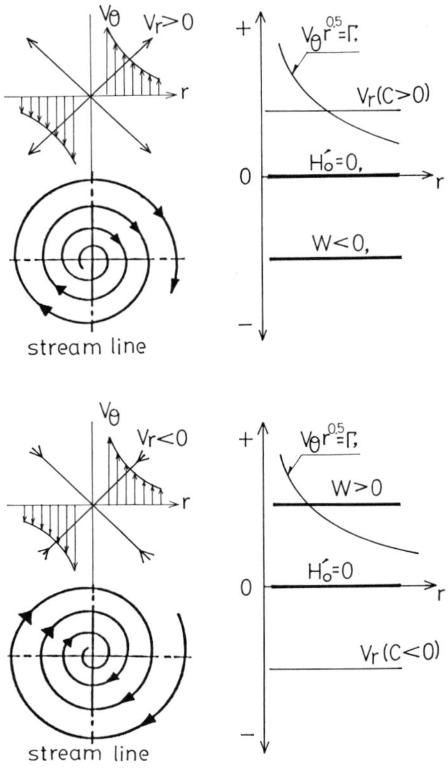

FIGURE 18–5. Schematic illustrations for $r\rho_o V_r = Cr$.

D. IN THE CASE OF $r\rho_o V_r = C(a - r)$, $(r \leq a)$

In this case the radial velocity V_r at the radius $r = a$ goes to zero. Substituting the above relation into Equation 18–40 and integrating one, we obtain

$$C(a - r)\Gamma r^{0.5} + \frac{2}{3}C\Gamma r^{1.5} = -r^2\rho_o\overline{v_r v_\theta} \quad (18\text{–}61)$$

Therefore, W and H_o' can be expressed as

$$W = -C\Gamma^2\left(\frac{a}{r} - \frac{1}{3}\right) \quad (18\text{–}62)$$

$$H_o' = \rho_o\frac{DI_o}{Dt} \propto -\frac{C\Gamma^2 a}{r^3} \quad (18\text{–}63)$$

Then when C is positive value $(C > 0)$, H_o' becomes negative. Therefore the fluid is negative. Therefore fluid is cooled during fluid flowing toward the outward directon to the radius $r = a$. When C is negative $(C < 0)$, H_o' becomes positive. Consequently fluid is heated during fluid flowing from the radius $r = a$ toward the center. These schematic illustrations are shown in Figure 18–7.

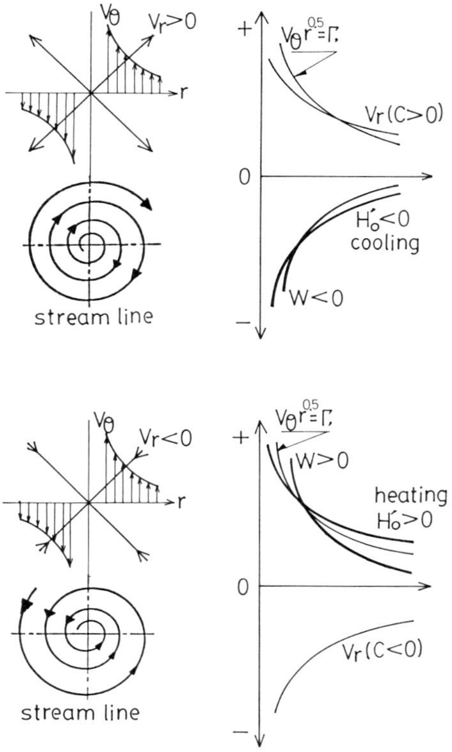

FIGURE 18–6. Schematic illustrations for $r\rho_o V_r = C$.

E. IN THE CASE OF $r\rho_o V_r = Cr^2(a - r)$, $(r \leq a)$

In this case the radial velocity V_r is zero at the two radial positions, $r = 0$ and $r = a$ as

$$V_r = \frac{C}{\rho_o} r(a - r) \tag{18–64}$$

Substituting this radial velocity equation into Equation 18–40 and then integrating once, we obtain

$$W = -C\Gamma^2 r\left(\frac{a}{5} - \frac{r}{7}\right) \tag{18–65}$$

$$H_o' = \rho_o \frac{DI_o}{Dt} \propto -C\Gamma^2\left(\frac{2}{7} - \frac{a}{5r}\right) \tag{18–66}$$

Then when C is positive, the radial velocity goes toward the outward direction between $r = 0$ and $r = a$. At the same time fluid is heated between $r = 0$ and $r = (7/10)a$, while fluid is cooled between $r = (7/10)a$ and $r = a$. On the other hand when C is negative, the fluid is cooled between $r = 0$ and $r = (7/10)a$, while fluid is heated between $r = (7/10)a$ and $r = a$. These schematic illustrations are shown in Figure 18–8.

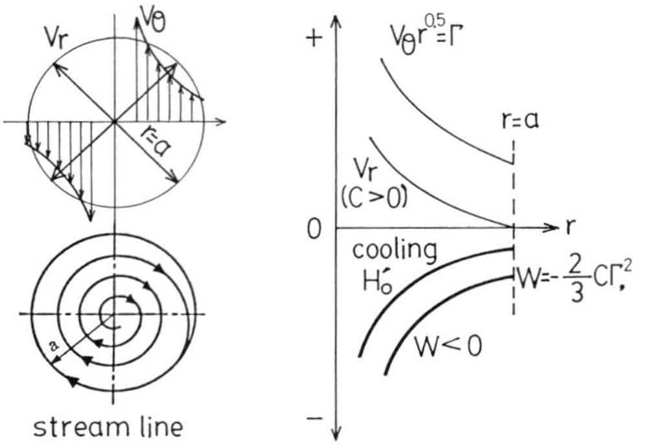

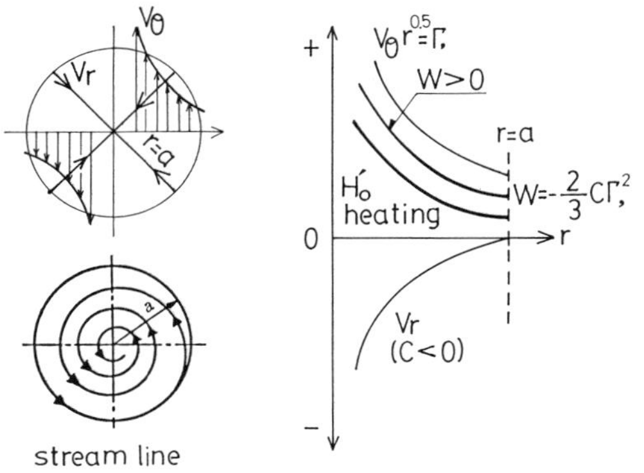

FIGURE 18–7. Schematic illustrations for $r\rho_o V_r = C(a - r)$.

F. IN THE CASE OF THE OGAWA COMBINED VORTEX MODEL (QUASI-FORCED VORTEX REGION, $0 \leq r \leq r_t$)

In the quasi-forced vortex region of the Ogawa combined vortex model, the tangential and radial velocities can be expressed as

$$V_\theta = Kr(1 - \Lambda r) \tag{18–67}$$

$$V_r = -fr \tag{18–68}$$

where the symbols K, Λ and f are the constants. Substituting the above equations into Equation 18–40 and after integrating once, W and H_o' can be written as

$$W = \frac{\rho_o f K^2}{2} r(1 - \Lambda r)\left(1 - \frac{6}{5}\Lambda r\right) \tag{18–69}$$

$$H_o' = \rho_o \frac{DI_o}{Dt} \propto -2\rho_o f K^2 r^2\left(1 - \frac{11}{4}\Lambda r + \frac{9}{5}\Lambda^2 r^2\right) \tag{18–70}$$

Now transforming the above equations into the dimensionless forms, the following equations are obtained:

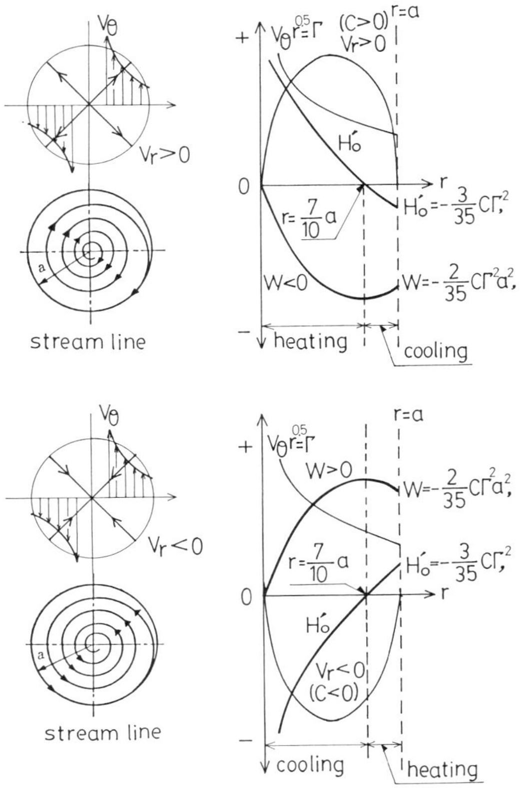

FIGURE 18–8. Schematic illustrations for $r\rho_o V_r = Cr^2(a - r)$.

$$\frac{2rV_\theta\rho_o\overline{v_r v_\theta}}{\rho_o f K^2 a^4} = \left(\frac{r}{a}\right)^4 \cdot \left(1 - \frac{r}{2a}\right) \cdot \left(1 - \frac{3}{5}\,\frac{a}{r}\right) \tag{18–71}$$

$$\frac{\rho_o(DI_o/Dt)}{2\rho_o f K^2 a^2} \propto -\left(\frac{r}{a}\right)^2 \cdot \left(1 - \frac{11}{8} \cdot \frac{r}{a} + \frac{9}{20} \cdot \frac{r^2}{a^2}\right) \tag{18–72}$$

These two relations are shown in Figure 18–9. From this figure it will be found that in the quasi-forced vortex region for the inward radial flow fluid is to be cooled, since in the quasi-forced vortex region H_0' becomes negative.

Accordingly from these various kinds of flow models we can obtain the following conclusions. The working direction of the work flux for the radial direction is, generally speaking, the opposite direction to the flow of the mass flux for the radial direction. Since the motion of fluid flow near the vortex core region in the vortex tube is governed by the inward radial flow ($C < 0$), the three-dimensional turbulent rotational flow near the vortex core will be cooled, while the rotating flow near the concave wall surface will be heated by the work flux of the tangential stress.

VI. WORK FLUX DUE TO THE REYNOLDS STRESS IN THE AXIAL DIRECTION

Here we consider simply the effect of the Reynolds stress in the axial direction. The momentum equation for the axial direction in the high turbulent rotational flow can be represented approximately as

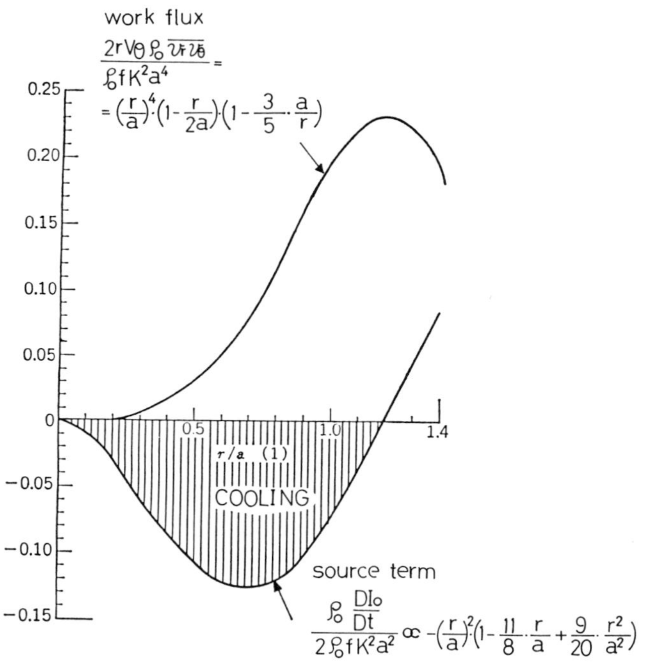

FIGURE 18–9. Numerical examples of the distributions of the work flux and of the source term by the Ogawa combined vortex model.

$$\frac{1}{r}\cdot\frac{d}{dr}(r\rho_o\overline{v_r v_z}) = -\frac{dp}{dz} \tag{18–73}$$

Integrating the above equation once, the following equation is obtained

$$r\rho_o\overline{v_r v_z} = -\int_0^r r\frac{dp}{dz}\,dr \tag{18–74}$$

Further multiplying both sides of Equation 18–74 by V_z, we obtain

$$rV_z\rho_o\overline{v_r v_z} = -V_z\int_0^r r\frac{dp}{dz}\,dr \tag{18–75}$$

Then we can suppose that the contribution of this term to $H_o' = \rho_o(DI_o/Dt)$ is smaller than that of the work flux of the tangential direction. The axial velocity V_z is smaller than the tangential velocity V_θ and the axial gradient of the static pressure is also a smaller value.

MAIN SYMBOLS

C_p(J/kg · K); specific heat of the constant pressure
H_o'(N/m² · s); contribution to the total enthalpy
i(J/kg); enthalpy per unit mass

i_o(J/kg);	total enthalpy per unit mass
k(1);	ratio of the specific heats
n(1);	velocity exponential of the quasi-free vortex
p(Pa);	static pressure
r(m);	radius
V(m/s);	velocity vector
V_r(m/s);	radial velocity
V_z(m/s);	axial velocity
V_θ(m/s);	tangential velocity
z(m);	axial distance
W(N/s);	circumferential work flux due to the Reynolds stress acting on the unit surface
ν_t(m^2/s);	eddy kinematic viscosity
ρ(kg/m^3);	density
ω(rad/s);	angular velocity

REFERENCES

1. **Reynolds, A. J.,** On the dynamics of turbulent vortical flow, *Z. A. M. P.*, 12, 149, 1961.
2. **Reynolds, A. J.,** Energy flows in a vortex tube, *Z. A. M. P.*, 12, 343, 1961.
3. **Stewart, H.,** The energy equation for a viscous compressible fluid, *Proc. Nat. Acad. Sci.*, 28, 161, 1942.
4. **Hartnett, J. P. and Eckert, E. R. G.,** Experimental study of the velocity and temperature distribution in a high-velocity vortex tube flow, *Trans. Am. Soc. Mech. Eng.*, May, 751, 1957.
5. **Hilsch, R.,** Die Expansion von Gasen im Zentrifugalfeld als Kälteprozeß, *Z. Naturforsch.*, 1, 208, 1946.
6. **Schmidt, K.,** Experimentelle Untersuchungen am Ranque-Wirbelrohr, *Z. Naturforsch.*, 7a, 480, 1952.
7. **Van Deemter, J. J.,** On the theory of the Ranque-Hilsch cooling effect, *Appl. Sci. Res.*, 3-A, 174, 1952.
8. **Hikichi, T. and Ogawa, A.,** On the characteristics of the cyclone dust collectors which are used for the Diesel engine, *Nihon Univ. J. Coll. Eng.*, 18-A (March), 99, 1977.

Chapter 19

INERTIA SEPARATION OF THE SMALL PARTICLES IN BAFFLED CONDUITS

I. INTRODUCTION

Calvert and Hodous[1] made an inertia collector for the collection of small particles in the baffled conduits. This mechanical mechanism of inertia collection is generally effective for the particles larger than about one micron diameter.

This research is concerned with the collection of the small particles in the baffled conduits as shown in Figure 19–1. Eddies are formed behind each baffle which are perpendicular to the major fluid flow direction and act as a series of the cyclone dust collectors.[2]

The major collection mechanism in the baffled conduit is that due to the centrifugal effect of a series of the rotating eddies. The particles which are transported from the main stream of fluid to an eddy in the baffles will move radially outward at velocities such that the centrifugal force is equal to the opposing fluid drag force (Stokes drag force). These particles which move far enough will strike the baffled walls and be collected.

II. THEORETICAL CONSIDERATION

In order to derive the mathematical model, the following assumptions are made:

1. There is instantaneous mixing of particles within the main fluid flow and between the main flow and the adjacent eddy.
2. The eddy is circular and its diameter is equal to the distance between the baffles.
3. The particles reach their terminal velocities in the centrifugal force field instantaneously.
4. All particles which strike a solid surface will stick to it.
5. Terminal velocities may be estimated by Stokes law.

Therefore the terminal settling velocity W_t in a gravitational force field can be described as

$$W_t = \frac{\rho_p \cdot g \cdot X_p^2}{18\eta} \tag{19–1}$$

Next, when the particle is now considered to be in a centrifugal force field at a distance r_c from the center of the eddy, and has a tangential velocity U_θ, the radial settling velocity U_r can be written as follows:

$$U_r = \frac{W_t U_\theta^2}{g r_c} \tag{19–2}$$

In addition to this, it will be assumed that the particles will be collected from the eddy only about 3/4 of its circumference with transfer of the particles from the main stream to the eddy occurring in the remaining 1/4 of its circumference. Then the required time T_c for the eddy making 3/4 of a revolution may be represented by

$$T_c = \frac{3}{4}\left(\frac{2\pi r_c}{U_\theta}\right) \tag{19–3}$$

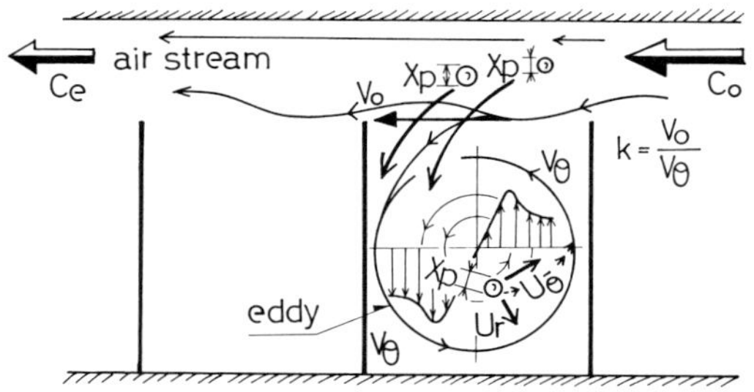

FIGURE 19–1. Schematic illustration of the separation mechanisms of the fine solid particles in the baffled conduits.

The radial distance δ of a moved particle in time T_c is

$$\delta = T_c U_r = \frac{3}{4}\left(\frac{2\pi r_c}{U\theta}\right) \cdot \frac{W_t U_\theta^2}{g r_c} = \frac{3\pi W_t U_\theta}{2g} \qquad (19\text{–}4)$$

Now the radial U_r and tangential U_0 velocities of the particle are assumed to be independent of radius, and it will be seen that this is valid because the small radial distance traveled by a particle during 3/4 spacing has no direct effect of δ.

The volume Q_c of the eddy fluid per unit time can be estimated as

$$Q_c = \delta U_\theta B = \frac{3\pi W_t U_\theta^2 B}{2g} \qquad (19\text{–}5)$$

where B is the width of the collector. The collection efficiency of an individual eddy will be defined as the ratio of the rate of the particle collection by the eddy to the rate of the particle flow through the total cross-section at this point in the collector, since the concentration in the eddy and that in the main stream are assumed to be equal. Therefore the total fluid flow rate Q_o can be represented as

$$Q_o = H\ B\ V_o \qquad (19\text{–}6)$$

where V_o is the mean stream velocity. Therefore the collection efficiency η_c' can be defined as

$$\eta_c' = \frac{Q_c}{Q_o} = \frac{3\pi W_t U_\theta^2}{2gHV_o} \qquad (19\text{–}7)$$

Here let us introduce the dimensionless factor $K = (V_o/V_\theta)$ which is obtained as the ratio of the mean air velocity V_o of the main stream to the tangential velocity $V_\theta \risingdotseq U_\theta$ of the eddy. Equation 19–7 can be transformed to

$$\eta_c' = \frac{3\pi W_t V_\theta^2 V_o}{2g \cdot HV_o^2} = \frac{3\pi V_o W_t}{2g \cdot HK^2} \qquad (19\text{–}8)$$

In this equation substituting Equation 19–1 of the terminal velocity into Equation 19–8, Equation 19–8 becomes

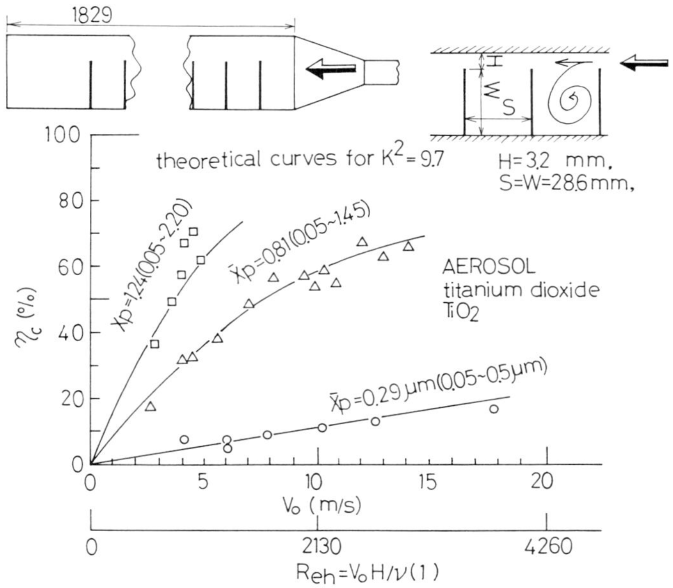

FIGURE 19–2. Relationship of the collection efficiency between the theoretical and experimental results.

$$\eta_c' = \frac{3\pi V_o \cdot \rho_p X_p^2}{2 \cdot H \cdot K^2 \cdot 18\eta} = \frac{\pi X_p^2 \rho_p V_o}{12\eta \cdot H K^2} \qquad (19\text{–}9)$$

Therefore, the collection efficiency η_c' increases with increase of mean air velocity V_o and also with decreasing height H.

Next we must define the relationship between η_c' and the overall collection efficiency η_c of the series of eddies for the particle size X_p. The overall collection efficiency for n baffles can be defined as

$$\eta_c = \frac{C_o - C_e}{C_o} \qquad (19\text{–}10)$$

where C_o is the number of the particles per unit volume entering the collector and C_e is the number of particles per unit volume leaving the collector, and also C_e can be calculated by

$$C_e = C_o(1 - \eta_c')^n \qquad (19\text{–}11)$$

Consequently, the overall collection efficiency η_c can be written as

$$\eta_c = 1 - (1 - \eta_c')^n \qquad (19\text{–}12)$$

One experimental result is shown in Figure 19–2. The test aerosol is a titanium dioxide (TiO_2) which has the mean diameter $\bar{X}_p$ = 0.29 μm (0.05 ~ 0.5 μm), $\bar{X}_p$ = 0.81 μm (0.05 ~ 1.45 μm), and $\bar{X}_p$ = 1.24 μm (0.05 ~ 2.20 μm). The height H is 3.2 mm and the sizes of S and W are the same (S = W = 28.6 mm), respectively. The mean air stream velocity V_o is V_o = 0 ~ 20 m/s and Reynolds number Re = $H \cdot V_o/\nu$ = 0 ~ 4260. The collection efficiency increases with increasing the mean velocity V_o and with increasing the particle diameter X_p.

MAIN SYMBOLS

$Q_c(m^3/s)$;	volume of the eddy fluid per unit time
$Q_o(m^3/s)$;	total fluid flow rate
$r_c(m)$;	distance from a center of the eddy
$T_c(s)$;	required time
$U_r(m/s)$;	radial settling velocity of the particles
$U_0(m/s)$;	tangential velocity of the particles
$V_o(m/s)$;	mean air velocity
$W_t(m/s)$;	terminal velocity
$X_p(\mu m)$;	particle size
$\delta(m)$;	radial distance
$\eta(Pa \cdot s)$;	viscosity
$\eta_c(1)$;	total collection efficiency
$\rho_p(kg/m^3)$;	density of the particle
$\nu(m^2/s)$;	kinematic viscosity

REFERENCES

1. **Hodous, R. and Calvert, S.,** Collection of small particles in baffled conduits, *J. Air Pollution Control Assoc.*, 12(7), 326, 1962.
2. **McCreary, F. J.,** Particle trajectories in free vortex flow models, Ph.D. thesis, University of Washington, Seattle, 1974.

Chapter 20

BEHAVIOR OF DISPERSED PARTICLES IN A SPIRAL-SHAPED AIR DRYER

I. INTRODUCTION

In this chapter, the mechanical behavior of dispersed particles in a spiral-shaped air dryer by Planobskiy et al.[1] is introduced for one of the applications of vortex flow to particle handling engineering. The author especially feels that this idea can be applied to the design of cyclone dust collectors.[1]

II. FUNDAMENTAL IDEA

One of the most effective methods for the dispersed particles in a convective dryer is an application of the curved or rotational motion of the dispersed particle-gas mixture. The powerful convective heat-mass exchange and the promotion of the relative velocity between the dispersed particles moves in the motion of the mixture along the curved locus, a factor which shows the motion of the particle to be stagnated. This is an effect of centrifugal force. An important problem is the aerodynamic investigation of the dispersed particles in the curved air flow. However, the research work clarifying this problem is sparse except for the study of radial particle sedimentation in the cyclone dust collectors. Therefore, in this section the interesting experimental study for the dispersed particle-gas mixture in the curved flow in the spiral-shaped apparatus by the above-stated authors will be described.

III. EXPERIMENTAL METHOD AND EMPIRICAL EQUATION

Let us consider the motion of particles in the curved locus. The following forces are considered:

1. An aerodynamic force on the particle of the diameter X_p

$$F_a = \zeta_d \frac{\pi}{4} X_p^2 \frac{\rho(V - U_p)^2}{2} \tag{20–1}$$

2. The particle mass

$$G = mg \tag{20–2}$$

3. A centrifugal force

$$F_c = m\frac{U_p^2}{R} \tag{20–3}$$

4. A reaction force from the concave wall surface of the slit as shown in Figure 20–1,

$$N \tag{20–4}$$

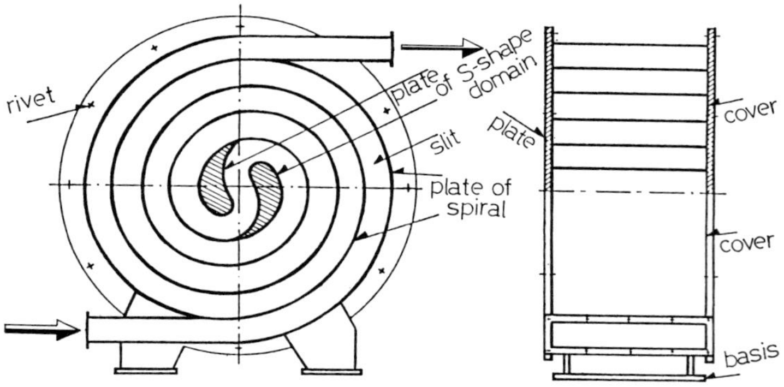

FIGURE 20–1. Schematic illustration of the spiral-shaped air dryer.

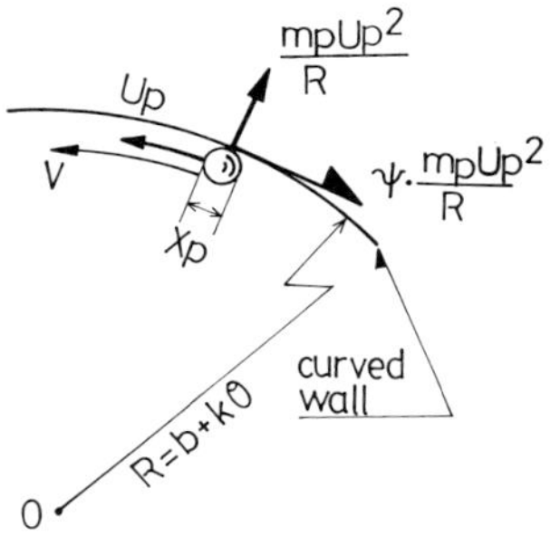

FIGURE 20–2. Fluid and particle velocities on the concave wall surface.

where m is a mass of the particle, R is a radius of curvature, V is the gas velocity, and U_p is the particle velocity as shown in Figure 20–2.

First we assume that the mass of the particle can be neglected in comparison with the centrifugal force F_c. Therefore the frictional force F_f on the curved wall surface of the slit can be written as

$$F_f = -\Psi N = \Psi F_c \tag{20–5}$$

where Ψ shows the effective frictional coefficient. So the differential equation of the motion of the particle of mass m along the curved surface is

$$m\frac{dU_p}{d\tau} = \zeta_d \frac{\pi X_p^2}{4} \cdot \frac{\rho(V - U_p)^2}{2} - \Psi\frac{mU_p^2}{R} \tag{20–6}$$

$$m\frac{U_p^2}{R} = -N \tag{20–7}$$

where τ means a time. In addition to this, we assume that the radial motion of the particle which is restricted by the curved surface in the spiral-shaped slit is followed up to the spiral equation as

$$R = b + k\theta \tag{20–8}$$

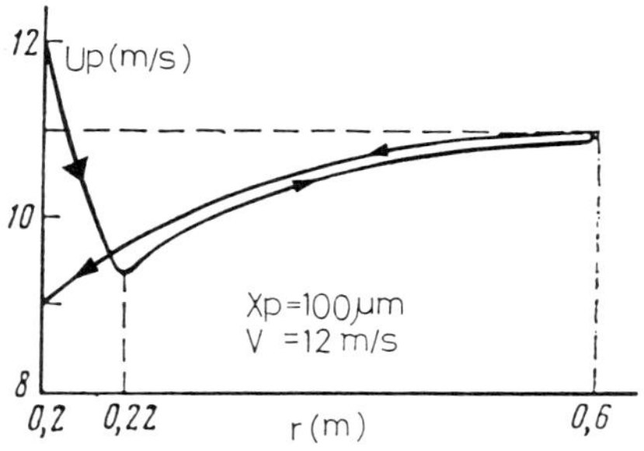

FIGURE 20–3. Calculated result of the particle velocity U_p depending on the radius r for V = 12 m/s and X_p = 100 μm.

Here transforming Equation 20–6 with $dR = k \cdot d\theta$ of Equation 20–8, we obtain

$$U_p \frac{dU_p}{dR} = B \cdot R \cdot U_p^2 - C \cdot U_p^2 - D \cdot R \cdot U_p + E \cdot R \qquad (20\text{–}9)$$

where these symbols B, C, D, and E which depend on the parameters ζ_d and Ψ and depend on the flow condition are represented as

$$B = \frac{\rho \zeta_d \pi X_p^2}{8mk}, \quad C = \frac{\Psi}{k}, \quad D = \frac{\rho \zeta_d \pi X_p^2 V}{4mk}, \quad E = \frac{\rho \zeta_d \pi X_p^2 V^2}{8mk} \qquad (20\text{–}10)$$

Planobskiy et al.[1] calculated the motion of the particle in the spiral-shaped slit by the electronic computer for the following numerical example: the air velocity, V = 12.0 m/s; the pitch of the spiral, t = 0.008 m/grad; the minimum radius of the spiral, R_{min} = 0.2 m; and the maximum radius of the spiral, R_{max} = 0.6 m. The values of the coefficients applied the mean values: the drag coefficient, ζ_d = 0.3; the frictional coefficient, Ψ = 0.4. And also the density and the diameter of the particle were ρ_p = 1100 kg/m^3 and X_p = 100 μm. The calculated results are shown in Figure 20–3, where from the stagnation point near the center of the spiral to the downward direction, the equation of motion must be transformed to

$$U_p \frac{dU_p}{dR} = -B \cdot R \cdot U_p^2 + C \cdot U_p^2 + D \cdot R \cdot U_p - E \cdot R \qquad (20\text{–}11)$$

As the results of Figure 20–3 and Equations 20–9 and 20–11 indicate the particle velocity increases with increasing radius, while the particle velocity decreases with decreasing radius. Further, the violent decreasing velocity of the particle near the inlet region is based upon the occurrence of the centrifugal force during flowing through from the straight slit into the curved slit.

Moreover, in order to study the aerodynamic research, it is necessary to establish the empirical relationship between ζ_d and Ψ. In connection with the above problem and with the engineering application, they proposed the estimating method of the relative velocity between gas and the particle in the curved slit as follows: on the mechanically equilibrium state Equation 20–6 can be transformed to

$$\zeta_d \frac{\pi X_p^2}{4} \cdot \frac{\rho (V - U_p)^2}{2} = \Psi \frac{\pi X_p^3}{6} \rho_p \frac{U_p^2}{R} \qquad (20\text{–}12)$$

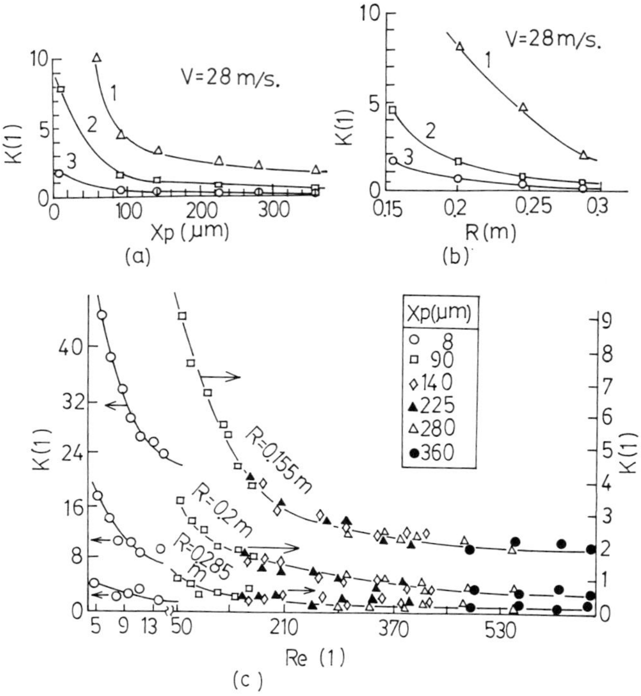

FIGURE 20–4. Experimental results of K depending on the particle diameter X_p, the radius R, and the Reynolds number R_e.

Introducing a coefficient $K = \Psi/\zeta_d$ and also putting $V_o = V - U_p$, the equation describing the relative velocity between them in the spiral slit can be obtained as

$$V_o = \frac{V}{1 + \sqrt{\frac{3}{4} \cdot \frac{1}{K} \cdot \frac{R}{X_p} \cdot \frac{\rho}{\rho_p}}} \tag{20–13}$$

in the above equation one of the unknown quantities is K. Then Planobskiy et al. suggested the functional relationship from the theoretical analysis as follows:

$$K = \frac{\Psi}{\zeta_d} = A \cdot R_{ep}^{n} \cdot F_{rp}^{m} \left(\frac{X_p}{R}\right)^{p} \tag{20–14}$$

where R_{ep} is the Reynolds number defined as $R_{ep} = V \cdot X_p/\nu$ and F_r is the Froude number defined as $F_{rp} = V^2/gX_p$. The experimental apparatus of the spiral-shaped air dryer is shown in Figure 20–1. The test particle was the polyvinyl-chloride (PVC). The diameter of the test particle was $X_p = 8$, 90, 140, 225, 280, and 360 μm, respectively. And the mean air velocity in this test was $V = 10 \sim 28$ m/s. The experimental results are shown in Figure 20–4. Figure 20–4(a) shows the relationship between K and X_p for $V = 28$ m/s and for the parameters $R = 0.155$ m (curve 1), 0.2 m (curve 2), and 0.285 m (curve 3), respectively. From these results the empirical relationships between K and X_p and between K and R are

$$K \propto X_p^{-0.7} \quad \text{and} \quad K \propto R^{-4} \tag{20–15}$$

Figure 20–4(b) shows the relationship between K and R for V = 28 m/s and for the parameters X_p = 8 μm (curve 1), X_p = 90 μm (curve 2), and X_p = 360 μm (curve 3), respectively. Figure 20–4(c) shows the relationship between K and R_{ep} for the parameters R = 0.155 m (curve 1), R = 0.2 m (curve 2), and R = 0.285 m (curve 3). From these experimental results the value of K for $R_{ep} > 500$ becomes constant, which corresponds to the rectilinear motion.

Consequently an empirical equation of K for the curvilinear motion of the dispersed particles in the spiral-shaped slit is

$$K = 1.76 \times 10^{13} \cdot R_{ep}^{-3.4} \cdot F_{rp}^{1.37} (X_p/R)^{4.1} \tag{20–16}$$

This result can be applied to the cyclone dust collectors.

MAIN SYMBOLS

$F_{rp}(1)$;	Froude number
k(m/rad);	constant value
m(kg);	mass of a solid particle
$R_{ep}(1)$;	Reynolds number about a solid particle
U_p(m/s);	particle velocity
V(m/s);	fluid velocity
X_p(μm);	particle diameter
$\zeta_d(1)$;	drag coefficient
$\rho(kg/m^3)$;	density of fluid
$\rho_p(kg/m^3)$;	density of the solid particle
τ(s);	time
$\Psi(1)$;	effective frictional coefficient

REFERENCE

1. **Planobskiy, A. N., Mushtaeb, I. M., Dedkob, B. V., and Uljianob, V. M.,** Aerodynamic study of the dispersed particle-gas mixture in the spiral shaped air dryer, *Teor. Osn., Khim. Tekhnol.*, 8(3), 407, 1974.

INDEX

E

F

G

H

I

J

K

L

M

N

O

P

Q

R

S

T

V

W

Y